LABORATORY METHODS
OF
ORGANIC CHEMISTRY

MACMILLAN AND CO., Limited
LONDON · BOMBAY · CALCUTTA · MADRAS
MELBOURNE

THE MACMILLAN COMPANY
NEW YORK · BOSTON · CHICAGO
DALLAS · ATLANTA · SAN FRANCISCO

THE MACMILLAN COMPANY
OF CANADA, LIMITED
TORONTO

LABORATORY METHODS
OF
ORGANIC CHEMISTRY

BY

L. GATTERMANN

COMPLETELY REVISED BY

HEINRICH WIELAND

TRANSLATED FROM
THE TWENTY-SECOND GERMAN EDITION
BY W. McCARTNEY, PH.D.(EDIN.), A.I.C.
ASSISTANT IN THE DEPARTMENT OF MEDICAL CHEMISTRY, UNIVERSITY OF EDINBURGH

WITH 55 ILLUSTRATIONS IN THE TEXT

MACMILLAN AND CO., LIMITED
ST. MARTIN'S STREET, LONDON
1932

PRINTED IN GREAT BRITAIN
BY R. & R. CLARK, LIMITED, EDINBURGH

PREFACE TO THE FIRST EDITION

THIS book was primarily written in order to satisfy the author's private requirements. When a large class starts practical work in organic chemistry it is, with the best will in the world, often impossible to teach each individual the manipulative details which abound in this work. In order that the student may have some help even in the absence of his teacher when carrying out general operations, the special instructions for the preparations are preceded by a general section on crystallisation, distillation, drying, analyses, etc. In writing this section less attention has been given to the complete enumeration of the numerous modifications of individual operations than to an attempt to describe the most important operations in such a way that the beginner can carry them out independently in the absence of the demonstrator.

In the second, special part each separate preparation is followed by general considerations which concern the nature and the significance of the reactions involved and aim at imparting to the student, already during his practical work, the widest possible range of theoretical knowledge. As is well known, the knowledge acquired under these conditions is more firmly fixed than when it is exclusively derived from a purely theoretical book. Thus the author hopes that, along with the excellent manuals of E. Fischer and Levy, his book also may meet with some acceptance. He will always be grateful to colleagues who point out defects.

L. GATTERMANN

HEIDELBERG, *August* 1894.

PREFACE TO THE REVISED (NINETEENTH) EDITION

It is rather more than thirty years ago since Ludwig Gattermann published the first edition of his *Anleitung für das organisch-chemische Praktikum.* The plan of providing the preparative directions with theoretical explanations has certainly proved satisfactory. That is already shown by the wide circulation of the book, of which eighteen editions have appeared. Methodology and technique are undoubtedly the chief objects of the practical course, but aiming merely at culinary art and technical achievement such a course does not accomplish enough. A command of methods implies above all an understanding of their rationale and a power of adapting their numerous modifications to particular requirements; the architect is more important than the mason. We demand that the student should be conversant with the theory of the transformations which he carries out practically. The comments made on the individual preparations are intended to facilitate a survey of the subject in hand, and to encourage the use of text-books and journals by further reading. Now that a knowledge of the principles of organic chemistry may be assumed during the preparative work in German universities, the danger of such comments becoming a *pons asinorum* is remote.

In rewriting the book the theoretical and practical requirements have been deliberately increased. The equipment which sufficed during the last three decades has now become insufficient for those who desire to work at present-day problems, where difficulties have been accentuated alike in pure science and in technology.

The idea of making the preparative work at once an explanation and a living experience of the science has demanded a rearrangement of the subject matter in accordance with systematic relationships. It will be seen that the new arrangement does not depart seriously from the path which ascends from the more simple to the more difficult. In any case a considerable educational advantage may be expected to result from the rounding off of successive subjects.

The general part as well as the analytical have been completely revised and greatly shortened in order to make more space for the preparative part. The increase in the number of preparations is intended to provide variety, and to counteract a tendency towards stereotyped routine in the organic practical course.

I am greatly indebted to my assistants, especially to Drs. Franz Bergel and F. Gottwalt Fischer, for untiring co-operation in carrying out numerous experiments. Dr. Fischer has, moreover, drawn the new diagrams for this edition and has prepared the index.

HEINRICH WIELAND

FREIBURG I. B., *Easter* 1925.

PREFACE TO THE TWENTY-SECOND EDITION

THE new edition, which has become necessary after two years, shows greater changes than do the two preceding ones. In the course of laboratory instruction new experience is constantly being gained, which leads to improvements in the directions and the explanations. On this occasion an opportunity occurred to make such changes in many parts of the book. The theoretical discussions have been made more complete by taking account of important new researches (*e.g.* Diels' diene synthesis, polyenes, haemin).

With a view to stimulating the interest of the student in biochemistry, the most important branch of the subject, two enzymic processes have been included. The saccharification of potato-starch is carried out, with reference to the quantitative aspects, and the mash is subsequently fermented. In addition, the preparation of invertase from yeast and the polarimetric observation of the inversion of cane sugar are described.

Alkaloids have, so far, not been represented. Now nicotine has been admitted because the starting material is readily obtainable everywhere.

It seems desirable to point out here that the preparations given in this book are intended to provide a selection, and that the inclusion of all of them in the course is by no means contemplated. In the Munich laboratory a suitable selection is made by the assistants, and, according to his capacity, the student carries out fifty to sixty preparations to which are added six to eight preparations from the original literature. The organic elementary analysis can be conveniently completed in one to two weeks, so that the practical work of the chemistry student in the organic laboratory requires less than one academic year. In our opinion a further limitation would be fatal to chemical education in our universities.

We also consider it unadvisable to reduce the requirements for those students who intend later to graduate in inorganic or physical chemistry; rather do we think that the thorough study of the varied methods peculiar to preparative organic chemistry provides a basis of experimental skill which should not be withheld from those who have aptitude for the other divisions of the subject. I have to thank sincerely Prof. G. Fischer, Freiburg, and Dr. A. Bertho, Munich, for their active co-operation. My thanks are also due to several other colleagues for many hints.

HEINRICH WIELAND

MUNICH, *July* 1930.

NOTE ON THE TRANSLATION

THE present translation has been prepared by Dr. McCartney while working in a German laboratory. I have read both the manuscript and the proofs. Although here and there the text may still savour of the German, I am confident that it is substantially free from error, and trust that it avoids obscurity. Colleagues and students who point out defects will earn my gratitude. The labour expended on the translation will be fully repaid if it enables English-speaking students to attain the standard of practical work demanded in the Munich laboratory.

G. BARGER

EDINBURGH, *April* 1932.

CONTENTS

A. SOME GENERAL LABORATORY RULES

B. ORGANIC ANALYTICAL METHODS

C. PREPARATIVE PART

B. Aromatic Diazo-Compounds

VIII. QUINONOID COMPOUNDS

IX. THE GRIGNARD AND FRIEDEL-CRAFTS SYNTHESES. ORGANIC RADICLES

The Grignard Reaction

ABBREVIATIONS

The abbreviations of the titles of journals are those employed in *British Chemical Abstracts.* The abbreviated title is followed by the year, volume number (in heavy type), and page.

Annalen	= Justus Liebig's Annalen der Chemie.
Ann. Chim.	= Annales de Chimie [et de Physique].
Ber.	= Berichte der Deutschen Chemischen Gesellschaft.
Bull. Soc. chim.	= Bulletin de la Société chimique de France.
Chem. Zentr.	= Chemisches Zentralblatt.
Helv. Chim. Acta	= Helvetica Chimica Acta.
J. Amer. Chem. Soc.	= Journal of the American Chemical Society.
J.C.S.	= Journal of the Chemical Society.
J. pr. Chem.	= Journal für praktische Chemie.
Monatsh.	= Monatshefte für Chemie und verwandte Teile anderer Wissenschaften.
Rec. trav. chim.	= Recueil des travaux chimiques des Pays-Bas.
Z. angew. Chem.	= Zeitschrift für angewandte Chemie.
Z. physiol. Chem.	= Hoppe Seyler's Zeitschrift für physiologische Chemie.

A. SOME GENERAL LABORATORY RULES

Reaction Velocity and Temperature.—Reactions with organic substances take place much more slowly than those which form the subject matter of a course of practical inorganic and analytical chemistry. The latter are nearly always *ionic reactions*, which proceed with immeasurable rapidity, but organic substances usually react much more slowly and therefore their preparation requires to be accelerated by increased temperature. A rise in the temperature of 10° doubles or trebles the velocity of most reactions. If the velocity at 20° is represented by v, then on the average that at 80° is $v \times 2{\cdot}5^6$. Consequently reactions will proceed in boiling alcohol about 250 times as fast as at room temperature.

For this reason many reactions of organic substances are brought about in heated solvents, usually at the boiling point.

The vapour of the solvent is cooled in a condenser fixed on the reaction vessel in such a way that the evaporated solvent continuously flows back again. Tap water is passed through the condenser.

In order to concentrate a solution the solvent is distilled "through a downward condenser". For this purpose various forms of coil condenser are more convenient than the Liebig pattern. For working "under reflux" such coil condensers are less suitable because of the layers of liquid which form in the coil between the vapour and the external atmosphere. A condenser designed by Dimroth has proved suitable for both types of work. In it the cooling water passes through the coil (Fig. 1). In order to prevent condensation of water vapour on the coil it is advisable to fix a calcium chloride tube into the upper opening of the condenser.

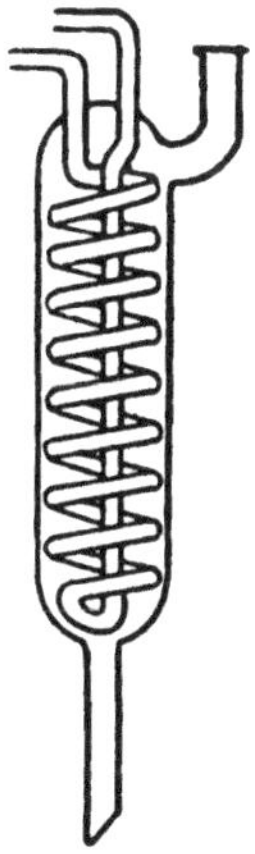

FIG. 1.

If a solvent which boils above 100° is used, the water-cooled condenser can be replaced by a long, wide glass tube (air condenser).

The condenser is attached to the reaction vessel by means of a tightly fitting *cork*, which is softened in a cork-squeezer before being bored. The diameter of the cork-borer chosen should be less than that of the glass tube for which the hole is made. The borer is heated in the flame of a Bunsen burner and is driven in a strictly vertical position through the cork, which stands on the bench with the narrow end upwards. As far as is practicable, collodion should not be used for making stoppers gas-tight. In general, rubber stoppers should not be used in experiments in which they are exposed to the vapours of boiling organic solvents, since they swell greatly and also give off soluble constituents which contaminate the reaction solution.

External Cooling.—Many reactions which occur with great evolution of heat require to be moderated. Further, in the preparation of labile substances which might be damaged by a high temperature, it is often necessary to provide for the cooling of the reaction mixture. The degree of cooling varies and, depending on the amount of heat to be removed and on the reaction temperature necessary, is produced by running tap water (8°–12°), by ice, which is finely crushed and mixed with a little water, by an ice and salt mixture (0° to $-20°$), or by a mixture of solid carbon dioxide with ether or acetone (down to $-80°$). Liquid air is generally not required in preparative organic work. To prepare freezing mixtures such as are often required, ice, well crushed in an ice mill or metal mortar, is thoroughly mixed by means of a small wooden shovel with about one third of its weight of rock-salt, preferably in a low, flat-bottomed glass jar or in a low enamelled saucepan.

In order to keep a freezing mixture cold for hours (or even over night) it is transferred to a " thermos " flask in which the contents of test tubes pushed into the freezing mixture can be maintained at low temperatures for a long time. For keeping larger vessels cold in this way, Piccard has indicated an arrangement easily constructed from two filter jars placed one inside the other. The bottom of the outer vessel is covered with kieselguhr until the rim of the centrally placed smaller jar is level with that of the outer jar. Then the annular space between the jars is likewise packed with pressed down kieselguhr and its upper portion between the rims is tightly closed with pitch.

Too little attention is generally paid to the *concentrations* of the reactants in preparative organic work. With the exception of rare cases (*e.g.* in intramolecular rearrangements) we are concerned with reactions of orders higher than the first, and in these several kinds of molecules—usually two—are involved. Since, according to the kinetic molecular theory, the velocity of bimolecular reactions is proportional to the number of collisions between the various dissolved molecules and therefore to the product of the concentrations,

$$v = C_A . C_B . K \text{ (K = velocity constant)},$$

it is advisable in all cases where there is no special reason to the contrary to choose the highest possible concentration for a reacting solution.

It should always be borne in mind that reduction of the concentration to one-half, one-quarter, or one-tenth makes the reaction four, sixteen, or one hundred times as slow.

Purification of Organic Substances

The substances which form the object of preparative work are usually solid crystalline materials or liquids—occasionally also gases. Because of the multiplicity of reactions in which organic substances can take part, and in pronounced contrast to most reactions of inorganic chemistry, it is rare for an organic reaction to proceed strictly in *one* direction only and to yield a *single* end-product. Almost always secondary reactions occur and greatly complicate the isolation of pure homogeneous substances from a reaction mixture; this isolation constitutes the chief aim of preparative exercises. Sometimes several well-defined chemical compounds are produced at the same time and must be separated; sometimes it is a question of separating the required substance with as little loss as possible from undesirable products which accompany it—the so-called resins or tars. These terms are used for by-products the origin and nature of which have usually not been investigated; sometimes they unfortunately become the main product. As regards classical organic chemistry, they have awakened interest only in so far as they are regarded as an unmitigated nuisance.

The substances to be prepared must be freed very carefully from all these undesirable admixtures. For this purpose two methods are in principle available :

1. **Crystallisation**
2. **Distillation**

1. Crystallisation

General Considerations.—Solid crystallisable substances are usually obtained at the end of a reaction in the form of a crude product which separates in more or less pure form from the solvent on cooling, either directly or after concentration. The rate at which organic substances crystallise varies within very wide limits, and their tendency to form supersaturated solutions is extraordinarily great. But even when supersaturation is counteracted by dropping a crystal into the solution—by "seeding"—the attainment of equilibrium in the cold saturated solution is often exceedingly slow. The cause is indeed the slow rate of crystallisation. Hence the full yield of crude product is often obtained only after the solution has been left for many hours.

The process of recrystallisation is most simply (and most frequently) carried out as follows: A hot saturated solution of the crude product in a suitable solvent is prepared, and from this solution the substance crystallises again in a purer condition. If the procedure is to succeed it is essential that the impurities should be more soluble than the substance itself, and should consequently remain dissolved in the cooled solution (*the mother liquor*).

The principle of differential solubility is also applied conversely, namely, when the by-product can be separated from the just-saturated solution of the substance because of its low solubility in an appropriate solvent. Since, in this case, the solution always remains saturated with respect to the by-product, it is never possible by this method to obtain a substance in *one* operation, as may be possible by the first method.

It is also important for recrystallisation from hot saturated solution that the temperature-solubility curve should rise as steeply as possible, *i.e.* that the dissolving power of the solvent should increase greatly with increasing temperature. In that case only can the amount of substance taken be recovered from the solution in the highest possible yield.

The choice of the right solvent is therefore of great importance for the process of recrystallisation. The most commonly used solvents are the following: *water, ethyl alcohol, methyl alcohol, ether, acetone, glacial acetic acid, ethyl acetate, benzene, petrol ether, chloroform, carbon bisulphide.*

For quite sparingly soluble substances, formic acid, pyridine, bromobenzene, nitrobenzene, and occasionally also phenol, ethyl benzoate, aniline, and dioxan are used. A distinct relation exists between the constitution of solute and solvent, and is expressed by the old rule: *similia similibus solvuntur.* Thus, as is well known, substances containing hydroxyl (*e.g.* sugars, carboxylic acids) are soluble in water, whereas hydrocarbons are more soluble in benzene and petrol ether than, for example, in alcohols.

The above statements, however, generally hold with some degree of certainty for simple organic compounds only. With complicated substances the conditions are more involved, and unless the worker has long experience he is obliged to test the available solvents *seriatim.* Alcohol is used most, and with this one usually begins; then perhaps water, benzene, and petrol ether. It may be said that, on the whole, of the more usual solvents, benzene, chloroform, and ether have a very great, petrol ether and water a moderate solvent power for organic substances. Although the validity of this rule is contravened by many substances, it nevertheless gives some indication for testing purposes. Thus if the sample is too sparingly soluble in alcohol a solvent from the first group is chosen; if it is too soluble, one from the second. In the case of sparingly soluble substances a higher boiling homologue of the same class is often chosen —in place of the lower alcohol, propyl or amyl alcohol, instead of benzene, toluene or xylene—because the higher boiling point brings about increased solvent power.

It very often happens that the preparation of a substance leads to an *amorphous* crude product, resinous or flocculent, which becomes crystalline on digestion with a suitable solvent or else by direct recrystallisation. It must be remembered that the solubilities of the amorphous and crystalline forms of the same substance are altogether different, and that the amorphous preparation is always much the more soluble.

Salts dissolve quite generally with ease in water, and often also in the alcohols, acetone, and chloroform, but they are not dissolved by ether, benzene, or petrol ether. Consequently organic acids can be extracted by aqueous solutions of alkali, and organic bases by aqueous solutions of acid, from a mixture of neutral substances in a solvent like ether.

When a substance has not the necessary moderate solubility in any solvent but is either too readily or too sparingly soluble, the combination of different solvents is a useful expedient. The solvents which are used together must be miscible with each other. The following are most frequently used :

Alcohol, glacial acetic acid, acetone with water.

Ether, acetone, benzene, chloroform with petrol ether.

Pyridine with water, ether, or alcohol.

The method of procedure is to add the solvent used as diluent drop by drop to the cold or hot concentrated solution until turbidity is just produced ; crystallisation is then induced by leaving the liquid to stand or by scratching with a sharp-edged glass rod. When crystallisation has begun the solution is cautiously diluted further. It is a mistake to precipitate the dissolved substance at one stroke with large amounts of the diluent.

In the case of all operations which are not yet under control, preliminary test tube experiments should be carried out. The student should acquire the habit of doing this from the very beginning.

Aqueous filtrates should be collected in beakers, but organic solvents in conical flasks, which prevent evaporation and so check the formation of crusts. Already in order to obtain some idea of the degree of purity from the appearance of the crystals, the crystallisation should be left to go on undisturbed so that crystals may separate in the best possible form. It is an error to assume that fine crystals produced by immediate strong cooling of a solution constitute an especially pure substance ; on the contrary, the large surface of the deposit favours the adsorption of by-products. Moreover, with well-formed crystals it is much easier for the organic chemist to meet the imperative requirement that he should check the homogeneity of a substance. The examination of the preparation with a lens or under the microscope should not be neglected ; 50- to 100-fold magnification is sufficient.

If a solution has become saturated at room temperature the yield of crystals can be increased by placing the vessel in ice-water or in a freezing mixture.

Substances of low melting point occasionally separate as oils when their hot saturated solutions are cooled. The solution must then be diluted somewhat. Moreover, in such cases provision is made for slow cooling by standing the flask containing the solution in a large beaker of water at the same temperature and leaving till cold. Of substances which crystallise with difficulty a small sample should always be retained for use as " seeding " crystals. Separation as an oil may then be obviated by dropping these crystals into the solution before it has become quite cold and rubbing with a glass rod.

Procedure.—In order to prepare a hot saturated solution the substance to be purified is covered, preferably in a short-necked, round-bottomed flask, with a little solvent which is then heated to boiling. More solvent is gradually added in portions until all the substance has dissolved. Since crude substances frequently contain insoluble impurities, the process of dissolution is carefully watched to see exactly if and when the compound to be recrystallised has completely dissolved. On account of the lability of many substances prolonged boiling is to be avoided. Solutions made with solvents which boil under 80° are prepared on the boiling water bath under reflux condenser; the solvent to be added may be poured into the flask through a funnel placed in the top of the condenser. It is better, however, at least when using large quantities, to fit a two-neck attachment (Anschütz tube, Fig. 32, p. 38) to the flask, since in this way it is possible to add the solvent conveniently and, in other cases, to drop in solid substances also. The condenser is fixed in an oblique position to the oblique tube of the attachment, whilst the vertical tube, through which substances are added, is closed with a cork.

Water and other solvents which boil above 80° are most suitably heated on an asbestos support in an air (Babo) oven or on an asbestos-wire gauze. If the boiling point lies considerably above that of water (more than 20°) the danger of cracking the condenser must be avoided by circulating warm water through it, or else the water condenser must be replaced by a long, wide glass tube (air condenser), which may be wrapped in moist filter paper if necessary. For test-tube experiments under reflux the so-called "*cold finger*" is exceedingly convenient. It consists of a glass tube about 15 cm. long and from 6 to 8 mm. wide sealed at one end. About 3 cm. from the other end a narrow tube 3 cm. long is attached at a right angle and bent downwards so that the apparatus can be hung on an iron ring. Cooling water is led away through this side tube to which thin rubber tubing is attached. The water is led into the "cold finger" through a bent glass tube which reaches to the bottom and is fixed in a small piece of rubber tubing which acts as a stopper. This handy condenser is fitted into the test tube by means of a notched cork.

The very troublesome " bumping " is avoided by adding porous pot, in pieces about half the size of a pea, before the boiling begins. When the pieces of pot become inactive they are replaced by new ones (do not drop them into superheated solutions !). When violent bumping occurs in large volumes of solution the addition of wooden rods is to be recommended.

In order to remove coloured impurities which often adhere tenaciously to a colourless substance, the hot saturated solution is boiled for a short time with a few knife-points of animal charcoal or specially prepared wood charcoal. Since the air which escapes from the charcoal causes copious frothing the adsorbent must be added carefully and with shaking. On account of their colloidal nature the coloured impurities are most easily adsorbed from aqueous solutions.

Filtration.—Solutions from which crystals are to be obtained are not completely clear, even in the absence of charcoal, and they must therefore be filtered. A filter paper in the ordinary conical form is generally to be preferred to the " folded " paper. The angle of glass funnels is usually not quite correct, and allowance can be made for this by making the second fold in such a way that the straight edges of the paper do not quite coincide and then using the larger cone for the filtration.

In preparative organic work readily permeable " grained " filter paper is alone of use.

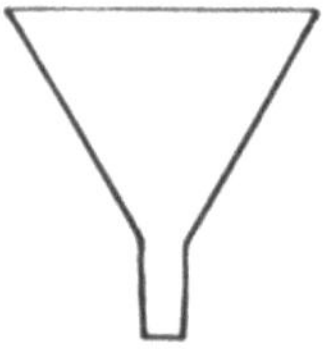

Fig. 2.

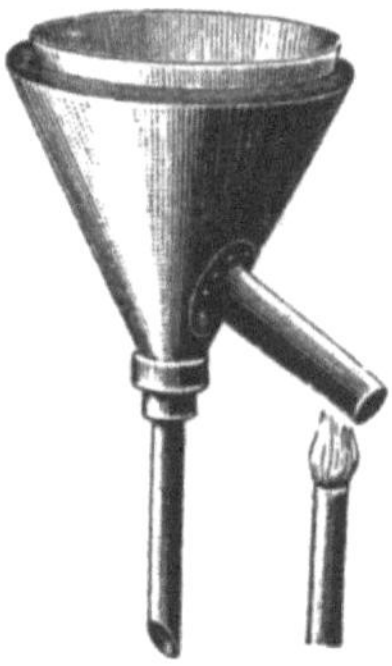

Fig. 3.

Fig 4.

Fig. 5.

The dissolved substance (especially if the solution is very concentrated) often crystallises in the funnel on account of local cooling and in this way filtration is hindered. This trouble can be partially met by using a funnel (Fig. 2) with the delivery tube cut short (0·5–1·0 cm.). But it is much more satisfactory to use a so-called hot water funnel (Fig. 3) in which the filtering surface of the funnel is heated with boiling water in a metal jacket. When inflammable solvents are used, the flame must be extinguished before filtration. The steam-heated funnel (as in Fig. 4) is likewise very useful. When small amounts of liquid are to be filtered, the empty glass funnel may be heated over a naked flame before use, or the paper may be fixed in the funnel, moistened with alcohol, ignited, and allowed to burn till it begins to char, while the funnel is held in a horizontal position and rotated.

It is often advisable, especially in the case of aqueous solutions which are difficult to filter, to use a porcelain Büchner funnel and apply suction. A well-fitting filter paper is required and the filter flask must be cautiously warmed before use, preferably by standing it in an enamelled pail of warm water and heating to boiling.

If the filter paper becomes choked by crystallisation of the substance, it is best not to push a hole through it. The paper should rather be held upright in a small beaker in which fresh solvent is kept boiling, and the more dilute solution thus obtained is poured through the same paper. In such cases the whole solution must generally be concentrated by evaporation.

If it is desired to produce well-developed crystals when re-crystallising, the filtrate, in which separation of crystals often occurs even during filtration, must be reheated till a clear solution is obtained and then allowed to cool slowly without being disturbed.

The **isolation of the crystals** is never accomplished by ordinary filtration, but always by collecting them at the pump on a filter paper, or, in the case of concentrated acids and alkalis, on glass wool, asbestos, or, best of all, on Schott filters of sintered glass. Large amounts of substance are collected on Büchner funnels (Fig. 5) of size appropriate to the quantity of the material to be separated. It is quite wrong to collect a few grammes of substance on a funnel six or more centimetres in diameter. For very large quantities the filter jar (Fig. 6, overleaf) is used. In many cases, especially for small amounts (5 g. or less), the Witt *filter plate* (Fig. 7) is to be preferred. It presents the advantage that its cleanliness can be checked much more readily than that of an opaque porcelain funnel, and, especially, that much less solvent is required to wash the more compact solid.

In order to prepare the filter paper a small piece of the paper is folded over the upper edge of the filter plate and then a piece having a radius 2–3 mm. greater is cut out with scissors. This piece is moistened with the solvent and fitted closely to the funnel by pressing, rubbing out small folds with a rounded glass rod or, in the case of larger plates, with the finger-nail.

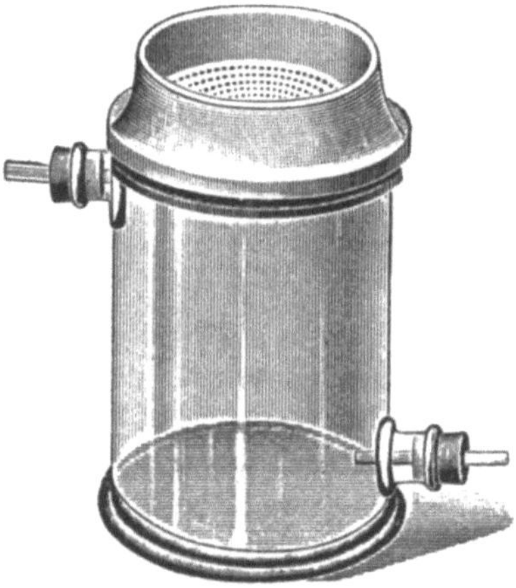

Fig. 6.

When minute amounts of substance (a few hundred milligrammes or less) have to be filtered, small glass plates 0·5–1·0 cm. in diameter are used as supports for the filter paper. These plates are made from thin glass rods by heating one end in the blow-pipe till soft and then pressing out flat on an iron plate (Diepolder). The glass rods must be long enough and thin enough to pass through the delivery tube of a quite small funnel and to project beyond its end. The pieces of filter paper which rest on the glass plates are cut somewhat larger than the plates themselves and are made to fit closely (Fig. 8).

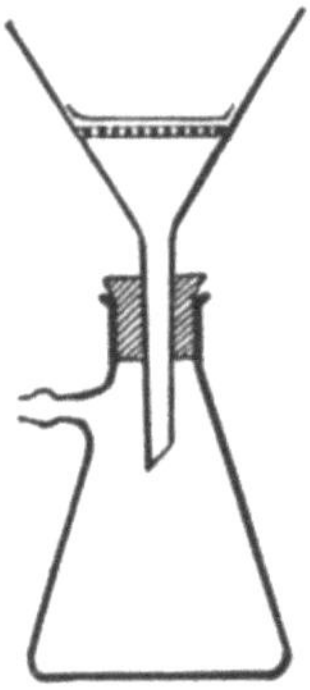

Fig. 7.

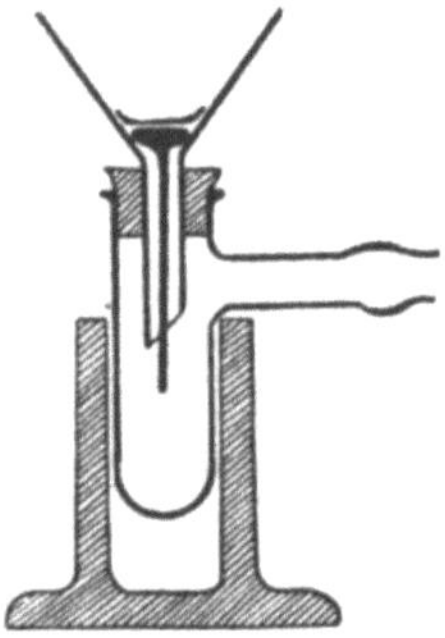

Fig. 8.

In order to remove the substance from the filter plate after filtration the funnel is inverted over a basin or watch-glass and all the material is transferred to the latter with the help of a thin glass rod or copper wire; the "glass button" is pushed out from its lower end. The plate is removed with forceps, but the paper not before it is dry. The material which adheres to the funnel is removed without loss by scraping with a small, obliquely cut piece of thin cardboard.

The filtrate is collected in a *filter flask* of a size appropriate to the volume of the solution. The very useful *filter tube* (Fig. 8) in its various sizes is also employed when filtering on a small scale. Such tubes stand in a lead support or in a wooden block bored with holes of various diameters.

In view of its great importance as a method for preparing analytically pure substances, the technique of filtration deserves the special attention of the student of practical organic chemistry.

The process of pouring the crystals along with the mother liquor on to porous plate and subsequently washing is emphatically to be rejected. Already in the preparation of organic substances the mind of the beginner should, above all, be directed to working as much as possible in quantitative fashion. It is not the number of preparations which indicates success, but the care and thoroughness with which each separate reaction is carried out.

For these reasons the " mother liquor " must not be treated as waste and neglected. Its importance will indeed only become clear to the research worker, but the beginner at preparative work should extract from it whatever is to be extracted for his purposes.

Filtrates are therefore reconverted into (cold) supersaturated solutions by evaporation of part of the solvent, and so a second crop of crystals is obtained. Occasionally yet another crop may be produced. As a rule the crops so prepared must be recrystallised once again from fresh solvent (check by melting-point determination!).

A few words should be added about the *washing* of crystalline precipitates with the object of freeing them from adherent mother liquor. The same solvent as was used for crystallisation must always be employed and, since its solvent power for the substance, even in the cold, leads to more or less appreciable loss, it must be used in the smallest possible amounts. Suction should not be applied while washing; the precipitate is saturated with the solvent and then the pump is turned on.

It is desirable to provide the Woulf bottle or filter flask, which should be connected to every water-jet pump, with a regulating cock which not only enables the suction to be cut off conveniently, but also allows of changes in the partial vacuum, which are necessary in many cases.

FIG. 9.

In the case of substances which are readily soluble even in the cold, the solvent used for washing must previously be cooled in a freezing mixture.

As long as mother liquor adheres to the crystals, although it no longer drains from them, no air should be drawn through the material if volatile solvents are used. Otherwise the impurities in the mother liquor are also deposited, and, especially in the case of easily soluble substances, there is no certainty that these impurities can again be completely removed by washing.

Small amounts of substances are washed with drops of the solvent. A so-called *dropping tube* (Fig. 9), *i.e.* a glass tube drawn out to a not too narrow capillary, is used. Such tubes are also very convenient for carrying out many reactions and they promote cleanliness in working.

The practice, which may often be observed, of "purifying" substances by evaporating their solutions to dryness in a crystallising basin, or of leaving them till the solvent has evaporated, naturally does not achieve its purpose because, of course, the impurities are not removed in this way.

Drying of the Substances.—A pure preparation must be completely freed from adherent solvent. Stable substances are most conveniently dried at room temperature by exposure to the air for one or two days between sheets of filter paper laid on a clean support. Substances of high melting point are more rapidly freed from solvent in a drying oven or on the water bath; some care is, however, always indicated.

The method which is most certain, and the sole applicable to analytically pure preparations, consists in drying in a *vacuum desiccator* containing sulphuric acid. The old type of Scheibler is probably the most suitable.

The consistency of the grease used to make the cover of the desiccator air-tight is very important; the most suitable grease is dry lanoline (*adeps lanae anhydricus*) or a mixture of equal parts of beef suet and vaseline. The tube carrying the stop-cock is moistened with glycerol and pushed through the rubber stopper previously fixed in the opening of the desiccator. Sharp edges on the tube must be rounded off and care taken that an air-tight closure is made. The support inside the desiccator consists of a porcelain plate having three short legs and several circular openings into which small basins, watch-glasses, and the like can be fitted. In order to prevent the support from sliding to and fro, three suitably cut pieces of cork are firmly fixed between the rim and the walls of the desiccator. When air is admitted to the desiccator there is a danger that substances which are being dried may be blown about. To prevent this a stiff strip of cardboard, or similar object, is fixed against the inner opening of the tube. In addition, before the cock is opened, a strip of filter paper is held before the outer opening. This paper is then sucked in against the tube and sufficiently moderates the current of air.

In order to dry the air which enters, a straight calcium chloride tube is attached to the tube of the desiccator. The calcium chloride must be firmly held in position at both ends by means of glass wool or, better, cotton wool. In desiccators which are often carried about, the container for the sulphuric acid is filled to the acid level with pieces of glass —small pieces of broken tube, stoppers, and the like—or with pieces of pumice which have previously been boiled with dilute hydrochloric acid and then dried. Splashing is thus avoided. From time to time the concentrated sulphuric acid is renewed. A special vacuum desiccator must be kept for analytical work.

For intensifying the drying effect, especially in respect of water, a small basin filled with solid technical potassium hydroxide is laid on the support. Most solvents, with the exception of chloroform, benzene, petrol ether, and carbon bisulphide, are absorbed by this combination. In order to free substances from these four solvents, thin slices of paraffin wax in a shallow basin are placed in the desiccator beside the substance, if its properties are such as to preclude drying in the air.

The rule should be adopted that any vacuum desiccator which does not fully keep its vacuum over night (test with a gauge) should be discarded. Thus it is sufficient to evacuate once and to leave over night. To continue suction at the pump for hours is to waste water.

Many substances contain water or other solvent so firmly bound that it cannot be removed in a vacuum at room temperature. These are dried in a vacuum at a higher temperature: they are heated in a small round flask on the water bath or oil bath until they cease to lose weight. The so-called "*drying pistol*" (Fig. 10) is especially convenient for this purpose. The vapours from the liquid boiling in A heat the wide inner tube B, in which the substance is exposed in a porcelain boat. C contains a drying agent—for water and alcohol, phosphorus pentoxide, for other vapours, paraffin wax. According to the temperature desired, the heating liquid chosen is *chloroform* (66°), *water* (100°), *toluene* (111°), *xylene* (140°). C is connected to the pump.

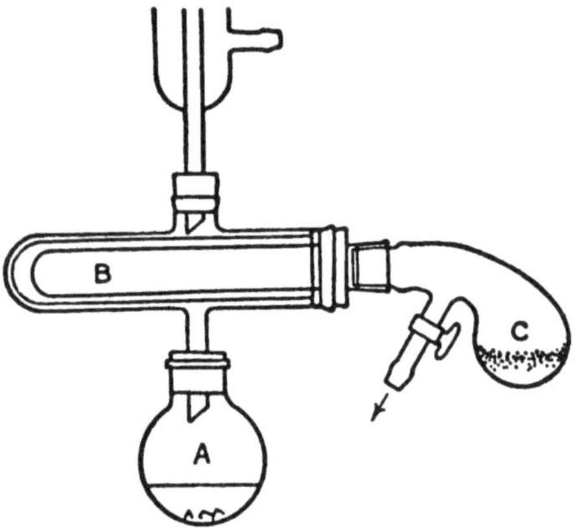

Fig. 10.

If a non-volatile solvent, such as glacial acetic acid, xylene, high-boiling petrol ether, or nitrobenzene, has been used for recrystallisation, it is washed out before drying by means of one which is more easily removed, *e.g.* ether, benzene, petrol. In general a substance which is sparingly soluble in glacial acetic acid or nitrobenzene is so also in ether.

2. Distillation

Purification by distillation consists in transferring the substance in the gaseous state to another place where it is again liquefied or solidified. Where this method of purification is used it is essential that the substance be stable at its boiling point. The latter can be lowered by distillation in a vacuum—at the pressure usually produced by the water-pump (12 mm.) the boiling point is on the average 100°–120° below that at atmospheric pressure. This difference is greater in the case of substances which boil above 250° at ordinary pressure. Very often, therefore, substances which do not distil unchanged at atmospheric pressure can be purified by distillation in a vacuum because they are thus exposed to a much lower temperature.

Simple substances, and in particular those which are also readily volatile, such as hydrocarbons, alcohols, esters, the lower acids and amines, are distilled under atmospheric pressure. All substances which decompose easily, and those which have especially high boiling points, are distilled under reduced pressure. In general solid substances should only be distilled when purification by crystallisation has been unsuccessful on account of too great solubility or for other reasons. Naturally in each case the possibility of distillation (without decomposition) must be established in advance.

Distillation, whether at atmospheric pressure or *in vacuo*, serves not only to separate the product from non-volatile impurities, but also to fractionate mixtures of volatile substances having different boiling points (fractional distillation).

Distillation at Atmospheric Pressure.—The simple distilling flask with side tube sloping downward (Fig. 11) serves exclusively as the distilling vessel. In general the side tube should be attached high in the case of low-boiling liquids and nearer the bulb in that of less volatile liquids.

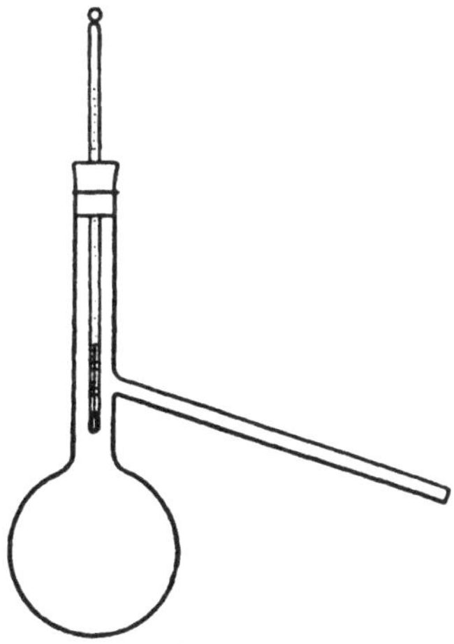

Fig. 11.

The thermometer is held in the flask by means of a clean bored cork; the bulb of the thermometer must be completely immersed in the vapour and hence must be below the junction of neck and side tube.

Since the ordinary laboratory thermometers are often inaccurate, they must be compared with a standard thermometer before use. The most accurate method of standardisation is to hang the two thermometers side by side with the bulbs dipping in concentrated sulphuric acid or paraffin at 250°, and then to observe the temperatures during cooling at intervals of 10° and record the deviations. Thermometers for distillations should have small bulbs so that they record rapidly.

Distilling flasks should be chosen of such a size that the bulb is half or two-thirds full of liquid. In order to avoid bumping and superheating a few pieces of porous plate (pot) half as large as peas are dropped into the flask before each distillation. If boiling is again delayed fresh pieces of pot must be added, not to the superheated liquid, however, but after brief cooling.

The flask is fixed above the side tube in a clamp lined with cork.

Sources of Heat.—Liquids which do not boil above 80° are heated in the water bath (enamelled jar or beaker); the temperature of the bath should be about 20° above the boiling point of the substance. The maintenance of the correct heating temperature is of the greatest importance, since if it is raised too much the boiling point of the distillate will be found too high in consequence of superheating.

In the case of substances of high boiling point where, for preparative purposes, a margin of a few degrees in the boiling point may be allowed, a naked smoky gas flame may generally be used, which is at first cautiously made to play round the flask. A conical air bath (Babo) or wire gauze may likewise be used. When the substance is valuable, when attention must be paid to analytical purity, or also when, on account of the degree of stability of the substance, superheating should be avoided, it is preferable to use an *oil* or *paraffin bath.* For temperatures greater than 220° a *bath* of Wood's or Rose's *metal* or a molten mixture of equal parts of potassium and sodium nitrates, both in an iron crucible, is to be preferred.

Substances of low boiling point are condensed in a Liebig condenser attached to the side tube of the flask by means of a cork. If it is desired to avoid all loss by volatilisation, the condenser is connected to the filter flask serving as receiver by means of a so-called adapter and the receiver is cooled in ice, or else in a freezing mixture.

For liquids which boil at about 100° a shorter condenser suffices, and when small amounts are being distilled it is especially advisable to use a small cooling jacket slipped firmly over the side tube so that loss of material is minimised. Such a device is illustrated in Figs. 19 and 25.

In the case of boiling points above 120° cooling with running water is generally not practicable, because the condenser tube, being in contact with the hot vapour, may easily crack; instead, the jacket contains as cooling agent standing water which gradually grows warm. When the boiling point exceeds 150° simple air cooling (wide condenser tube without jacket) suffices.

Substances which solidify rapidly after condensation should never be distilled from a flask having a narrow side tube; the distillate in the tube can indeed be reliquefied by warming with the flame, but often it is hardly possible to clear away the material which blocks parts covered by corks or other connections, so that time is lost and annoyance caused.

The *sausage flask* (Fig. 12) is therefore at once chosen. It has a wide side tube from which the product can be removed without trouble when distillation is complete, preferably by melting.

Normally distillation is carried out as follows. After the contents of the flask have been gradually heated, the visible signs of boiling appear, the mercury column of the thermometer rises rapidly and becomes steady at a definite temperature—the boiling point. When this temperature has been attained to within one degree the receiver (a small wide tube or similar vessel) for the first runnings is replaced by one suited to the amount of distillate expected (a conical flask or a narrow-necked bottle in which a small funnel is placed), and heating is continued at such a rate that one drop distils every second or every other second. The thermometer must be watched all the time. In general the boiling point range should not exceed 1°–2°; in the case of analytically pure preparations the limits should be even closer. If a naked flame is used for distillation the boiling point rises a few degrees as a rule towards the end of the process on account of superheating, although pure substance is still passing over. If the boiling point rises even earlier, beyond the limit given, the receiver must be changed again and the distillation continued so that a third fraction, the "last runnings", is collected.

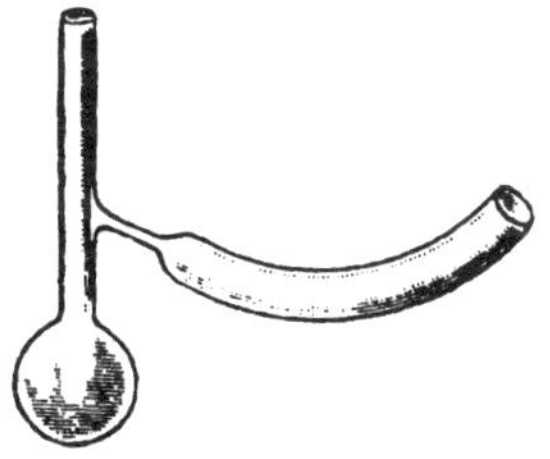

FIG. 12.

It should be borne in mind that both in the first and last runnings there is some of the main product. The vapour pressure of a distillable substance is so considerable even below the boiling point that its vapours already pass over along with the more volatile constituents (usually residues of solvent) of the original material. On the other hand the boiling point of a substance rises when it is mixed with higher boiling substances.

Thus ether, which is very extensively used to take up organic preparations, is not completely removed from a much less volatile substance even on the boiling water bath, although the boiling point of this solvent is 35°. The benzene wash of coke ovens is another well-known example which cannot be discussed in detail here.

Hence it is evident that the last runnings also are not free from the main product, and when first and last runnings form an important amount it is worth while to redistil these two portions separately according to the rules given.

Fractional Distillation.—When several volatile reaction products are to be separated from one another the procedure is not so simple as described above. In proportion as the boiling points of the various constituents approach each other the separation becomes more difficult, and it is not easy with the help of the usual laboratory apparatus to separate with any degree of precision substances which differ in boiling point by 10°.

The nearest approach to the desired end is attained by repeating the process of distillation. In the case of low-boiling substances it can be done in *one* operation with the help of so-called *fractionating columns*, which are devices introduced into the gaseous phase before final condensation occurs. In the several divisions of these columns, which can be constructed in various forms (Fig. 13), vapours are liquefied by air cooling, and the vapours which are formed later must pass through this liquid which lies in their path. In this way the less volatile constituents of the vapour are condensed, while the more volatile are exposed to the same treatment in the next division. Thus a number of separate distillations corresponding to the number of bulbs in the column occurs, and if the operation is carried out carefully and slowly a far-reaching separation is made possible. Cylindrical columns irregularly filled with glass Raschig rings are especially suitable.

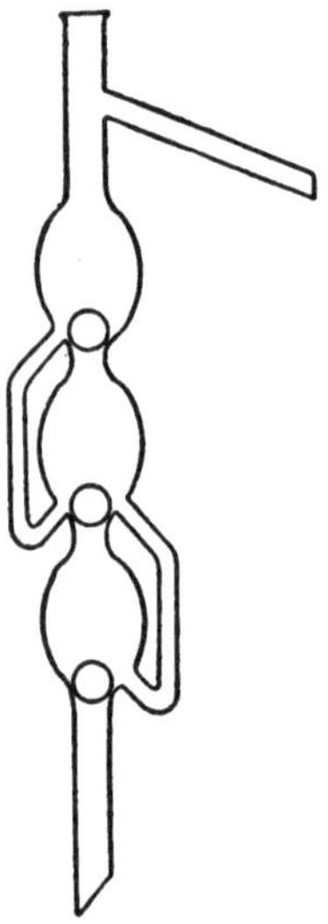

Fig. 13.

If only small amounts of substance are available, glass spirals[1] inserted into the lumen of the distilling flask yield quite excellent results.

Technically the principle of fractional distillation is applied in the manufacture of spirit and in the isolation of aromatic hydrocarbons from the light oil of coal-tar.

Mixtures of liquids of higher boiling point (above 120°) are first separated by distillation into several fractions of about the same boiling-point interval; the separate distillates (in smaller flasks) are again divided by distillation into fractions, and the fractions of adjacent boiling points are then fractionally redistilled, while the boiling-point ranges are continually reduced. If, as is very advisable, the above-mentioned Widmer spiral is also used here, the column in which it is placed must be well lagged with asbestos.

[1] Widmer, *Helv. Chim. Acta*, **1924**, **7**, 59.

Not all mixtures are separable by distillation; occasionally substances which boil at different temperatures form constant boiling mixtures.

Detailed information about the theory of fractional distillation will be found in Nernst's *Theoretical Chemistry*, 1923, p. 118.

Vacuum Distillation.—The organic chemist must continually bear in mind that almost all substances with which he has to deal are, from the thermodynamical standpoint, metastable. In all cases, however, increased temperature favours the setting up of the true equilibrium —here decomposition—and hence it is appropriate to adopt the rule: Never endanger substances unnecessarily.

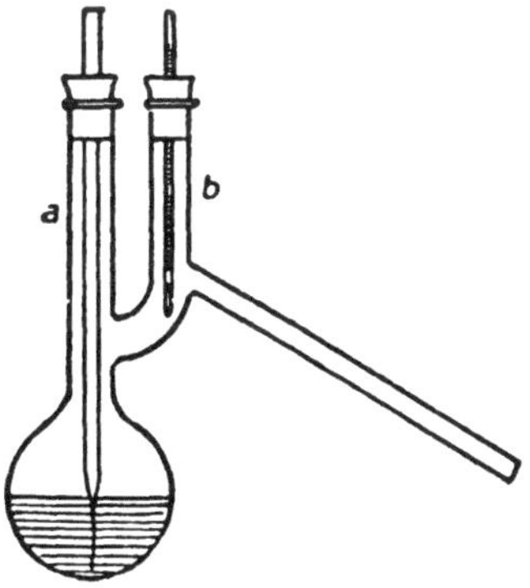

FIG. 14.

Hence distillation under reduced pressure, whereby the boiling point can be reduced by one hundred and more degrees, acquires great importance in organic work. The organic chemist who is doing preparative work must quickly learn how to apply the method, and should early accustom himself to regard vacuum distillation not as in any way extraordinary, but as one of the most elementary operations of laboratory practice.

The appropriate vessel for the distillation is the *Claisen flask* (Fig. 14). The very practical division of the neck into branches minimises the spurting of the boiling liquid, which is especially dangerous in this process. In order that the bumping, which very easily occurs during vacuum distillation, may be avoided, minute bubbles of air, or in the case of substances sensitive to the action of air, bubbles of hydrogen or carbon dioxide, are drawn continuously through the boiling liquid by means of a fine capillary tube.

This tube is drawn out in the flame of a blow-pipe from glass tubing 4–8 mm. wide and is then drawn out again to sufficient fineness in a micro-flame. To make sure before use that the capillary is not closed, the tip is submerged in ether in a small test tube and air is blown in from the mouth. The bubbles should emerge separately and slowly. Capillaries for distillation in a high vacuum should emit air bubbles only on powerful blowing, and then with difficulty.

Occasionally, and especially when liquids which foam are being distilled, it is necessary to regulate the amount of air passing through the capillary. To do this the capillary must not be too fine, and a small piece of thick-walled rubber tubing is fixed to the upper end of the tube and a screw clip is so attached that its jaws grip the rubber immediately above the glass. It must be borne in mind, however, that if the distillation is interrupted the liquid in the flask will be driven back by the external air pressure into the still evacuated capillary tube, and sometimes into the rubber tubing. This is avoided by cautiously unscrewing the clip before the interruption.

When obstinate foaming occurs, the thermometer is discarded in favour of a second capillary tube in the front neck of the Claisen flask (*b* in Fig. 14). The fine current of air drawn in through this tube bursts the bubbles of foam before they can pass over.[1]

The capillary tube is inserted (with a little glycerol as lubricant), tip first, into a narrow-bored, undamaged rubber stopper which fits tightly into the neck *a* of the Claisen flask. The correct position of the tip is immediately above the deepest part of the bulb of the flask. A thermometer, likewise pushed through a rubber stopper, is inserted into the neck *b*. If it is desired to prevent contact of the substance with rubber, Claisen flasks with

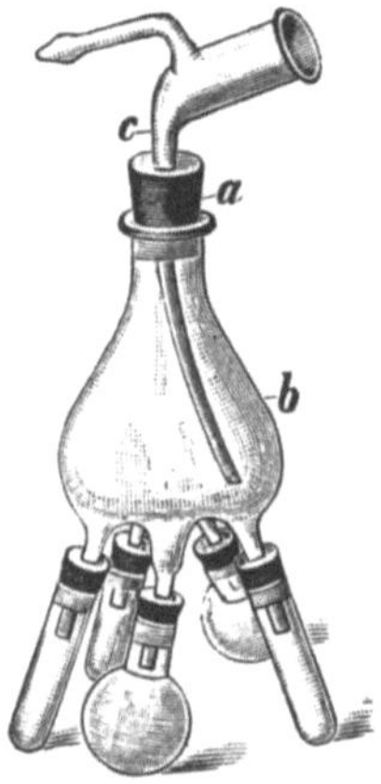

FIG. 15.

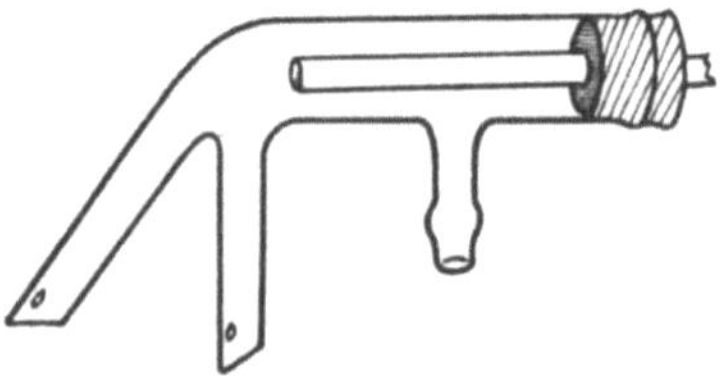

FIG. 16.

constricted necks are used. The capillary tube and thermometer are held in position in these necks by means of small pieces of rubber tubing drawn over each neck and its capillary or thermometer. The proper use of cork stoppers in vacuum distillations requires much practice.

[1] E. Dorrer, *Dissertation*, Munich, 1926.

The vapours are cooled in the way described ; here the use of the small water-cooling jacket drawn over the side tube is especially to be recommended.

Receivers.—When only one or two fractions are expected, small filter tubes of suitable size, as illustrated in Fig. 8, are used (the smallest for the first runnings), or, in the case of larger amounts of substance, small filter flasks. The rubber stoppers for attaching them should be tested in advance as to fit. When the receiver is being changed, the distillation must, naturally, be interrupted.

If this is to be avoided and several fractions are expected, it is preferable to use the so-called *cow's udder*, an apparatus which is illustrated in Fig. 15. By holding the glass tube *c* with one hand and turning the vessel *b* with the other the various receivers can successively be brought under the mouth of the delivery tube. (The tube *c* is lubricated with glycerol.) The form shown in Fig. 16 is handier and more suitable for the distillation of small amounts. According to its shape it is known in laboratory slang as a " spider ", " frog ", or " pigling ".

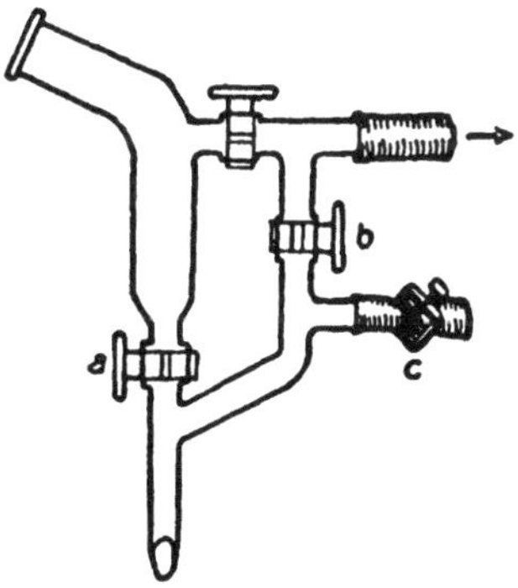

FIG. 17.

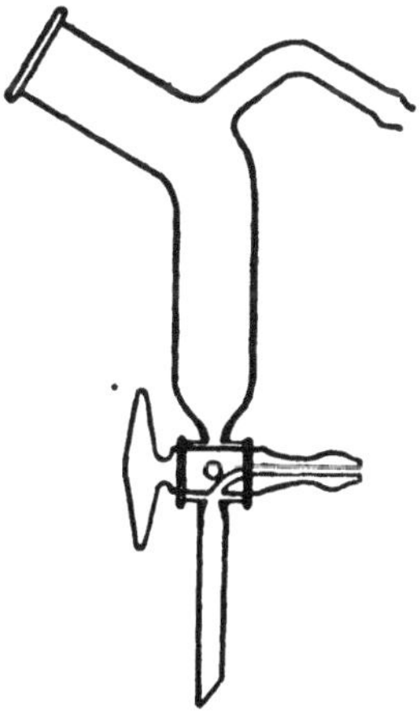

FIG. 18.

Finally, Anschütz and Thiele's adapter with stop-cock (Fig. 17) may be mentioned; it is very useful, in particular in the distillation of large amounts of substance. After the cocks *a* and *b* of this adapter have been closed the pressure in the receiver can be allowed to rise by unscrewing the clip *c* and the receiver can be changed.

Then after *c* has again been closed and a vacuum has been reestablished throughout by opening *b*, the cock *a* may also be opened and distillation continued. The third cock is not required. Of still simpler construction is the adapter (Fig. 18) with a three-way cock. By means of a boring in the cock the receiver can be opened to the atmosphere while the vacuum in the apparatus is maintained. After the receiver has been changed the cock must, however, be turned very cautiously, so that the distillate which has collected above is not blown out by air drawn in from below.

The two pieces of apparatus just described have the great advantage that the various fractions are at once completely isolated and do not come into contact with the vapours; on the other hand, they cannot be used for thick viscous liquids which do not pass through the boring of the cock.

They are therefore of special use for the distillation of relatively low-boiling substances the vapour pressure of which cannot be neglected.

When substances which solidify rapidly are distilled in a vacuum, the side tube of the Claisen flask should be wide, as described for ordinary distillation.

Heating.—Only after much practice can a vacuum distillation be carried out with a naked flame. Indirect heating in a bath is much more reliable. Here also the temperature of the bath should be adjusted with the greatest care to suit the boiling point of the substance (about 20° higher; when the side tube is attached high up, the difference must be increased); when the boiling point of a fraction has been reached, the temperature of the bath should be kept constant.

The flask is immersed in the bath to such a depth that the surface of the material to be distilled lies below that of the heating liquid. The bulb of the flask should not be more than half filled with substance.

When high-boiling substances are distilled, the flask is immersed as deeply as possible and the portion above the bath to the junction with the side tube is wrapped in asbestos paper held in position by a thin wire or by a string.

Sensitive substances which, as such, can be distilled in a vacuum, occasionally decompose when subjected abruptly while hot to large changes of pressure. In such cases the pressure is allowed to rise only after the contents of the flask have cooled. To proceed in this way is quite generally appropriate because thereby the very common oxidising action of hot air is also avoided.

In all distillations under reduced pressure a small gauge (Fig. 19), introduced between pump and apparatus, is indispensable, since the pressure must be controlled throughout, in view of its effect on the boiling point. Inconstant boiling points are very often due to variations in pressure. In order to hold back vapours which would contaminate the gauge by condensing in it, the cock is kept closed during the distillation, and only opened from time to time to test the pressure.

Before every vacuum distillation the whole apparatus must be tested for tightness with the gauge, i.e. *it must be shown to keep up a satisfactory vacuum.*

The heating of the bath is only begun after the vacuum has been produced. If the pressure over the *already warmed* liquid is reduced,

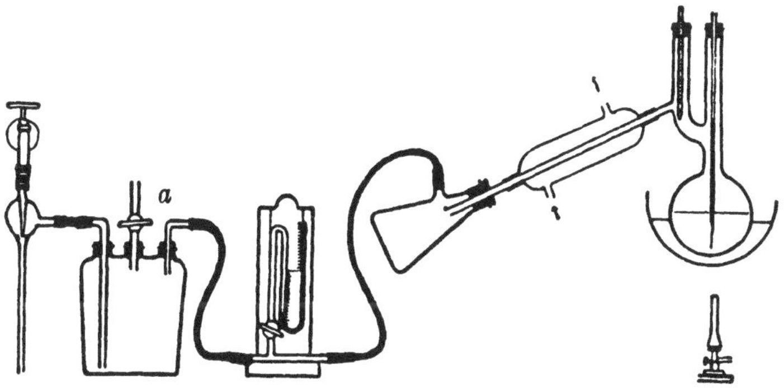

FIG. 19.

it often froths over, owing to superheating. This may happen before the boiling point of the substance is reached: it is enough if the material still contains a little solvent, *e.g.* ether, the removal of which on the water bath is never quite complete because of the greatly reduced vapour pressure.

In many cases when readily volatile low-boiling substances are distilled in a vacuum it is necessary to reduce the volatility by raising the pressure. The full vacuum of the water pump, which, depending on the pressure and temperature of the tap water, amounts to 10-20 mm. of mercury, is then not used, but, instead, pressures of 20–100 mm. Since the action of the pump cannot be regulated, a cock attached to the Woulf bottle (*a*, Fig. 19) is used, so that with the help of the gauge any desired pressure can be obtained. For substances which boil above 150° under atmospheric pressure the maximum effect of the water pump is employed.

The extent to which the reduction of pressure in a vacuum distillation lowers the boiling point can be seen in Fig. 20, in which *nitrobenzene*, boiling point 208°/760 mm. (curve I), and *benzaldehyde* (II), boiling point 179°/760 mm., are the examples chosen. The importance of a "good vacuum" in preparative work is expressed in the steep slope of the curves in the low-pressure region. When the distillation is carried out at 20 mm. the boiling point is about 15° higher than when the pressure is 10 mm. As the pressure rises this effect is reduced, as the curve III in the upper part of the figure makes clear. This curve, which is on a different scale, represents the region of pressure from 760 mm. downwards. At Munich a reduction of the sea-level barometer to 720 mm. only lowers the boiling point of water by 1·5°.

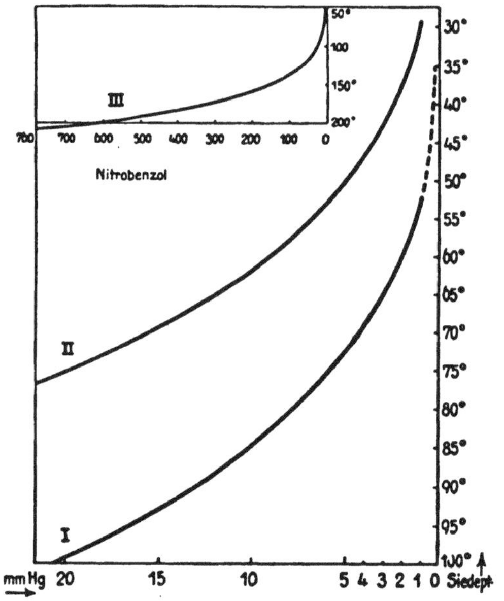

FIG. 20.

The quantitative relations between pressure and boiling point vary from substance to substance, but among organic compounds the variation is not very great, so that the curves here reproduced can in practice very well be used as a guide.

For example, if according to the literature a substance boils at 96°/12 mm., it will boil at 104–105° at 18 mm.

Substances which boil at too high a temperature, even in the vacuum of the water pump, can often be distilled without decomposition in a *high vacuum, i.e.* at pressures of 1 mm. or less. Reduction of pressure to this limit reduces the boiling point on the average by 150° as compared with the boiling point at atmospheric pressure, and by about 40° as compared with that in the vacuum of the water pump. The dotted continuation of the nitrobenzene curve I illustrates this—it is not based on actual measurements.

Since the introduction of the so-called mercury vapour-jet pumps, which are to-day available in almost every university laboratory, distillation in a high vacuum presents no difficulty, and whoever has mastered ordinary vacuum distillation will also be able to distil in a high vacuum when the occasion arises, possibly when working according to directions from original literature. Because the apparatus is easily damaged, at least in ordinary usage, it has not been applied to the practical exercises in this book, and is not further described. The excellent mercury vapour-jet pump of *Volmer* should be available to-day in every organic teaching laboratory.

Always protect the eyes during a vacuum distillation !

SUBLIMATION

Volatile substances of which the vapours, on cooling, condense directly to crystals without passing through the liquid phase are sometimes advantageously purified by sublimation, particularly when solubility relations render recrystallisation difficult. The purification of iodine is a well-known case in point. In organic chemistry this process is particularly suitable for quinones.

Small amounts of substance are conveniently sublimed between two watch-glasses of equal size. The substance is placed on the lower glass and covered with a round filter paper, having several perforations in the centre, and somewhat larger than the glass. The second watch-glass is laid with its convex side upwards on the lower glass, and the two are fixed together with a watch-glass clip. When now the lower glass is heated as slowly as possible on the sand bath with a small flame the vapour from the substance condenses in crystals on the cold upper glass; and the filter paper prevents the small crystals from falling back into the hot lower glass. The upper watch-glass can be kept cool by covering it with several layers of moist filter paper or with a small piece of wet cloth.

FIG. 21.

If it is desired to sublime larger amounts of substance, the upper watch-glass of the apparatus just described is replaced by a funnel, of which the diameter is somewhat less than that of the watch-glass (Fig. 21).

Sublimation can also be carried out in crucibles, flasks, beakers, retorts, tubes, etc. If the substance to be purified only sublimes at a high temperature, as is the case, for example, with indigo or alizarin, a vacuum is applied (small round-bottomed flask or retort). When carrying out sublimations the student should always make sure that the apparatus has cooled completely before taking it apart.

Distillation with Steam

This important method of purification is very extensively used, both in the laboratory and on a large scale in chemical industry. It is based on the fact that many substances, the boiling points of which may lie considerably above that of water, are volatilised by injected steam to an extent proportional to their vapour pressure at that temperature, and are then condensed along with the accompanying steam in a condenser. The most suitable and theoretically simplest case (see below) arises when the substance is sparingly soluble or practically insoluble in water.

To test for volatility in steam, heat a small sample of the substance to boiling (porous pot!) in a test tube with about 2 c.c. of water and hold the bottom of a second test tube containing some ice in the issuing vapours until a drop of water has formed on the cold surface. A turbidity in the drop indicates that the substance is volatile with steam.

The substance to be distilled is placed along with a little water in a capacious, long-necked, round-bottomed flask which should not be more than one-third filled, and is heated over a burner until near the boiling point (in order to avoid great increase in volume due to condensation of water). Then after the water has been turned on in the *long* condenser and the receiver has been placed in position, a rather vigorous current of steam is passed in. The wide steam-delivery tube should reach nearly to the bottom of the flask and should be bent slightly, as shown in Fig. 22. If steam is not laid on in the laboratory it is produced in a tin can rather more than half filled and provided with a long vertical safety tube. As a rule distillation is continued until the distillate passes over clear. If the substance crystallises in the condenser the jacket is partly emptied for a short time and the steam then melts the crystals, so that the substance flows away. Care must be taken, however, when this procedure is adopted, that there is no loss of uncondensed vapour carried away with the steam. The readmission of the cooling water to the hot tube must be done cautiously.

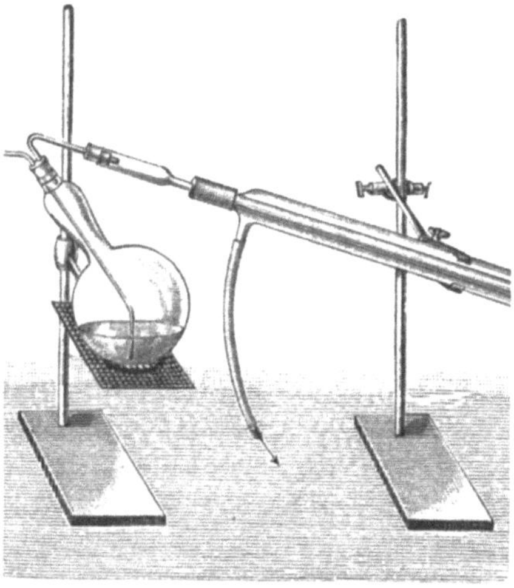

FIG. 22.

When distillation is complete, the connection between the steam tube and the flask is broken before shutting off the steam, because otherwise the residue in the flask might be sucked back through the inlet tube. Special attention must be paid to this point when the steam is taken from the laboratory supply.

Smaller amounts of substance can also be steam distilled from a distilling flask of sufficient size having the side tube attached high up on the neck, and specially volatile substances can also be distilled by simple heating with water, without a current of steam.

Substances which volatilise only with very great difficulty are distilled with *superheated* steam. The superheating takes place in a copper tube (Fig. 23) wound in a conical spiral, interposed between the steam generator and the flask and heated from below with a burner. The flask with the substance stands in an oil bath heated to a high temperature (about 150°).

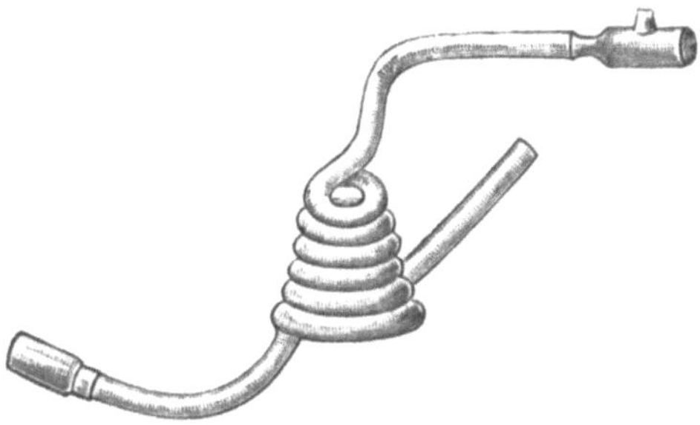

FIG. 23.

Sometimes the distillation can be performed without the superheater if the steam, dried as completely as possible, is passed, not too rapidly, into the distilling flask containing the dry substance, which is kept hot in the bath. Volatile substances, which tend to decompose, are occasionally distilled with steam under reduced pressure and consequently at reduced temperature.

On the Theory of Steam Distillation.—The ideal case occurs when the substance to be distilled is insoluble, or more accurately, sparingly soluble in water (examples: toluene, bromobenzene, nitrobenzene) so that the vapour pressures of water and the substance do not affect each other or hardly so. The case of substances which are miscible with water (alcohol, acetic acid) is quite different and involves the more complicated theory of fractional distillation. Let us consider the first case only and take as our example bromobenzene, which boils at 155°. If we warm this liquid with water, its vapour pressure will rise in the manner shown by its own vapour pressure curve and independently of that of water. Ebullition will begin when the sum of the vapour pressures of the two substances has become equal to the prevailing atmospheric pressure. This is the case, as we can find from the vapour pressure curves, at 95·25° under a pressure of 760 mm.

At this temperature the vapour pressure of bromobenzene amounts to 121 mm., that of water to 639 mm., and their sum consequently to 760 mm.

According to Avogadro's rule, therefore, the vapour phase will contain the two components in the molecular ratio 121 : 639, *i.e.* there will be 5·28 times as many water molecules in the mixed vapour as molecules of bromobenzene. The absolute ratio in which bromobenzene passes over with steam is simply determined by taking the molecular weights into consideration. To 1 mole of bromobenzene of molecular weight 157 there are 5·28 moles of water of molecular weight 18, or, with 157 parts by weight of the first there distil $5{\cdot}28 \times 18 = 95$ parts by weight of the second. This corresponds to a ratio bromobenzene : water of about 5:3.

Accordingly, if the vapour pressure curve of a substance not miscible with water is known, it is easy to calculate approximately its degree of volatility with steam. The calculation is approximate only because the condition of mutual insolubility is hardly ever fulfilled.

Evaporation of Solvents

Since, during preparative organic work, substances have very often to be isolated from dilute solution, this operation is one of the commonest. Ether is distilled from the steam or water bath through a downward condenser (coil condenser Fig. 24) and is used again, possibly after purification. If it contains volatile acids it is shaken with dilute sodium carbonate solution; if volatile bases are present it is shaken with dilute sulphuric acid.

In order to avoid loss and conflagration due to the volatility of the ether, a filter flask, connected to the condenser with a cork, is used as receiver, and a length of rubber tubing hanging down below the bench is attached to the side tube of the receiver for safety.

Naked flames must never be allowed on the bench when work with ether or any inflammable solvent is in progress.

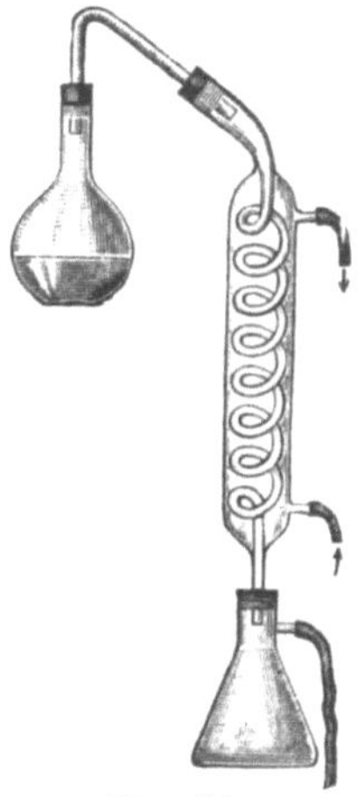

FIG. 24.

If large amounts of solvent have to be evaporated, and if it is also desired to distil the dissolved substance after removal of the solvent, the solution is admitted from a dropping funnel, in portions, to a suitable distilling flask at the rate at which the solvent distils (porous pot). In this way the use of too large a vessel is avoided. If a steam bath is not available and a water bath has to be used, the flame must be extinguished each time the vessel is refilled (funnel!). In this case the process is usually more rapidly carried out by distilling the whole of the solvent from a large round-bottomed or conical flask and then washing the residue (completely!) with a little solvent into the smaller vessel.

Small amounts of easily evaporated liquids can be removed directly from a test tube or a small flask heated on the water bath. On each occasion the test tube is filled to a depth of 2–3 cm. only, and is replenished from time to time; during the boiling on the water bath it must be shaken continuously. All preliminary tests with solutions are carried out in this simple way previous to an examination of the residues. For the latter purpose solutions of substances liable to decompose are left exposed to the air in watch-glasses or small crystallising basins so that the solvent may evaporate.

When it is necessary to remove *completely* such solvents as alcohol or benzene it is not sufficient to use the steam or water bath alone, because the boiling point rises higher and higher as the concentration of the solution increases; even ether causes difficulties.

In this case, when the solvent ceases to distil, an oil bath or, more frequently, a vacuum is used. It is sufficient to mount a capillary tube, place the flask in a deep porcelain basin or in an enamelled basin maintained at a moderate temperature, and connect directly to the pump; this removes most solvents, water included, rapidly and completely.

Thin-walled glass apparatus such as conical flasks, flat-bottomed flasks, and test tubes should never be evacuated.

Round-bottomed flasks should always be employed, or, in certain cases, filter flasks, but the latter should only be warmed cautiously. If, as so often happens in the case of sensitive substances, large amounts of solvent have to be removed by distillation in a vacuum, condensation is accelerated by the use of a fairly large condenser and, where necessary, by cooling the receiver with ice.

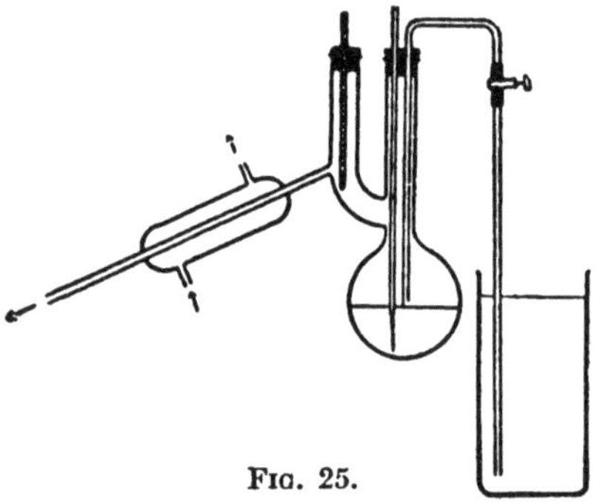

Fig. 25.

The condenser can be dispensed with if a simple distilling flask supported on a large funnel having a wide rubber tube attached to the stem is used as receiver. Water from the tap is kept pouring on the flask from above. The end of the side tube of the flask containing the material to be distilled must reach to the middle of the bulb of the receiving flask. This arrangement is especially suitable for concentrating aqueous solutions.

The arrangement illustrated in Fig. 25 permits of the continuous distillation in a vacuum of large amounts of liquid, in particular water. By operating the cock from time to time the liquid distilled can be replaced by suction from the storage vessel.

The persistent foaming of aqueous solutions during distillation frequently causes annoyance and loss of time. By adding to the solution about 3 per cent of its volume of isoamyl alcohol the trouble can be overcome. The aim is attained more certainly by drawing the solution into the empty flask, fixed ready for distillation, at the same rate as that at which the water distils. If this procedure is adopted it is advisable to draw out the end of the delivery tube to a narrow bore and to adjust the rate of flow from the storage vessel exactly by means of a screw clip (Fig. 25).

Extraction

In order to extract from water a dissolved reaction product, or one which is not sufficiently solid or crystalline to be removed by filtration, or else to separate a substance from insoluble material which accompanies it, it is taken up in a suitable solvent. *Ether* is most commonly used for this purpose. Thus, for example, the distillate from a steam distillation is treated in this way unless it spontaneously separates into two layers.

Fig. 26.

For the above purpose separating funnels are used. For small volumes (25 c.c. and less) funnels of the type shown in Fig. 26 are employed. The stems of these are at most 5 cm. long and are cut obliquely so as to facilitate delivery. A filter funnel is always used when pouring liquids into the separator. After the layers have separated, the lower is run off through the cock and the upper through the upper opening. The cock is not opened until the heavier liquid has settled to the bottom, and, especially when ether is the liquid used for extraction, the pouring away of part of the aqueous solution along with the ether is to be avoided. Small preliminary samples are separated after being withdrawn from the main solution by means of a dropping tube (Fig. 9).

Occasionally when aqueous solutions, and particularly suspensions, are extracted with an organic solvent very unpleasant emulsions make a clean separation impossible. The most effective way of avoiding this difficulty is to mix the liquids carefully. Further remedies are : creation of a vacuum in the separating funnel, addition of a few drops of alcohol, saturation of the aqueous phase with common salt, . . . and patience.

If a substance is soluble in both water and the organic solvent, the result of the extraction depends on the ratio of solubilities ; if the "partition coefficient", *e.g.* the ratio of the solubility in water to that in ether, is large, correspondingly more ether must be used or the number of extractions must be increased. For this coefficient determines how a substance soluble in two immiscible solvents will distribute itself between them. Whether an aqueous solution should be extracted with a certain amount of ether in one portion or whether it is better to extract several times with smaller portions is determined by the following simple consideration. Let us assume that the distribution coefficient is equal to 1, and that we have one volume of water and two of ether, which in the one case are used in one lot and in the other in two equal portions. Suppose that the amount of substance in solution is a grammes. Then in the first case the amount which passes into the ether is $\frac{2a}{3}$, whilst in the second the first half of the total volume of ether takes up $\frac{a}{2}$, and the second once again takes half of the remaining $\frac{a}{2}$ g., *i.e.* $\frac{a}{4}$, making a total of $\frac{3a}{4}$ g. In order to extract this amount from water in *one* operation three volumes of ether would be necessary. This means that two lots of one litre used separately perform the same service as three litres used in one lot. The practical conclusion is obvious.

The coefficient of partition of organic substances between water and lipins (these are fat-like constituents of the cell wall) is of great importance in biological processes (H. H. Meyer and Overton's theory of narcosis).

For the extraction from water of a dissolved substance, ethyl acetate, chloroform, benzene, amyl alcohol are also used occasionally instead of ether. Since water dissolves about 10 per cent of its volume of ether, unnecessary dilution is to be avoided, already for the sake of economy.

The Drying of Solutions.—After a substance has been extracted from aqueous solution or suspension with an organic solvent its solution is saturated with water and must therefore be dried; if drying were omitted, the dissolved water would mostly remain behind with the substance to be isolated, after the solvent had been evaporated. In choosing a drying agent it should be borne in mind that it must not react with the solvent nor with the solute, and must be completely insoluble in the former. The ethereal solution of a base may well be dried with solid potassium hydroxide, but, of course, not that of an acid. The most active and most commonly used drying agent is *calcium chloride*, which is used in either granulated or (previously) fused form; ethereal solutions are almost exclusively dried with it, except when they contain substances, such as alcohols and amines, which yield addition compounds with the salt. Ethereal solutions containing alcohol should, therefore, never be dried with calcium chloride; the alcohol must first be removed by repeated extraction with water. As a rule much too much drying agent is used. For ordinary purposes it is sufficient to use so much calcium chloride that after some time there remains, along with saturated calcium chloride solution, about an equal amount of the solid salt.

Anhydrous sodium sulphate, even when freshly ignited before use, is much less active than calcium chloride. It is used when a substitute for calcium chloride is indicated for the reasons given above. For solutions of basic substances ignited potassium carbonate, solid potassium hydroxide, and barium oxide are much used as drying agents.

In order to free the usual solvents completely from water the following agents are used:

For ether, benzene, and its homologues and petrol ether: sodium.

For acetone, chloroform, ethyl acetate, carbon bisulphide: calcium chloride.

The alcohols are freed from water by boiling under reflux for several hours with freshly ignited quick-lime and are then distilled.

Solvents such as chloroform and carbon tetrachloride, which contain chlorine, should on no account be dried with sodium, because of the danger of explosion.

Extraction Apparatus.—Very often an organic substance is so much more soluble in water than in ether and other solvents that shaking with a solvent even when repeated is ineffective. In such cases an apparatus for continuous extraction of solutions is used; it should not be wanting in any organic laboratory. The apparatus of Schacherl (Fig. 27 *a*), which can be constructed from simple laboratory materials, indicates the principle involved. Still more convenient is the apparatus shown in Fig. 27 *b*. It likewise can be constructed in all dimensions from easily obtainable materials.

This leads to a consideration of extraction apparatus for solid substances. The best known is that of *Soxhlet,* which is much used, especially for analytical purposes. For preparative work the simplified extractor (Fig. 28) is to be preferred, for it is cheaper and works more quickly. In order to prevent the formation of channels by the drops of solvent in the material to be extracted, a thin porcelain sieve-plate (filter plate) is laid on top.

The extraction apparatus is used chiefly for dissolving out sparingly soluble constituents from mixtures, and for isolating natural products from (dry) vegetable or animal material. Occasionally it is very useful for "recrystallising" sparingly soluble substances from the extraction thimble by means of an appropriate solvent (especially ether). As a rule, crystals of the dissolved material separate already from the solution in the flask during the

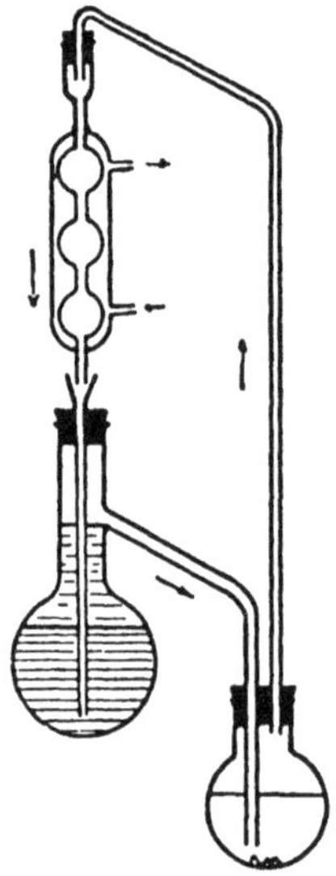

FIG. 27 *a*.

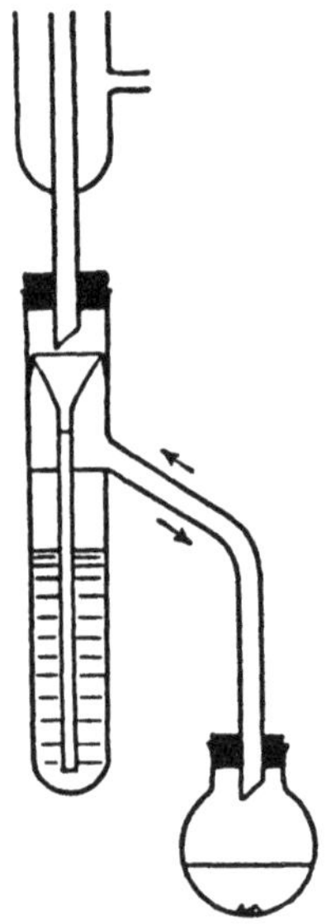

FIG. 27 *b*.

FIG. 28.

process of extraction. This solution soon becomes supersaturated even while hot.

When solvents of high boiling point are used the extraction thimble is directly suspended by a thin wire in the round-bottomed flask, but should not dip into the liquid.

WORKING WITH COMPRESSED GASES

Nowadays every university laboratory is doubtless supplied with steel cylinders in which the most important gases in common use are stored in the compressed state.

These gases are:

(1) Oxygen, hydrogen, nitrogen.

(2) Carbon dioxide, chlorine, ammonia, sulphur dioxide.

The elements under (1), with very low critical temperatures, are in the gaseous state in the cylinders, but the substances under (2) are present in liquid form. Oxygen, hydrogen, and nitrogen are usually under a pressure of 150 atmospheres in cylinders having a capacity of 10 litres; after filling, therefore, the cylinders contain gas sufficient to occupy 1·5 cubic metres at atmospheric pressure. The taps of the hydrogen cylinders have reversed screw threads so as to prevent accidental charging with oxygen.

In the laboratory all cylinders should be fitted with reducing valves, for the maintenance of which an assistant should be responsible. The use by itself of the simple valve in the head of the cylinder makes regulation of the gas stream difficult and invariably leads to waste.

For all gases (chlorine included) the so-called conical valve of aluminium bronze can be used. It can be made cheaply by any skilled mechanic (Fig. 29).

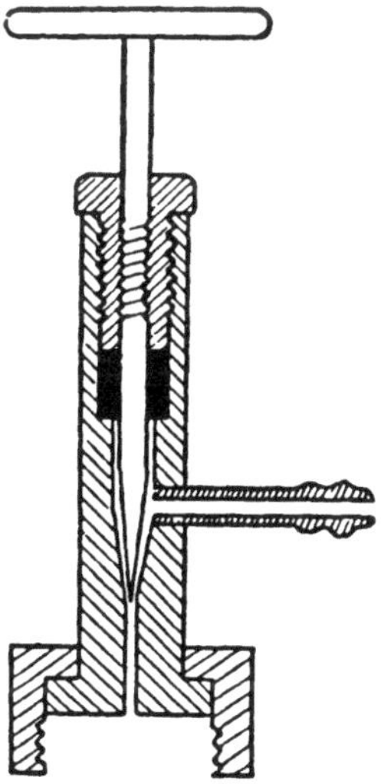

FIG. 29.

In all work with gases—whether from cylinders or from Kipp's apparatus—*a means of gauging the velocity of the current must be used.* Except with ammonia, a small bubbler containing concentrated sulphuric acid serves. It can be attached to the cylinder itself. As a rule, in order to dry at the same time, a wash-bottle is attached, preferably not one in two parts with a ground glass joint, because such bottles are often forced open by the slightest excess pressure.[1] If the gases have to be dried very thoroughly a wash-bottle is not sufficient, and one or two U-tubes containing phosphorus pentoxide spread on glass wool are added. Ammonia is passed through potassium hydroxide solution 1:1 and is then dried by passage through a tower containing potassium hydroxide and calcium oxide.

Ordinary laboratory apparatus cannot be used with gas from a cylinder for carrying out reactions under pressure in a closed system. For example, if it is desired to leave a reaction solution under pressure in hydrogen or carbon dioxide, the reaction vessel should not simply be left connected to the cylinder. In order to relieve the pressure in

[1] After use the connection between steel cylinder and wash-bottle should always be broken so as to prevent sucking back of the sulphuric acid.

the apparatus the circuit includes a T-tube, one branch of which is connected to a glass tube dipping into a cylinder with mercury or water.

In such cases it is more convenient to use a Kipp or, when nitrogen is employed, a gas-holder filled from a cylinder.

Experience shows that much gas is wasted because the beginner rarely considers how much is approximately required for his reaction. Nevertheless he should do so.

With the exception of nitrogen, all commonly used gases can, if necessary, be prepared by simple, well-known methods.

Heating under Pressure

When it is desired to accelerate a reaction in solutions or with undissolved substances at a temperature above their boiling point, the reactants must be shut off from the external atmosphere either by sealing in a glass tube in which they are then heated, or in a closed metallic vessel (autoclave). This is obviously necessary even when we wish a reaction to take place in alcoholic solution at 100° or in aqueous solution at about 120°. Thus the aim is entirely to increase the *temperature* of the reaction; the accompanying increase in pressure has no effect on the reaction velocity, for as a rule it is unaccompanied by any significant change in the concentration.

Since *solutions* are most frequently heated in sealed tubes in which the vapour pressure of the solvent determines the pressure, it is necessary to reckon with quite considerable pressures at temperatures which are appreciably above 100°. Gases which may possibly be formed during the reaction will add to the pressure. An estimate of the degree of pressure to be expected in a sealed tube reaction should be made with the help of the vapour pressure curve of the solvent used. In preparative reactions carried out in the heated tube, the pressure is always that of the saturated vapour. The pressure is therefore not dependent on the absolute amount of the solution taken. But since particularly water in the liquid state, and hence also solvents containing water, attack glass seriously at high temperatures, the tube is generally not more than half filled. Naturally when a gas is produced during the reaction the amount of free space available for it plays a part in determining the ultimate pressure.

The pressure tubes of Jena glass commonly used can be employed with some degree of certainty for pressures of 20 to 25 atmospheres if chemical action on the glass is excluded.

Tubes for sealing should always be filled through a funnel, and the inner walls in the neighbourhood of the part to be sealed must be kept clean. Concerning the handling of sealed tubes see pp. 65-67.

If the temperature is not to be raised above 100° the tube is wrapped in a towel and suspended in a specially protected water bath by means of a string or wire.

When no pressure or only slight pressure is developed a soda-water bottle with patent stopper (Fig. 30) is used instead of the sealed tube and is heated in the water bath.

For preparative work sealed tubes are unsatisfactory because of their small capacity, and therefore for larger quantities of material *autoclaves* are used. These are metal vessels capable of withstanding high pressures. The covers (see Fig. 31) are made tight with a lead ring, and are fastened by means of six to eight bolts, of which the nuts are gradually tightened in regular succession. Various types of autoclave are in use. That known as the Pfungst *tube* may be specially recommended. Autoclaves should always be heated in the oil bath.

FIG. 30.

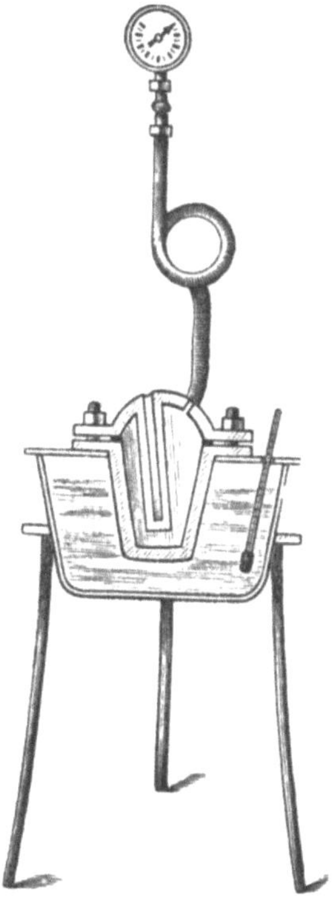

FIG. 31.

When carrying out pressure work always protect the eyes, and form in advance some physical idea of the strain imposed on the apparatus.

STIRRING AND SHAKING

In work with homogeneous solutions mechanical agitation is not necessary except when it is desired to add a substance in small portions or drop by drop, so as to bring it at once into a state of fine division, whether in solution or in suspension. This holds especially in cases where a sensitive preparation is endangered by the heat of reaction locally developed, *e.g.* when concentrated sulphuric acid is added.

In such cases the solution must be continuously agitated—by manual shaking or, better, by mechanical stirring. A glass rod, to the lower end of which propeller-like blades of glass are sealed, makes a suitable stirrer. Its upper end is slipped through a somewhat wider glass tube or cork-borer, which again is pushed through a cork, clamped so as to hold the stirrer vertically. To the upper end of the stirring rod a small pulley or a grooved cork or rubber stopper is firmly fixed. The rod should move with the least possible friction (in a narrow rubber ring smeared with glycerol). For driving, a Rabe water turbine is used or else a small suitably geared electro-motor ($\frac{1}{16}$th horse-power is sufficient). In places where there is no water supply the stirrer may be driven by a small hot-air engine, such as that made by Heinrici of Zwickau, which is also excellent for other purposes.

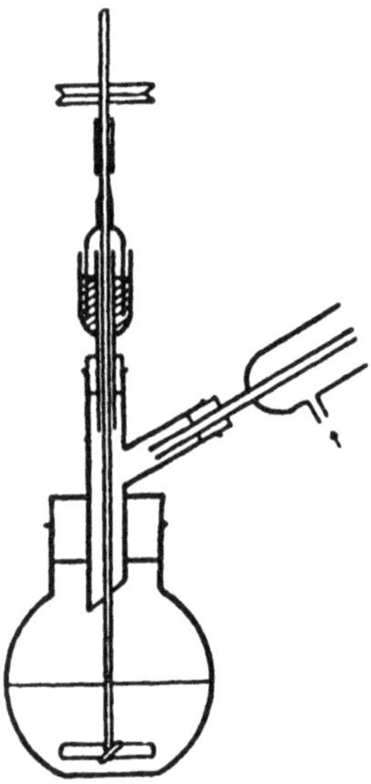

FIG. 32.

If the contents of a closed vessel have to be stirred or if the material is at the same time to be heated under reflux, a mercury seal must be provided on the stirrer, as shown in Fig. 32. Such a seal cannot, however, withstand pressure developed inside the flask.

When liquids consisting of non-miscible layers are to be stirred the blades of the stirrer must work at the interface. Heavy deposits, *e.g.* zinc dust, sodium amalgam, are generally not moved about adequately by small glass stirrers. In such cases mechanical stirring is often illusory, and a more powerful effect is obtained manually by stirring with a glass rod or strip of wood, or by frequent shaking.

The *shaking machine* must also be mentioned here ; it is used to produce the finest possible mechanical division in heterogeneous systems. Narrow-necked bottles with well-fitting ground glass stoppers are almost exclusively used as containers. The stopper is held down by means of a piece of rubber tube drawn over it and fastened to the neck with thin wire. Unless appropriate safeguards are provided, reactions in which gas or much heat is produced may not be performed in the shaking machine.

Determination of the Melting Point

The purity of a crystallised organic substance is checked by the *melting point*. This easily determined constant also serves for the identification of substances and, in the case of new compounds, for their characterisation. The apparatus is a long-necked, round-bottomed flask into which a tested thermometer is fixed by means of a cork; a sector is cut out of the cork with a sharp knife in order that the whole of the thermometer scale may be visible (Fig. 33). The heating liquid is pure concentrated sulphuric acid with which the bulb of the flask is three-quarters filled. The substance, in powder form, is introduced into a small, thin-walled capillary tube. Such tubes are made as follows from test tubes (preferably from damaged tubes which must, however, be clean and dry!). The tubes are rotated in the flame of the blow-pipe till soft and then drawn out rapidly; already after short practice the student can strike the correct diameter, which should be 1·0–1·5 mm. internally. Suitable portions of the drawn-out material are cut off with scissors. It is convenient to cut double lengths (about 12 cm.), so that by sealing each length in the middle (micro-burner) two melting-point tubes are obtained ready for use.

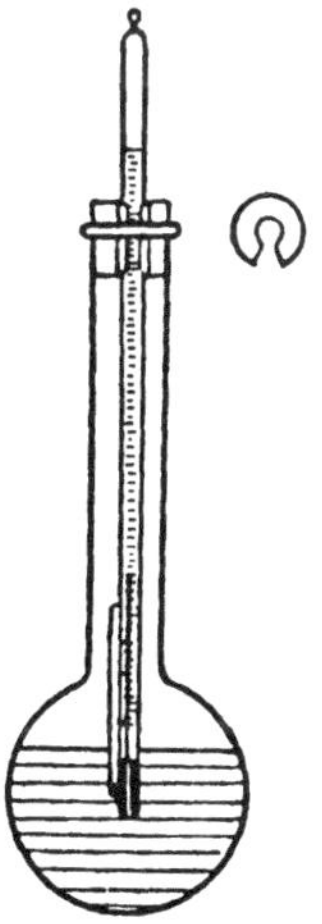

Fig. 33.

A small amount of the thoroughly dried substance is powdered on a watch-glass or on a small piece of porous plate with a spatula or pestle, and a column of the powder about 2 mm. long is formed at the bottom of the capillary tube by dipping the open end into it and tapping cautiously so as to shake the substance down from the mouth. If the substance sticks at any point above the bottom, the tube is allowed to fall several times through a long glass tube on to a glass plate or a watch-glass.

The melting-point tube is then attached to the lower end of the thermometer in such a position that the substance is at the level of the middle of the mercury bulb; this is done most conveniently by transferring a drop of concentrated sulphuric acid from the bulb to the upper end of the capillary. The bulb itself must be completely immersed during the determination. The apparatus is now heated with a moderate-sized flame held obliquely and moved slowly and regularly round the bath, which must be illuminated by reflected light. Substances having high melting points can be heated rapidly at first, but in the neighbourhood of the melting point the temperature should be raised slowly. As a rule, at this stage small particles of substance adhering to the upper part of the tube are softened by the hotter ascending sulphuric acid, and further heating should be carried out cautiously. The melting point is reached when the specimen, having collapsed, forms a clear liquid. In the case of substances of unknown melting point a preliminary determination is made.

Many organic compounds do not melt without decomposition, which often shows itself by change of colour and usually by evolution of gas, which can be very distinctly observed in the capillary. Such substances do not possess sharp melting points, but have decomposition points which are almost always dependent on the rate of heating; acceleration of the latter raises the decomposition point. The change produced by heat in these substances, below their point of decomposition, is also made evident by a shrinking and softening, which is called "*sintering*".

When the melting point of a decomposing substance is to be determined the bath is heated rather rapidly to within 10°–20° of the temperature of decomposition, and from that point onwards the rate of heating is reduced to about 5° per minute.

When stable substances sinter before melting it is a sign that they have not been completely purified, and should be recrystallised or redistilled. Some substances sinter, however, even when perfectly pure, *i.e.* they have no sharp melting point. In this connection the so-called "liquid crystals" may also be mentioned (Lehmann, Vorländer).

The rule should be adopted that a substance is only regarded as pure when its melting point no longer changes on repetition of the process of purification.

The reason why impurities lower the melting point is that they act to some extent as dissolved substances; now it is well known that the freezing point of a solution is always lower than that of the solvent (cryoscopy). An important method of identification is based on this fact. When the melting point of a compound, obtained by a new method, suggests identity with a known substance a certain decision can be reached by determining the melting point of an intimate mixture of the two compounds. If A differs from B the two substances will behave towards each other as impurities and the melting point of the mixture will be lowered ; on the other hand, if they are identical the melting point will remain unchanged. It is convenient, when determining a " mixed melting point ", to test the three samples (A, B, and A + B) simultaneously on the same thermometer. With a little practice this can be done with one sample on each side and one in front, or, if the thermometer is thick enough, with all three alongside each other in front (at the same level !).

In the single case of isomorphous substances the mixed melting point test fails.

There are also several methods for the determination of the *boiling point*, with small amounts of substance, in the melting-point apparatus.

The use of the sulphuric acid bath above 250° is not without danger ; as soon as signs of boiling appear heating is stopped and even before that temperature is reached the possibility that the flask may crack must be taken into consideration. Higher temperatures (up to 350°) are attained by using a bath of sulphuric acid in which potassium sulphate has been dissolved by heat. This heating bath solidifies on cooling, because potassium hydrogen sulphate crystallises from it ; it must therefore be just melted before the thermometer is introduced.

Here only a general review of methods and manipulations has been given, in so far as they are employed in the preparative exercises. The comprehensive and thorough works of Lassar-Cohn, Hans Meyer, and Houben-Weyl should be consulted for special requirements.

B. ORGANIC ANALYTICAL METHODS

DETECTION OF CARBON, HYDROGEN, NITROGEN, SULPHUR, AND THE HALOGENS

Testing for Carbon and Hydrogen.—If a substance burns with a flame (exceptions, *e.g.* sulphur) or decomposes with separation of black charcoal when heated on platinum foil, it is to be regarded as organic. The tests for carbon and hydrogen can be carried out simultaneously as follows: Mix a dry sample of the substance with several times its volume of fine ignited copper oxide in a small test tube and cover the mixture with a little more oxide. Stopper the test tube with a cork carrying a tube bent at right angles and heat strongly. If the gases evolved produce turbidity in clear baryta water (CO_2), the substance contains carbon, while the appearance of small drops of water in the upper part of the tube indicates the presence of hydrogen.

Testing for Nitrogen.—Press a clean piece of potassium or sodium half the size of a lentil between filter paper, transfer it to a small test tube about 5 mm. wide and 6 cm. long, add a small sample of the substance and heat in the flame of a Bunsen burner until decomposition occurs, usually accompanied by slight explosion and darkening. Then heat the tube to redness and dip it, while still hot, into a small beaker containing 5 c.c. of water. Uncombined metal may be ignited and the tube will crack (fume chamber!). Remove glass and carbon by filtering the solution, which contains cyanide if the substance contained nitrogen. Add two drops each of ferrous sulphate and ferric chloride solutions to the filtrate, make sure that it is alkaline by testing, and heat. If cyanide is present potassium ferrocyanide is formed in one to two minutes. Now cool and acidify the alkaline liquid with hydrochloric acid. The ferric and ferrous hydroxides which were precipitated go into solution and the potassium ferrocyanide reacts with the ferric chloride in the usual way, yielding

Prussian blue. Hence if the substance contained nitrogen a blue precipitate occurs. If the proportion of nitrogen in the substance is only small it occasionally happens that, at first, no precipitate is obtained, but only a bluish-green solution. If this is kept, for instance over night, a precipitate is deposited. When readily volatile substances are tested for nitrogen a longer tube is used and the material which condenses on the cold parts is allowed to flow back repeatedly on to the hot metal. Substances which lose their nitrogen already at moderate temperatures, *e.g.* diazo-compounds, cannot be tested in this way. When such substances are burned with copper oxide in a tube filled with carbon dioxide, a gas (nitrogen), which is not absorbed by potassium hydroxide solution, is produced (cf. quantitative determination of nitrogen).

Testing for Sulphur.—The qualitative test for sulphur is carried out in the same way as that for nitrogen. Ignite the substance in a small tube with sodium, dissolve the product in water, and add to one half of the cooled solution a few drops of sodium nitroprusside solution freshly prepared by shaking a few particles of the solid salt with cold water. A violet colour indicates the presence of sulphur. Since the nitroprusside reaction is extremely sensitive and does not allow any estimate of the amount of sulphur to be made, filter the second half of the liquid, add lead acetate solution to the filtrate, and acidify with acetic acid. According as the amount of sulphur is small or large, a dark turbidity or a more or less heavy precipitate will form.

Readily volatile compounds cannot, as a rule, be tested in this way. As described in the method for the quantitative determination of sulphur, they are heated in a sealed tube to about 200°–300° with fuming nitric acid and the solution produced is diluted with water and tested for sulphuric acid with barium chloride.

Testing for the Halogens.—Chlorine, bromine, and iodine can but rarely be detected in organic compounds by direct precipitation with silver nitrate, since the halogen usually does not form ions.

In order to test for un-ionised halogen ignite the substance with an excess of chemically pure lime in a moderately wide test tube in a Bunsen flame, dip the tube while still hot in a little water so that the glass is shattered, acidify with nitric acid, filter, and add silver nitrate solution.

In substances which contain no nitrogen the halogens can be detected by ignition with sodium as in the test for nitrogen. In this case remove glass and decomposition products by filtration, acidify the filtrate with pure nitric acid, and add silver nitrate solution.

Halogen can be detected very rapidly and conveniently by Beilstein's test. Twist a platinum wire round a small piece of copper oxide the size of a lentil or round a small rod of the oxide 0·5 cm. long and heat in the Bunsen flame until the latter appears colourless. Now cool the oxide, place a minute portion of the substance containing halogen on it, and heat again in the outer part of the flame. The carbon first burns with a luminous flame which soon disappears, leaving a green or blue-green region produced by the copper halide vapour. It is possible to deduce from the time during which the colour persists whether the substance contains only traces of halogen or more. The Beilstein test can also be made with a copper wire fixed in a cork.

Other elements which occur in organic compounds, such as phosphorus, arsenic, other non-metals, and metals in organic combination, are detected by destroying the organic material by oxidation (with nitric acid in a sealed tube or by fusion with potassium nitrate or sodium peroxide) and then applying the usual tests.

The determination of the elements which make up a substance is but a small part of its qualitative investigation. The next and more difficult task is to classify it, to show to what class of compounds it belongs in view of its chemical and physical properties and reactions. The ease with which it is possible to show that a polar compound is an acid or a base contrasts with the difficulties encountered in assigning a neutral substance of unknown constitution to the correct class. The recognition of the characters of the most important organic groups (of alcohols, aldehydes, ketones, esters, amides, nitriles and nitro-compounds), the differentiation of saturated, unsaturated, and aromatic substances by their reactions—the experimental solution of these and many other questions constitute an indispensable secondary object of a course of preparative organic chemistry. The student should not only acquire practice in the synthesis of representative compounds; he should also become conversant with the properties of the substances he prepares, he should become thoroughly acquainted with their characteristic reactions, and by close scrutiny and experimental observation he should fix their individuality in his mind. Hence the illustrative experiments included in the following course of preparative work are intended to further the educational object above indicated, and should be taken seriously. The carrying out of these reactions should be considered just as important as the purely preparative work.

In the course of the scientific work which succeeds the preparative course, and indeed whenever the chemist has to deal independently with a problem, questions requiring a knowledge of qualitative analysis continually arise. Because of the limited time available we have unfortunately hitherto not attained the ideal of following up the preparative work with a comprehensive course in the qualitative identification of organic substances, but nevertheless the closest attention should be given to this branch of study.

General aids to the study here outlined should be obtained already during the practical course from Hans Meyer's thorough work, *Analyse und Konstitutionsermittlung organischer Verbindungen.* Formal schemes of analysis, leading to the classification of organic substances, have also been worked out, *e.g.* by H. Staudinger in his *Anleitung zur organischen qualitativen Analyse*, 2 Aufl., Berlin, 1929.

ORGANIC ELEMENTARY ANALYSIS[1]

The quantitative determination of the elements in an organic substance is known as elementary analysis. In this process carbon and hydrogen are determined simultaneously, whilst a separate analysis must be carried out for the determination of each of the other elements.

I. DETERMINATION OF NITROGEN BY DUMAS' METHOD

A weighed quantity of substance is ignited with copper oxide in an atmosphere of carbon dioxide. The carbon is thus oxidised to carbon dioxide and the hydrogen to water, whilst the nitrogen is liberated as such and is determined volumetrically after collection over potassium hydroxide solution. Any oxides of nitrogen which may be formed are reduced to nitrogen by means of a glowing copper

[1] In the main the combustion methods described here are those which have been evolved in the course of years in the Staatslaboratorium at Munich and described in the directions prepared for use in this institute by F. Henle. Cf. also the directions for elementary analysis in F. Henle, *Organ.-chem. Praktikum*, Leipzig, 1921.

spiral. For the determination of nitrogen the following apparatus and chemicals are required :

A hard glass combustion tube of such length that it extends about 8 cm. beyond the furnace at each end.

An air-free carbon dioxide Kipp ; a vacuum desiccator ; a glass stop-cock.

A mixing tube with ground glass stopper (the opening of this tube must be capable of being introduced into the combustion tube).

Several small weighing tubes (the openings of these must be capable of being introduced into the mixing tube).

A porcelain boat.

Several small glass dishes with ground glass covers for substances for analysis.

A small mortar with glazed surface.

A bulb containing copper oxide, with rubber stopper.

One-holed rubber stoppers which fit the openings of the combustion tube ; rubber tubing for connections.

500 g. of wire-form copper oxide ; a copper spiral 10 cm. long.[1]

Three copper spirals each 2 cm. long.[1]

Pure potassium hydroxide in sticks.

Stout iron wire about 75 cm. long with hooked ends (for moving the spirals).

Methyl alcohol, two asbestos sheets, glass wool.

Nitrometer, gasholder, and furnace are supplied by the laboratory

Preparing the Apparatus

Air-free Kipp for Carbon Dioxide.—Cover small pieces of marble in a basin with dilute hydrochloric acid, pour the acid away after a few seconds, and then boil the pieces for about two hours with distilled water, replacing from time to time what is evaporated. Transfer the marble to the Kipp, fill the two lower bulbs with air-free boiled water, and then fill the top bulb with pure concentrated hydrochloric acid. Remove any air bubbles which may appear in the apparatus by opening the cock for a short time or by raising and lowering the funnel tube. Then open the cock and allow water to escape until the

[1] Strips of copper gauze 10 and 2 cm. wide are rolled firmly round straight pieces of thick copper wire 14 and 6 cm. long. The long spiral should slip easily into the tube but the short ones should fit tightly. The ends of the wires are bent into hooks by means of a pair of pliers.

acid reaches the marble. Mix the acid and water well by opening and closing the cock until a strong current of carbon dioxide is obtained. Attach a tall safety tube to the Kipp, connect it to a wash-bottle containing lead acetate solution in order to remove hydrogen sulphide and allow a moderately strong current of gas to pass for one to two hours. Then close the wash-bottle by means of a rubber tube carrying a screw clip, since the stop-cock on the Kipp must always be left open.

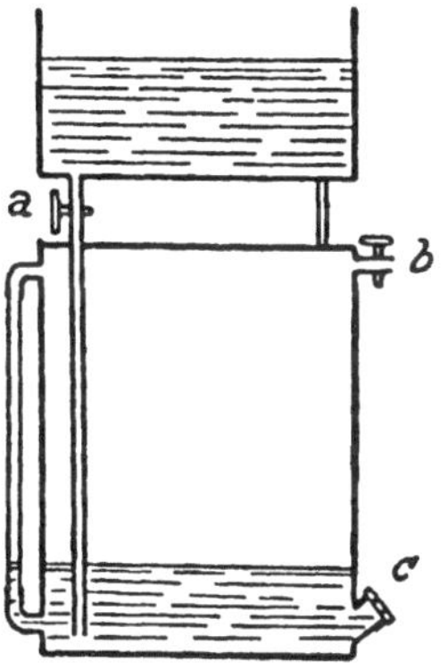

FIG. 34.

Filling the Oxygen Gas-holder.—Open the cocks *a* and *b* (Fig. 34) and so fill the lower container completely with water. Now close both cocks, unscrew the cap *c*, introduce a rubber tube connected with an oxygen cylinder through the opening *c* and fill the holder with oxygen. Screw on the cap and fill the upper container with water. When the cocks *a* and *b* are opened oxygen flows out at *b*.

Cleaning and Filling the Combustion Tube

Clean the combustion tube by pushing a roll of cotton wool or of filter paper several times through it. Then push in a 2-cm. copper spiral to a point one third from one end. The spiral should be of such thickness that it only slides under gentle pressure. Fill into the longer division of the tube copper oxide wire which has been freed from dust by means of a sieve. Shake the copper oxide together by striking the tube with the palm of the hand and insert a second short copper spiral of the same thickness as the first to hold the column in position. At this end of the tube the free space should now be about 20 cm., and is reserved for the reduced spiral. Fill the other (shorter) portion with finer[1] copper oxide prepared from the wire oxide by grinding. In order to hold this fine oxide in position insert a third short copper spiral, which should slide with moderate ease. The free space which still remains should be about 8 cm. long (cf. Fig. 35). Ensure the purity of the spirals by heating them in the flame of the blow-pipe until the greenish colour vanishes.

Burning Out the Filled Tube.—Attach a glass stop-cock to the end of the tube which contains the fine copper oxide and lay the tube, with the cock towards the gas-holder, on a strip of asbestos paper which rests on the iron gutter in the combustion furnace.

[1] Pieces of wire about 2 mm. long.

Then light all the flames together and regulate individual burners so that the flames are of equal height; adjust the main gas tap so that the tips of the flames just touch the iron gutter and close the tiles. After a few minutes turn on the main tap further from time to time, until the flames separate at the gutter and ascend its sides without, however, meeting above the tube. In this way the tube is raised to a moderate red heat in the course of half an hour.

FIG. 35.

FIG. 36.

Avoid too rapid heating since it easily cracks the tube. In order to protect the ends of the tube from too strong cooling, attach asbestos shields (Fig. 36), which at the same time prevent burning of the rubber stoppers. As soon as red heat is attained, connect the tube to the oxygen holder and pass a slow current of the gas, using a wash-bottle containing water as a bubble counter.[1] When oxygen can be detected at the forward end of the tube turn off the current of gas, extinguish the flames, and allow the tube to cool.

Preparation of the Reduced Copper Spiral.—Pour 1 c.c. of methyl alcohol into a test tube in the bottom of which is a roll of glass wool. Then heat a copper spiral 10 cm. long to bright redness in a blowpipe flame, drop it into the tube, and allow the ignited gases to burn away.

Preparation of the Nitrometer.—Clean the nitrometer to be used for collecting and measuring the nitrogen by washing with dilute hydrochloric acid and distilled water. The mercury which forms the seal must cut off the delivery tube from the measuring tube. Close the screw clip, open the stop-cock, lower the pear-shaped levelling bulb as much as possible and fill it with 50 per cent potassium hydroxide solution. Exclude air until the nitrometer is to be used.

Drying of the Substance for Analysis.—If the substance is a solid, grind it to a fine powder and free it from adherent moisture by drying to constant weight in an evacuated desiccator containing sulphuric acid. Keep liquids in a non-evacuated desiccator. Substances which are sensitive towards acids are dried in a vacuum over

[1] For the determination of nitrogen the tube does not require to be kept free from water.

potassium hydroxide, and those which are stable to heat and have high melting points are dried in an air bath at about 110°. Analyse substances which contain water or alcohol of crystallisation either after drying in the air or after bringing to constant weight in the desiccator. If the water of crystallisation is very firmly bound, dry in a high vacuum at elevated temperature (cf. foot of p. 13).

Carrying Out the Combustion

Weighing.—All objects to be weighed must be brought into the balance room at least one hour before weighing, so that temperature equilibrium may be attained. When possible hold the objects with forceps or at the narrowest part. Before weighing determine the zero point of the balance. Weigh accurately to 0·1 mg.

Determine first the approximate weight of the weighing tube, then introduce the substance, wipe the tube with a cloth, and weigh. Now empty the substance into the mixing tube, which should already contain a little copper oxide. By re-weighing the tube with the residue of adherent substance determine the exact weight of the substance to be burned. For nitrogen determinations take 0·1–0·16 g. of substance.

Filling the Combustion Tube.—Remove the copper spiral which holds the fine copper oxide (rear end of the combustion tube) and empty the fine oxide into the copper oxide bulb. Then cover the substance in the mixing tube with a layer about 5 cm. deep of fine oxide from the bulb, close the mixing tube with the glass stopper and mix well by thorough shaking. Then pour the contents into the combustion tube. Now wash out the mixing tube several times with fine copper oxide and pour the separate portions into the combustion tube until all the fine copper oxide has been transferred. After the copper spiral has been slipped in again, close the rear end of the tube with a one-holed rubber stopper carrying a glass stopcock. Into the front end, after the reduced spiral has been inserted, fix a glass tube by means of a rubber stopper—the tube is bent downwards at an obtuse angle. The stoppers must be pushed in to half their length and their narrow end should still be in contact with the walls of the tube. The ends of the glass tubing should not project more than 1 mm. inwards from the stopper.

Setting up the Apparatus.—After the tube has been placed in the cold furnace with the tiles turned back, connect it with the air-free Kipp by means of the rubber tubing attached to the wash-bottle and carrying a screw clip. At the other end of the tube attach the nitrometer. To test the apparatus for gas-tightness (*this must be done each time before a combustion is carried out*), close the screw clip on the nitrometer and open the connections to the Kipp apparatus. After pressure equilibrium has been attained no bubbles should pass through the wash-bottle during a period of several minutes.

The Combustion.—First open the cock of the nitrometer and lower the levelling bulb until all the potassium hydroxide solution runs into it. Then open the screw clip and pass a vigorous stream of carbon dioxide through the combustion tube for about five minutes, regulating the rate by means of the clip on the wash-bottle. In order to test whether there is still air in the tube, close the clip at the nitrometer, raise the bulb until the solution rises a little above the boring of the stop-cock, close the latter, and again lower the bulb to its original position. When the screw clip is opened the bubbles which rise should be practically completely absorbed. During a period of three minutes no measurable volume (less than 0·1 c.c.) of unabsorbed gas should collect. If this is not the case, open the stop-cock so that the solution runs back into the bulb, and continue passing carbon dioxide through the tube for some little time longer. Then test again in the manner just described. When all the air has been expelled cut off the carbon dioxide current by screwing down the clip on the wash-bottle and also close the stop-cock at the rear end of the tube. Now light the burners under the reduced copper spiral and under part of the permanent column of copper oxide, and at the same time light also one or two burners at the rear end of the tube. Cover the heated portions with tiles, attach the asbestos sheets to the ends of the tube, and also protect, by small sheets of asbestos, those portions of the tube which are to be kept cold. Heat as cautiously as when burning out the tube. As soon as the copper oxide has reached dull red heat and the reduced spiral moderate red heat (as judged in the absence of direct sunlight) begin heating the substance from both ends, first warming a short length by lowering the tile, then lighting the burners below and gradually increasing the size of the flames, but still protecting the cold parts of the tube by asbestos sheets. Continue in this way until the whole tube has been brought to dull red heat. Keep cool the part in which the substance lies until a length of the copper oxide column sufficient for combustion is glowing. Remember that the effect of lowered tiles and newly lighted burners only shows itself gradually.

The application of heat causes carbon dioxide to pass from the tube into the nitrometer. As soon as the combustion begins, bubbles of nitrogen appear and increase more and more in number and size. Regulate the rate of combustion so that not more than one or two bubbles per second appear. If they come too rapidly moderate the speed by temporarily turning back the tiles and extinguishing the flames in the neighbourhood of the substance. When combustion is over only a few bubbles appear, and these are all absorbed by the potassium hydroxide solution. Drive out the nitrogen which remains in the tube by passing a moderate current of carbon dioxide until the bubbles become as small as at the beginning of the analysis, and no further increase in the volume of the gas collected takes place (five to ten minutes). Then screw down the clip on the nitrometer, detach the combustion tube by withdrawing the stopper carrying the glass tube, extinguish the flames beneath the reduced spiral, remove the Kipp apparatus, connect the tube with the oxygen gas-holder, and pass oxygen through until it can be detected at the front end of the tube. Then extinguish the flames. When the tube has cooled and the copper spiral has again been reduced, the apparatus is once more ready for use.

Taking the Reading.—Transfer the nitrometer to the room in which the barometer is kept, hang a thermometer on the nitrometer (bulb of the thermometer alongside the middle of the gas column), and leave for *at least* one hour. In order to ensure that the stop-cock is gas-tight always have a layer of potassium hydroxide solution a few millimetres deep in the upper part of the nitrometer. Remove bubbles from the surface of the column of potassium hydroxide by raising and lowering the levelling bulb and so causing movement. Then place the bulb in such a position that the level of the liquid in it is the same as that of the liquid in the measuring tube. Read the scale-division which is in the same horizontal plane as the lower edge of the meniscus. Also note the temperature and the barometric pressure.

Calculation.—The percentage of nitrogen is given by the following formula :

$$\% \text{ N} = \left(1{\cdot}2507 \cdot \frac{1}{1+\alpha t} \cdot \frac{b-\delta-e}{760}\right) \cdot \frac{v}{s},$$

in which v is the volume of nitrogen read, $a = \frac{1}{273} = 0{\cdot}003663$, b is the barometric height, δ is the correction of the barometric height to 0°, and e is the vapour pressure of potassium hydroxide at t°.[1]

Limits of error of the determination: 0·3 per cent high, 0·1 per cent low.

The apparatus can at once be used again for a further determination. After being used 4–5 times the potassium hydroxide solution in the nitrometer must be renewed.

II. DETERMINATION OF CARBON AND HYDROGEN BY LIEBIG'S METHOD

1. In the Absence of Elements other than Carbon, Hydrogen, and Oxygen

The method is essentially as follows: A weighed amount of substance is oxidised in a combustion tube with red-hot copper oxide, and the oxidation products are absorbed, *carbon dioxide* by potassium hydroxide solution, *water* by calcium chloride, and are weighed. The combustion is carried out in a current of oxygen, but it may also be performed in a closed tube.

In addition to the apparatus used for the determination of nitrogen the following are required:

A gas-holder for air; a drying apparatus for air and oxygen.

3 m. of clean, thick-walled, narrow-bored rubber tubing for the drying apparatus and the gas-holder.

An ordinary straight calcium chloride tube.

Granulated calcium chloride (fine); a potash bulb.

A calcium chloride tube for the determination of water.

PREPARING THE APPARATUS

Combustion Tube.—Prepare the copper oxide[2] and fill the tube in the same way as for the determination of nitrogen, but instead of the reduced copper spiral use an oxidised one. It is absolutely essential that the tube and its contents should be perfectly dry after

[1] The values of the expression between brackets are given for various values of $b - \delta - e$ and t in the table on pp. 408, 409. With sufficient accuracy δ may be replaced by $\frac{t}{8}$ and e by $\frac{t}{5}$. For example: readings, $b = 738$ mm., $t = 20°$; look up in the table the value for $p = 738 - 2{\cdot}5 - 4 = 731{\cdot}5$. When calculating logarithmically use Küster's tables.

[2] Prepare a special lot of copper oxide for the C-, H-determination, and keep this lot separate from that used for the determination of nitrogen.

filling. Before each combustion heat the tube at about 200° for ten to fifteen minutes while a slow current of air passes through, and carefully expel the water, which is deposited in the cold end of the tube, by means of a hot tile which, however, must not touch the tube. After the flames have been extinguished close the tube with a rubber stopper and allow to cool while the pressure inside is maintained by connecting to the gas-holder.

Filling the Air Gas-Holder (cf. Fig. 34).—Unscrew the cap *c*, open the cock *b*, and let the water in the lower container flow away. Replace the cap and fill the upper container with water. The gas-holder is now ready for use.

The Drying Apparatus.—In order to free the air and oxygen from carbon dioxide and moisture, both gases are passed through the drying apparatus. This consists of wash-bottles, one for each gas, containing potassium hydroxide solution (1 : 1) and a system of tubes, filled with soda-lime and calcium chloride. By means of glass stop-cocks and a three-way tube attached to the system it is possible to pass at will air or oxygen through a single length of rubber tubing which carries a screw clip and, at its free end, a calcium chloride tube. The clip regulates the current of gas. The tubing must not have been used for any other purpose.

THE ABSORPTION APPARATUS

(*a*) **The Potash Bulb.**—The "screw" potash apparatus of Greiner and Friedrichs, in which the gas passes through a spiral, and the three-bulb apparatus of the firm of Dr. Bender und Dr. Hobein, Munich (Fig. 37), are both to be recommended. The apparatus is filled with 50 per cent potassium hydroxide solution, of which a stock should be kept. Potassium hydroxide used for this purpose should not have been purified with alcohol. One filling suffices for two combustions (about 1 g. of CO_2). The outside of the potash apparatus is cleaned with a dry cloth. The small tube sealed on

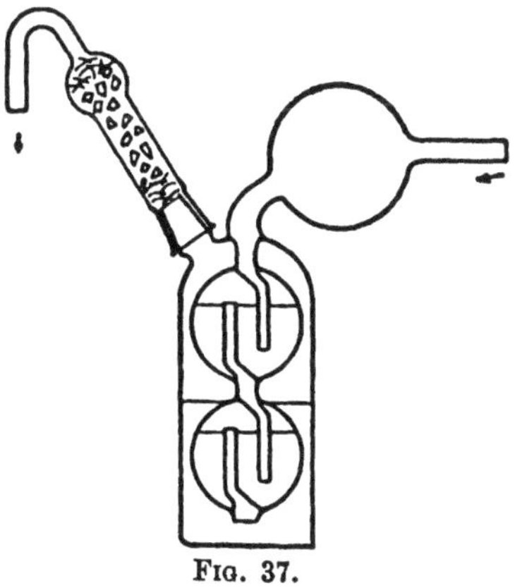

FIG. 37.

should also be freed internally from solution by means of a roll of filter paper. The small tube attached by means of a ground joint is filled with finely granulated calcium chloride held in position by glass wool; after about eight combustions the calcium chloride should be renewed. The ground joint should be slightly greased. When not in use the apparatus must always be closed by means of caps made from short pieces of rubber tubing tightly closed at one end by short lengths of glass rod.

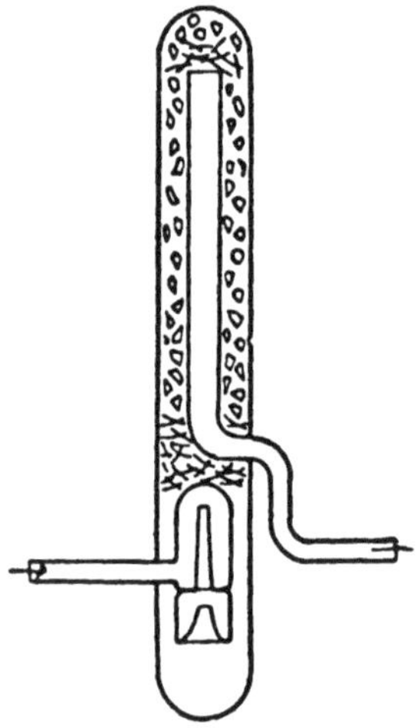

FIG. 38.

(*b*) **The Calcium Chloride Tube.**—The pattern of Dr. Bender and Dr. Hobein with the bottom cap attached by a ground joint is to be recommended. The lower half is filled with fine calcium chloride held in position by glass wool. Before the bottom cap is placed in position the ground joint is carefully freed from glass wool and greased slightly. This cap is kept in place by means of a stout rubber band. The one-piece calcium chloride tube (Fig. 38) is equally useful. Since the calcium chloride contains basic constituents a strong current of carbon dioxide should be passed through it for about half an hour before it is used for the first time. It should then be closed at one side with a cap and left over night in connection with the Kipp apparatus, so that the internal pressure is that of the Kipp. Then the carbon dioxide is expelled with air. The calcium chloride tube is cleaned externally in the same way as the potash apparatus. For hanging the tube on the balance an aluminium wire is used. When not in use the calcium chloride tube must always be kept closed with caps.

The remainder of the apparatus required for the carbon and hydrogen determination is prepared in the same way as that described for the determination of nitrogen. What was said in that connection about the drying of the substance to be analysed holds here also.

CARRYING OUT THE COMBUSTION

Weighing.—At least one hour before weighing, bring the absorption apparatus into the balance room and again wipe off adherent dust with a cloth. Do not wipe immediately before weighing. Weigh the apparatus without caps or stand. Weigh the substance in the same way as for the nitrogen determination. Take about 0·12–0·16 g. of substance. Immediately after weighing the substance fill the combustion tube in the manner described in detail for the determination of nitrogen.

Setting up the Apparatus.—Into the rear end of the tube (the end nearer to the substance to be analysed) fix a rubber stopper carrying a glass stop-cock to which a straight calcium chloride tube is attached. Connect this tube by means of rubber tubing to the three-way tube of the drying apparatus. At the front end of the combustion tube attach the calcium chloride absorption tube directly by pushing the small tube which leads to the water trap through a rubber stopper. First insert the stopper in the combustion tube and insert the absorption tube by turning it to and fro. Attach the potash apparatus, supported on a stand or on a small wooden block, to the calcium chloride tube by means of a piece of seamless rubber tubing 3–4 cm. long, taking care that the ends of the glass tubes are in contact. Reduce the friction between rubber and glass by blowing several times through the holes of the stoppers and the rubber tubing. Before a C- and H-determination is made also test the air-tightness of the apparatus by closing the exit tube of the potash apparatus with a cap and opening the cock which makes connection with the drying apparatus and the way to one of the gas-holders. No bubbles should pass through the wash-bottle of the drying apparatus during a period of several minutes after pressure equilibrium has been attained. As an additional check close one of the stop-cocks. If there is the slightest leak in the apparatus the level in the potash bulb changes. As a rule, leakage takes place in the rubber connections or at the cap. When the test has been made close the glass stop-cock nearest to the combustion tube and then cautiously remove the cap.

The Actual Combustion is carried out in a closed tube or in a current of oxygen. In the latter case regulate the current so that one bubble escapes from the potash apparatus approximately every two seconds. Light the burners under the front portion of the fixed column of copper oxide, later the last two burners at the rear end of the tube, and cover the corresponding parts with tiles. Attach the asbestos sheets to the ends of the tube and protect the part of the tube which is still to be kept cold with strips of asbestos. After the tube has been brought to moderate red heat during a quarter to half an hour, as described for the nitrogen determination (for substances which burn with difficulty heat more strongly), gradually extend the heating towards the substance, first warming the tube by lowering the tiles and then lighting the burners below. At what rate this can be done can be decided by watching the bubbles in the potash apparatus. Not more than one or two bubbles per second should escape from it. When the substance begins to burn the stream of bubbles in the potash apparatus becomes more rapid, and in the small tube leading from the front end of the furnace there appears a film of moisture which gradually collects into small droplets. Very soon also the bubbles in the potash apparatus are observed to be partly, and later completely, absorbed. The rate of burning should not be so great that the bubbles which appear in the potash apparatus cannot conveniently be counted. By adjusting the tiles and regulating the size of the flames it is possible to control this rate.

When the whole tube has attained red heat and the evolution of carbon dioxide is slackening, the current of oxygen is increased. If the combustion has been carried out in a closed tube the water which condenses in the cold rear end of the tube is driven forward by means of a hot tile. Since the oxygen is powerfully absorbed by the metallic copper produced during the combustion, the current of the gas must be allowed to flow rather strongly for some time. In proportion as the stream of bubbles in the potash apparatus becomes rapid again, slow down the current of oxygen. If the latter is not sufficient to prevent the potassium hydroxide solution from running backwards, pass air instead of oxygen through the tube. As soon as oxygen can be detected at the front end of the tube, cut off this gas and pass a current of air for about ten minutes. Regulate the air current likewise, so that at most two bubbles per second appear in the potash bulb. While the air is passing through, lower the flames and drive the condensed water out of the front end of the tube with a hot tile.

After the combustion is complete, detach the absorption apparatus, close it with caps, and transfer it to the balance room. Weigh the apparatus on the same day after the lapse of one hour.

Calculation.—The percentages of carbon and hydrogen can be calculated by means of the following formulae:

$$\% \ C = \frac{CO_2 \text{ found}}{\text{substance}} \times \frac{300}{11},$$

$$\% \ H = \frac{H_2O \text{ found}}{\text{substance}} \times \frac{201 \cdot 6}{18 \cdot 016}.$$

Use Küster's logarithm tables. The limits of error are: for carbon $\pm 0 \cdot 3$ per cent, for hydrogen $+0 \cdot 3$ and $-0 \cdot 1$ per cent. Good analyses give about 0·1 per cent C too little and 0·1 per cent H too much.

Subsequent Operations.—After the absorption apparatus has been detached, the flames are extinguished and the closed tube is allowed to cool while the pressure of the gas-holder is maintained. What was said with regard to the further use of the tube in connection with the nitrogen determination holds here also. When a tube is once filled there is no sense in spending a long time burning it out before each combustion, as is often done. It is sufficient to pass air through the moderately warmed tube for ten to fifteen minutes between combustions in order to dry the contents completely.

If, after a few combustions, a large amount of water has collected in the water trap of the calcium chloride tube, it should be removed by shaking and the exit tube dried with filter paper. The potassium hydroxide solution in the potash apparatus should be renewed after two combustions.

2. Procedure when Nitrogen is Present

If the substance to be analysed contains nitrogen use a tube filled as for a nitrogen determination, but in this case make certain that the filled tube and the reduced spiral are carefully freed from water and carbon dioxide. In order to do this, dry the filled tube as described under 1. Prepare the reduced copper spiral as follows: heat it to redness in a blow-pipe flame and drop it into a hard glass tube, in the bottom of which are a wad of glass wool and about 1 c.c. of methyl alcohol. After the flame has burnt out, close the tube with a rubber stopper carrying a glass stop-cock and evacuate with a water-jet pump. Then heat further for ten minutes with a powerful burner until the tube and also the stop-cock and rubber stopper (carefully avoid burning the latter) are completely free from moisture. Close the cock and leave the spiral in the evacuated tube until required for use. It is also sufficient to dry the reduced spiral in a test tube in a drying oven at 100°-105° for one hour. Slight tarnishing is harmless.

Carry out the combustion in a closed tube, and heat the reduced spiral to a bright red heat. Extinguish the flames under this spiral before passing the oxygen. Use two reduced spirals in the combustion of nitro-compounds rich in nitrogen.

3. Procedure when Halogen or Sulphur is Present

If the substance to be analysed contains halogen or sulphur, then copper halide or sulphur dioxide is formed. The former is rather volatile at red heat and would be deposited in the front part of the tube or in the calcium chloride tube. The sulphur dioxide, on the other hand, would be absorbed by the potassium hydroxide solution. Consequently a column about 15–20 cm. long of lead chromate is introduced into the foremost part of the tube and retains the halogen as lead halide and the sulphur as lead sulphate. But since these substances also are somewhat volatile when strongly heated, the part of the lead chromate column nearest to the absorption apparatus should be heated only just enough to prevent the retention of water. It is most convenient, therefore, to allow part of the lead chromate column to extend outside the furnace and also to heat the part in the heated zone less strongly than the copper oxide.

If the substance to be burned contains nitrogen in addition, a reduced copper spiral must further be inserted between the copper oxide and lead chromate columns. The length of the copper oxide column must then be correspondingly reduced. Copper oxide with which substances containing halogen or sulphur have been burned should not, in general, be used for the determination of carbon and hydrogen in substances free from halogen or sulphur.

4. Combustion in a Boat

Instead of burning the substance in admixture with fine copper oxide, it is often advantageous to carry out the combustion in a porcelain or platinum boat (contact action). The advantages of this procedure are :

1. Saving of time. While the substance and the absorption apparatus are being weighed the copper oxide wire can already be heated.

2. Incomplete combustion is shown by a residue of charcoal.
3. Inorganic residues can be detected and weighed.

The tube is filled in the same way as in the determination of carbon and hydrogen. In place of the fine copper oxide, there are the boat and a readily movable oxidised spiral about 10 cm. long. At the place where the boat rests the tube should be transparent, so that the course of the combustion can be observed. As usual, before combustion is begun the contents of the tube are dried by warming for a short time in a current of air. Then the rear end is allowed to cool while the front is heated to redness. In the interval the absorption apparatus is weighed, and the boat, after being ignited in the flame of the blow-pipe and left to cool in a desiccator, is likewise weighed, first empty, then with the substance.

The air current is next cut off, the calcium chloride tube and the potash apparatus are attached at the front end, the movable copper oxide spiral is pulled out with a hooked wire, the boat and the spiral are slipped in again, and the tube is connected to the drying apparatus. The combustion then proceeds in a slow current of oxygen in the manner described above.

Substances which decompose violently on warming cannot be burned in boats.

5. Procedure when Alkalis or Alkaline Earths are Present

If the substance contains an alkali or alkaline earth it is burned in a boat in which a layer of potassium dichromate covers the substance, so that the stable carbonate formed during the combustion is decomposed. The dichromate is previously fused, powdered, and dried for one day in the desiccator before use. Combustion is continued until the blistered mass has again completely fused to a compact substance.

6. Combustion of Liquids

Before combustion, liquids are filled into small glass bulbs provided with a capillary tube. These *small bulbs are made* as follows : draw out a clean thin glass tube at two adjacent points to not very narrow capillaries and seal one of them. Then, using the glass tube as a mouthpiece, rotate the *ampoule*-shaped part while heating it to red heat and blow a bulb (diameter of about 5–7 mm.) of a size which can be introduced into the combustion tube while lying in the porcelain boat. Make a number of these bulbs and keep them in a desiccator.

Introducing the Substance.—In order to fill a (previously weighed) small glass bulb with liquid, place the bulb with the capillary downwards in the vessel containing the liquid and by evacuating the desiccator remove the air from the bulb. Then allow air to re-enter the desiccator and the bulb will fill with liquid. The amount of liquid which it is desired to have in the bulb can be regulated by partial evacuation. Clean the outside of the bulb, remove liquid from the capillary with a small flame, and, after cooling, weigh the filled bulb. Do not touch the bulb, but only the capillary.

Filling the Combustion Tube.—For the combustion use a tube filled as for combustion in a boat. The tube should be transparent at the rear end and for substances containing nitrogen, sulphur, or halogen it must be filled in the manner described above for these cases. Lay the small glass bulb in the boat and push the latter into the combustion tube in such a way that the capillary is directed towards the absorption apparatus. Behind the boat introduce an easily moved oxidised copper spiral about 10 cm. long. In the case of substances which are not volatile without decomposition and of substances which burn with great difficulty and leave a carbonaceous residue, the bulb must be smashed. To do this, place a layer of fine copper oxide 2–3 cm. long in the tube, let the bulb (with the capillary directed forwards) slide on to this layer by holding the tube in an oblique position, and break up the bulb with the oxidised copper spiral. Liquids which volatilise with very great difficulty can also be burned directly in the boat.

Combustion.—Liquids must be burned still more carefully than solid substances. They are burned in a slow current of oxygen. The front half of the copper oxide wire is heated to moderate red heat and the extreme rear end is also heated, while care is taken to keep the part of the tube containing the bulb cool, by excluding heat from the sides with strips of asbestos, and possibly by laying on moist asbestos. Next the liquid is expelled from the bulb by occasionally holding a hot tile near it. Then carefully extend the heating from both ends towards the substance, and always give a preliminary heating by occasionally lowering a hot tile. Light the burners under the bulb only at the end of this process.

To be completely conversant with the methods of organic elementary analysis is the first essential for successful scientific work. The candidate for the doctorate who cannot carry out a combustion greatly adds to the difficulty of his own work, since he cannot establish definitely and conclusively the composition of any new substances which he may prepare.

In the Munich laboratory the student is required to carry out the combustion of a substance, unknown to him, belonging to each group (nitrogen ; C- and H- determination in the absence of other elements ; the same with nitrogen ; the same with halogen or sulphur ; a liquid). He does not go on to the next group until he has correctly analysed, at the first attempt, an example from the previous group. A test analysis is given in completion of the course of elementary analysis.

III. DETERMINATION OF HALOGEN, SULPHUR, AND OTHER ELEMENTS

If a substance contains yet other elements besides carbon, hydrogen, oxygen, and nitrogen, the former are usually determined by oxidising the substance in a sealed tube by heating with red fuming nitric acid. Halogens are determined as silver halide, sulphur as barium sulphate, phosphorus as magnesium ammonium phosphate or as magnesium pyrophosphate (Carius method).

The halogens can also be determined by ignition with halogen-free lime, whereby calcium halide is produced. The lime method can always be used, whereas certain aromatic halogen compounds are only decomposed with very great difficulty by the Carius method. Organically bound halogen can also be determined very simply by the use of hydrazine hydrate and active palladium (process of M. Busch).

In the case of loosely bound halogen, such as that in not too volatile acyl halides and alkyl halides, the substance to be analysed is boiled for some time under reflux with a pure alcoholic solution of potassium hydroxide, the alcohol is evaporated, and the ionised halogen is determined in the usual way. For this purpose it is often sufficient to titrate with acid when standard alcoholic potassium hydroxide solution has been used. Jena flasks which have been well steamed are alone used for such determinations because of the alkalinity of ordinary glass. Compare the example on p. 90.

Halogen compounds which volatilise with difficulty can also be converted into halide by mixing them with sodium peroxide and igniting (H. Pringsheim).

Alkali and alkaline earth metals in salts are determined as sulphate after evaporation with concentrated sulphuric acid.

For these determinations the following are necessary:

Hard glass tubes at least 50 cm. long, as thick as a lead pencil and sealed at one end.

Carius tubes; small, wide, thick-walled specimen tubes; a long-stemmed funnel.

Glass crucibles with sintered glass filter plates or Gooch crucibles; filter flask with adapter and rubber ring.

Solid silver nitrate.

Pure lime (free from halogen, calcium sulphate, and heavy metals); filter papers of known ash content.

Purified Asbestos.—Free the asbestos mechanically from coarse impurities, and boil it for about one hour with much pure dilute nitric acid (1 part of acid to 5–10 parts of water); renew the liquid frequently, and each time pour off the finest particles with the liquid. Store the long fibres in a closed vessel under dilute nitric acid.

1. Determination of Halogen by the Lime Method

Filling the Tube.—Make a column of pure lime 3–5 cm. high in a narrow combustion tube 50 cm. long sealed at one end, remove adherent lime from the upper end of the tube with filter paper, pour the dried and weighed substance in from the weighing bottle, and cover the substance with a layer of lime 3 cm. deep, which washes down small particles of substance that remain adhering to the walls of the tube. Mix the substance with the lime by shaking and tapping the tube while it is held alternately in a horizontal and a vertical position. Then fill the tube to within 10 cm. from the open end with lime, and make a channel for air by tapping the tube while holding it horizontally.

Heating the Tube.—Lay the filled tube with the open end projecting from the furnace, heat the part nearer this end to red heat, and then extend the heating slowly towards the substance. Since no combustion occurs, more or less coloured vapours with a "burnt" smell escape from the tube. The halogen, however, is quantitatively retained. When the whole tube has reached red heat keep it in this condition for a quarter to half an hour longer, then allow it to cool and close the tube with a stopper.

Dissolving and Precipitating.—Pour the cooled contents of the tube carefully into a one-litre beaker containing some distilled water, and wash out the tube, first with distilled water then with pure dilute nitric acid (1:3). Dissolve the lime completely by gradual addition of pure dilute nitric acid, then filter from carbon and fragments of glass, wash out with quite dilute nitric acid, and precipitate the halogen in the filtrate with silver nitrate.

Filtration with the Gooch Crucible.—Prepare the Gooch crucible before and during the operations described above. Pour a few cubic centimetres of the asbestos " sludge " into a test tube along with distilled water, wait for a few seconds, and pour the liquid with the fine fibres into a second test tube. Repeat this operation, then pour the long and afterwards the fine fibres into the crucible, drop in the perforated plate, and wash several times with distilled water with good suction at the pump in order to draw the fine fibres through. Then dry the crucible to constant weight in an oven at 130°.

For filtration of silver halides the glass crucibles of Schott of Jena are especially to be recommended. The bottoms of these crucibles are made of sintered glass.

Warm the contents of the beaker until the silver halide has settled and the supernatant liquid is clear, pour the latter through the crucible and digest the precipitate 4–5 times with water containing nitric acid at 60°–80° until the filtrate is free from silver. Then transfer the silver halide quantitatively to the crucible with the help of a rubber-tipped glass rod. Do not wash the material in the crucible any further. If these precautions are taken and a permanently acid washing liquid is used and kept at approximately the same temperature throughout, the halide should not pass through and form a turbidity. If this should nevertheless happen, warm the filtrate and filter again after the precipitate settles. Dry the crucible to constant weight at 130° and weigh. During the whole operation protect the silver halide from light. The limits of error in halogen determinations amount to ±0·3 per cent.

2. Determination of Halogen by the Method of M. Busch[1]

This process depends on the fact that hydrazine hydrate in alkaline solution is decomposed by metals of the platinum group

[1] *Z. angew. Chem.*, 1925, 38, 519.

according to the equation: $N_2H_5.OH \longrightarrow N_2 + 2H_2 + H_2O$. The hydrogen produced is then applied to the catalytic conversion of the combined halogen into the ionised form. Palladium on barium sulphate (preparation, see pp. 369, 370), nickel (according to Kelber, *Ber.*, 1917, **50**, 305), or, best of all, *palladinised calcium carbonate*, serve as catalysts.

Suspend 50 g. of well washed calcium carbonate which has been precipitated while hot, in about 200 c.c. of water and mix the suspension by shaking and gentle warming with a solution of 1 g. of palladium chloride. When the liquid over the calcium carbonate is completely decolorised, pour off, wash the solid a few times with distilled water, and then filter at the pump. Wash the material on the funnel until the filtrate is free from chlorine ions, dry in a vacuum desiccator and store in a well-stoppered reagent bottle. The adsorbed palladium hydroxide is reduced during the determination.

Carrying out the Determination.—Dissolve 0·1–0·2 g. of the substance in 20–40 c.c. of alcohol in a round-bottomed flask (capacity 150 c.c.) and add 10 c.c. of a freshly-prepared, pure, colourless, 10 per cent alcoholic solution of potassium hydroxide and after that 3 g. of the palladium catalyst. Then add 10 drops of hydrazine hydrate, and heat the liquid under reflux (employ cork!) for half an hour on the boiling water bath. Next filter, wash the solid on the funnel with some alcohol, then with water, and remove most of the alcohol from the filtrate by evaporation. If this or the subsequent cooling causes separation of the hydrogenation product filter again and then acidify with nitric acid. If the substance taken is an organic acid its hydrogenation product is liable to be precipitated on acidification. In this case filter once again before adding silver nitrate. Collect the silver halide in a Gooch crucible at the pump and wash with hot alcohol and water.

On account of the danger (not considerable) that organic substances remain adsorbed on the silver halide, the volumetric method with thiocyanate is especially to be recommended here.

This elegant method is limited to the analysis of substances soluble in alcohol and water and to such as contain no sulphur, phosphorus, or arsenic, since these three elements inactivate the catalyst. The limits of error are $\pm 0{\cdot}3$ per cent.

The palladium can be recovered from the used catalyst by dissolving the calcium carbonate in acetic acid.

3. Determination of Halogen by the Carius Method

Preparing the Carius Tube.—Weigh out 0·1–0·2 g. of the dry substance in a small, thick-walled specimen tube. Then weigh out the amount of finely powdered silver nitrate required for the reaction (assume the highest possible halogen content) and drop it on to the bottom of the Carius tube through a long-stemmed funnel. After adding 20 to 30 drops (about 1 c.c.) of red fuming nitric acid (which must not moisten the upper part of the tube) let the small tube with the substance slide down beside the acid and the nitrate. Take care not to allow the substance to come into contact with the nitric acid before the tube is sealed and placed in the iron tube. Cover substances which are already decomposed by nitrous vapours, with a little silver nitrate. Begin the sealing of the tube immediately after its contents are in place.

The Sealing of the tube is done in the flame of the blow-pipe. Grasp the Carius tube just above the middle with the left hand, hold it as much inclined as possible and take care that the nitric acid does not run into the specimen tube. Rotate the tube uniformly during the whole course of the sealing operation. First hold the mouth of the tube in a luminous flame and gradually increase the air supply until the luminosity disappears. When the glass softens, seal a thin glass rod, red hot at the end, to one side of the interior of the tube, pull it towards the other side and after this side has also become attached to the rod, move the latter so that it lies in the axis of the tube. Now heat the tube a little further down where it is still cylindrical, first with a luminous, then in the tip of a non-luminous flame until softening takes place. At this stage continue the rotation but do not draw the tube out. When the glass walls at the heated part have collapsed and thickened greatly, remove the tube from the flame and draw it out slowly in a straight line. When solidification has taken place, seal off the capillary in a fine-pointed flame. The capillary should be straight, thick-walled, and of small internal diameter. Cool the tube at first slowly in a luminous flame, then completely in the air and place it in an iron case with the capillary projecting a few centimetres. If the Carius tube is too short, fill a corresponding part of the iron case with sand or iron filings. *As long as the tube remains sealed it must not be removed from the iron case or from the furnace room.*

Heating the Tube.—Place the iron case with its tube in a bomb or tube furnace in such a way that the end with the capillary is raised somewhat and directed towards the wall on which a shield is fixed, and close the furnace. Several tubes may be heated simultaneously. Light all the burners and heat gradually to the desired temperature, regulating the gas supply at the main tap. For aliphatic halogen compounds (and many sulphur compounds) this temperature is about 250°, for the aromatic compounds (and sulphonic acids) it is about 300°. Ordinary Carius tubes are unsuitable for use at still higher temperatures and it is then better to use tubes of hard glass (combustion tubing). Most substances are completely oxidised after three to four hours, but in the case of aromatic compounds the heating is continued for some hours longer. Turn off the gas and leave to cool. If it is necessary to heat still longer, open the capillary in the manner about to be described, seal again after the gases have escaped, and heat further.

Opening the Tube.—When the tube is quite cold, remove the iron case from the furnace, expel with a luminous flame any liquid which may have collected in the capillary, and hold the latter in the tip of a blow-pipe flame (goggles!). After the compressed gases have perforated the softened glass, remove the tube from the case, make sure that the capillary is not choked, and break off the top of the tube as follows. Make a scratch round the tube with a glass-knife or a triangular file and touch with a white-hot thin glass rod. Remove splinters of glass from the edge and clean the outside of the tube.

Washing Out the Material in the Tube.—Carefully dilute the contents of the tube with some distilled water and, by pouring and repeated washing out, transfer them quantitatively to a beaker. Remove any silver halide which may remain adhering to the tube with a long glass rod. Since the specimen tube might break the beaker in falling, it is held back with a glass rod, then grasped with the fingers or with bone forceps and washed.

Further Treatment.—Break up the lumps of silver halide with a glass rod so that any occluded silver nitrate may be easily washed out. In the case of silver iodide (also of the bromide) warm for two hours on the water bath, since silver iodide forms a solid compound with silver nitrate and this compound is only gradually decomposed by water. Further, when iodine is being determined, silver iodate which is formed must first be reduced by addition of pure sulphurous acid solution. The subsequent treatment is the same as that which has already been described above.

The Carius tubes may be used until they have become too short.

4. Determination of Sulphur by the Carius Method

The determination of sulphur by the Carius method is carried out in the same way as the determination of halogen but, of course, without silver nitrate. If, after heating and opening the capillary, the liquid is not yet clear, the heating must be repeated.

Precipitating.—Again dilute the contents of the tube with water, pour them into a porcelain basin, and evaporate completely to dryness on the water bath in order to remove the nitric acid and so prevent the precipitation by barium chloride of nitrate along with the sulphate. Take up the residue in water and hydrochloric acid, filter if necessary from splinters of glass, and heat to boiling in a beaker. Drop the barium chloride solution very slowly from a capillary funnel [1] into the boiling liquid.

If no further turbidity is produced on addition of more barium chloride solution after the precipitate has settled (on the water bath !), filter the precipitate and wash with hot water until the washings no longer contain the barium ion.

Drying and Igniting.—Fold the paper containing the precipitate, press it firmly into the bottom of a porcelain crucible (or, better, a platinum crucible) which has been ignited and weighed, set the crucible in a position as nearly horizontal as possible on a pipe-clay triangle, and incinerate by careful heating. After some time heat the crucible from below with a full flame until the contents are white. Leave the crucible to cool in a desiccator and weigh.

If the above directions are not followed during the precipitation one is apt to obtain a somewhat too high sulphur content because of the carrying down of barium chloride. Remove the latter by repeated extraction on the water bath with water containing a little hydrochloric acid and then repeat the treatment described above. If this procedure is not successful, decompose the barium sulphate by fusion with sodium and potassium carbonates and determine the sulphate in the melt in the usual way.

The limits of error for sulphur determinations are $\pm 0 \cdot 3$ per cent.

[1] Draw out a test tube to a capillary in the middle and use the upper half.

5. Simultaneous Determination of Halogen and Sulphur

Prepare the Carius tube as for the halogen determination. After heating, and washing the contents of the tube into a beaker, remove the silver halide by filtration as described above. The filtrate contains the sulphuric acid produced by the oxidation along with the excess silver nitrate. Add to the filtrate an aqueous solution of barium nitrate which has previously been tested and found to give no precipitate with silver nitrate. Carry out the precipitation in the hot solution which should be diluted as much as possible (to at least 300 c.c.). If halogen-free barium nitrate is not available, add silver nitrate to a solution of the ordinary quality of this salt until no further precipitate forms, remove the silver halide by filtration, and use the now halogen-free filtrate for the precipitation.

6. Determination of Other Elements

Most of the other elements are determined by the methods of inorganic analysis after organic material has been oxidised in nitric acid solution by the Carius procedure.

Alkali and Alkaline Earth Metals are determined as sulphate. Weigh the substance in a quartz or platinum crucible,[1] add a few drops of concentrated [2] sulphuric acid and evaporate carefully ; loss of spirting liquid is prevented by inclining the crucible, covering with a lid, and heating from the rim downwards ; ignite (in the case of alkali sulphates at not more than dull red heat) until the contents of the crucible are white. Occasionally evaporation with concentrated sulphuric acid must be repeated until the weight of the crucible does not change.

7. Quantitative Determination of the Methoxyl Group by Zeisel's Method

This method depends on the conversion of the methyl of the CH_3O-group into methyl iodide by boiling hydriodic acid and on the determination of the silver iodide, which is precipitated in a receiver, at first in the form of a double salt with silver nitrate.

[1] Preferably a " finger " crucible.

[2] In the case of sensitive or explosive substances use a corresponding amount of dilute acid (1 : 1 or 1 : 2).

The silver nitrate is used in aqueous-alcoholic solution. Any glass-blower can construct the necessary apparatus (Fig. 39).

Reagents.—*Hydriodic acid* of specific gravity 1·7, boiling point 127°, free from sulphur and phosphorus compounds and therefore not prepared from iodine and hydrogen sulphide.

Red Phosphorus.—Suspended in water in the washing vessel *b*, it serves to trap the iodine which passes over; the volatile hydrogen iodide is trapped in the water. The phosphorus must be thoroughly washed with water, alcohol, and ether, and finally again with *much* water.

Alcoholic Silver Nitrate Solution; prepared by dissolving 12 g. of silver nitrate in 10 c.c. of water and adding 90 c.c. of pure alcohol which has been distilled over potassium hydroxide. Keep this solution in the dark.

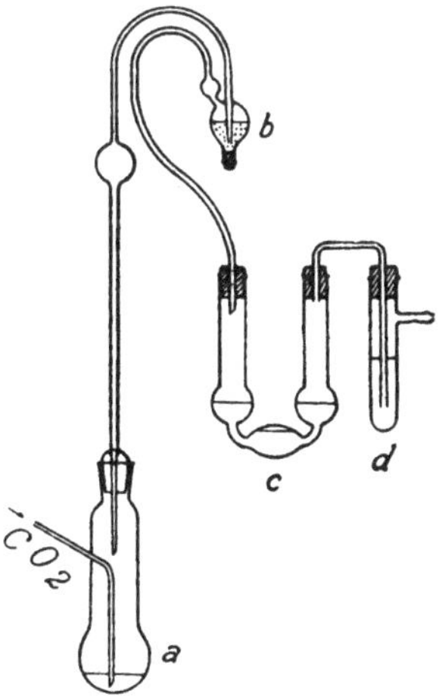

FIG. 39.

Carrying out the Determination.—Invert the apparatus and pour about 50 mg. of red phosphorus and some water into the washing vessel *b*, and close with a cork stopper. Pour 30 c.c. of the silver nitrate solution into the receiver *c* and 10 c.c. into the receiver *d*; into the small boiling flask *a* pour 10 c.c. of hydriodic acid along with 0·15–0·2 g. of substance. Fix the receiver *c* on a stand by means of a clamp. Attach the flask *a* to the apparatus and heat the contents to boiling with a micro-burner while passing carbon dioxide, washed by passage through lead acetate solution and concentrated sulphuric acid, through the apparatus at the rate of about three bubbles in two seconds.

In from fifteen to twenty minutes after the hydriodic acid begins to boil the silver nitrate solution becomes turbid and soon opaque as a result of the separation of the white double compound. When the liquid over the settled crystalline precipitate clears completely, the operation is at an end. Duration of the determination: one to two hours.

As a rule the contents of the tube *d* remain clear even when they are diluted with five volumes of water and left for ten minutes. Transfer the solution and precipitate quantitatively from *c* to a beaker, dilute to 500 c.c. with water and decompose the double compound on the water bath by concentration to about three quarters. Add a little pure dilute nitric acid, heat until the silver iodide has settled again and determine in the usual way. The ethoxyl group can also be determined in this way, one OC_2H_5 yielding as much silver iodide as one OCH_3.

Calculation.—Calculate the amount of OCH_3 or OC_2H_5 thus:

100 parts of AgI correspond to 13·20 parts of CH_3O or 19·21 parts of C_2H_5O.

Limits of error: ±0·5 per cent of the total alkoxyl content.

This method cannot be applied without modification to compounds containing sulphur and it must also be modified when applied to volatile substances.

8. Quantitative Determination of the Acetyl Group by Freudenberg's Method [1]

This process depends on the trans-esterification of the acetyl compound in alcoholic solution by *p*-toluenesulphonic acid and titration, after hydrolysis, of the ethyl acetate removed by distillation. The apparatus (Fig. 40) required for the acetyl determination can easily be constructed from articles to be found in the laboratory. The flask *a* should have a capacity of 100 c.c. and the flask *b* one of 150 c.c.; the side tubes (bent upwards in the one and down in the other) must have an internal diameter of at least 5 mm. The tube connecting *a* and *b* should extend sideways further than is shown in the figure.

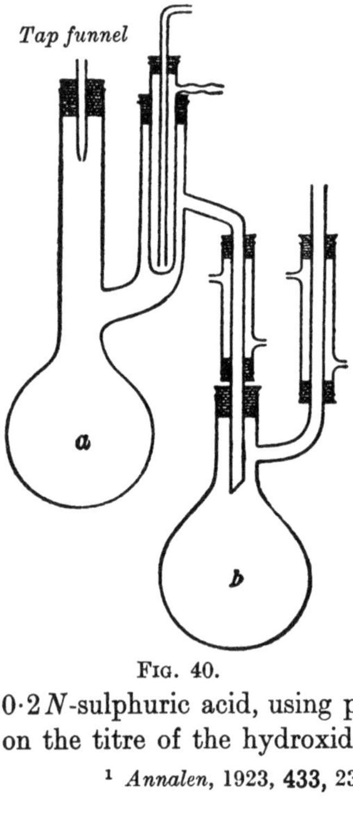

Fig. 40.

Reagents.—0·2 *N*-sulphuric acid; 0·2*N*-sodium hydroxide solution; for standardisation, dilute a sample of approximately 0·2 *N*-sodium hydroxide with one volume of water and two volumes of alcohol, and titrate with 0·2 *N*-sulphuric acid, using phenolphthalein as indicator (as a check on the titre of the hydroxide solution the blank experiment made

[1] *Annalen*, 1923, **433**, 230. Improved, *ibid.*, 1932, **494**, 68.

to test the purity of the reagents may also be used); *p*-toluenesulphonic acid.[1]

Carrying Out the Determination: pour 10 c.c. of absolute alcohol into the flask *b*, connect to the flask *a*, place 0·3 – 0·4 g. of the substance in the latter along with 30 c.c. of absolute alcohol, 5 g. of *p*-toluenesulphonic acid and a few pieces of porous pot, close the neck of the flask with a one-holed rubber stopper, carrying a dropping funnel, and turn on the water in the small condenser surrounding the distillation neck. Now dip the flask *a* into hot water, and cool the flask *b* in ice; the contents of the flask are thus removed by distillation, the condenser being kept empty; when distillation is over, add alcohol from the funnel and distil again. Repeat once more the addition of alcohol and distillation of the contents of the flask. When carrying out the determination, work to the following times:

Boil for ten minutes (in the case of N-acetyl, forty-five minutes) under reflux; bath temperature, 100°.

In the following operations maintain the bath: for O-acetyl at 95°, for N-acetyl at 100°.

Distil for fifteen minutes with condenser empty; after distillation is over, run in 20 c.c. of absolute alcohol and

Boil for ten minutes (in the case of N-acetyl, thirty minutes) under reflux; distil for ten minutes with condenser empty;

Allow 20 c.c. of absolute alcohol to drop in during fifteen minutes, continuing the distillation at the same time with condenser empty and then distil for ten minutes longer with empty condenser.

For the determination of acetyl in sugar derivatives the times and temperatures given must be strictly adhered to; in the case also of O-acetylated substances outside the sugar series the times and temperatures should not be below those given.

When the distillation is complete, pour 30 c.c. of 0·2 *N*-alkali hydroxide solution into the lower flask through its side tube, replace the cold bath by a hot one and keep the latter boiling for ten minutes; during the heating maintain the temperature of the upper flask at 80° C.

Remove the hot bath, cool the lower flask, pour 30 c.c. of water into it, disconnect from the flask *a* and titrate the contents (of *b*) with 0·2 *N*-sulphuric acid, using phenolphthalein as indicator.

[1] Preparation, see p. 183.

Calculation.—For alkali of normality N :

$$1 \text{ c.c. of alkali} \equiv N \times 43{\cdot}024 \text{ mg. of acetyl}$$

so that when v c.c. of alkali are used :

$$\textbf{Amount of Acetyl} = v \times N \times 43{\cdot}024 \text{ mg.}$$

Limits of error : $-0{\cdot}1$ per cent to $+0{\cdot}4$ per cent below and above the calculated value.

Duration of the determination : in the case of O-acetyl, two hours ; in that of N-acetyl, three hours.

The method can also be applied to substances containing halogen. In this case the reaction is carried out in the presence of silver toluenesulphonate ; the silver halide can be collected and weighed.

9. Determination of Active Hydrogen by the Method of Tschugaeff and Zerevitinoff [1]

Prepare a Grignard solution in a distilling flask held in an oblique position and having a small condenser on the side tube (this condenser

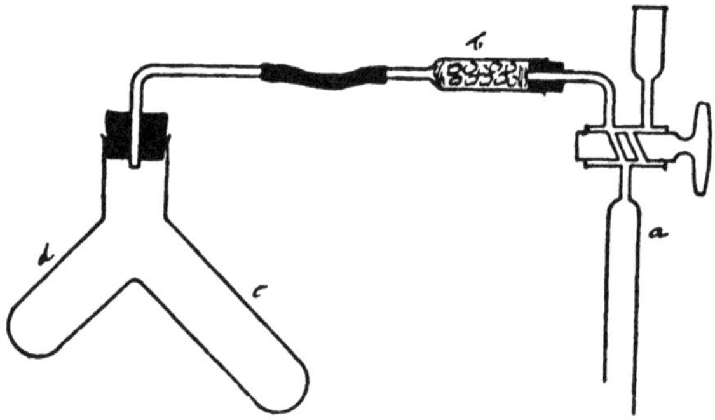

Fig. 41.

acts as reflux) ; use for this 20 c.c. of amyl ether which has been distilled over sodium,[2] 7 g. of methyl iodide and 2 g. of magnesium ; add also a few particles of iodine. If the reaction does not begin spontaneously, set it going by warming for a short time at 50° and complete by heating for one hour on the water bath. Then set the flask upright in the normal position and heat for a quarter of an hour longer so that the last traces of methyl iodide are removed. Pour off the Grignard solution so obtained from unused magnesium

[1] *Ber.*, 1907, **40**, 2027. [2] Preparation, see p. 104.

and store in a well-closed bottle. For each determination use about 5 c.c. of this solution.

The apparatus for the determination of active hydrogen is shown in Fig. 41. The Lunge nitrometer *a* of which the levelling bulb is not shown in the sketch is filled with saturated brine. By means of the short calcium chloride tube *b*, interposed between the reaction vessel and the nitrometer, the access of water vapour to that vessel is prevented. For the determination weigh accurately, according to the molecular weight and hydroxyl content, 0·1–0·2 g. of the substance [1] from a weighing bottle into the longer branch *c* of the well-dried reaction vessel, cover with amyl ether, and dissolve by careful shaking. Then pour 5 c.c. of the Grignard solution through a funnel into the other branch *d*. Attach the reaction vessel by means of a clean, tightly fitting rubber stopper and a piece of rubber tubing to the calcium chloride tube of the nitrometer from which the stop-cock has been removed, dip the reaction vessel into a beaker containing water at room temperature, wait for five minutes until temperature equilibrium has been attained, replace the stop-cock, and, by raising the levelling bulb, fill the nitrometer with brine. Then turn the cock through 90°, lower the levelling vessel, and put the reaction vessel into communication with the measuring tube by turning the cock again through 90°. Now remove the reaction vessel from the beaker, allow the solution of the substance to flow into the Grignard solution, pour the mixture backward and forward a few times and shake until the meniscus in the measuring tube no longer falls and the evolution of methane is therefore at an end. Put the vessel back into the beaker, wait for ten minutes until it has regained the temperature which prevailed at the beginning of the experiment (check with a thermometer), and read off in the usual way the amount of methane produced. At the same time take the temperature of the gas with a thermometer hung on the measuring tube and also read the barometric height. Reduce the volume to 0° and 760 mm.

Calculation.—According to the equation $RHn + nCH_3.MgI \longrightarrow R.(Mg.I)n + nCH_4$ one mole of substance liberates $n \times 22{\cdot}4$ litres of methane where n is the number of atoms of active H; a grammes of substance $= \frac{a}{M}$ moles liberate $\frac{n.22400.a}{M}$ c.c. of CH_4. The volume found and reduced (V_t) must be equal to volume

[1] Of the compounds prepared by the student triphenylcarbinol, β-naphthol, quinol, and benzoic acid may be used.

calculated ($V_{calc.}$) for one active H-atom ($n=1$), or, when there are several active H-atoms then V_f must be a simple multiple of $V_{calc.}$. It is convenient to express the result as the number of atoms of active H according to the ratio $V_f/V_{calc.}$. The limits of error amount to 5–10 per cent.

10. Determination of Molecular Weight

The various methods are not described here since, as a rule, they are learned in the practical course in physics or physical chemistry. The cryoscopic process is much to be preferred to the ebullioscopic. The most frequently used solvents are benzene and glacial acetic acid, and the best apparatus is the closed one of Beckmann, with electromagnetic stirring.

A very elegant and simple method by which the molecular weight of organic substances can be determined in a melting point apparatus was recently described by K. Rast.[1]

Camphor has a very high cryoscopic constant and its melting point is very greatly depressed by substances dissolved in it. The depression is about 8 times as great as that in benzene. $E_{benzene}=5{\cdot}1$, $E_{camphor}=40$. This means that a molar solution in camphor melts 40° lower than the solvent, *i.e.* than camphor itself. Accordingly, even with relatively dilute camphor solutions the depressions obtained are so great that the sensitivity of an ordinary thermometer (which can be read to 0·25°) suffices completely for the determination.[2]

On an ordinary analytical balance weigh out about 10 mg. of substance and 10–15 times as much camphor into a small, carefully cleaned test tube standing in a cork support; rapidly melt the two substances to a clear liquid in a sulphuric acid bath previously heated to 180°. Allow to solidify with shaking and thoroughly mix the mass, preferably while still warm, with a brass wire hammered flat at the end so as to form a spatula. Now melt the mixture again in the bath at a somewhat lower temperature and stir the mass once more after solidification.

Make the melting point tubes as described on p. 39 from a

[1] *Ber.*, 1922, 55, 1051, 3727. Abderhalden, *Arbeitsmethoden*, Abt. III, Teil A, p. 754.

[2] In the following description the procedure worked out by W. Münster is followed, by which, in contradistinction to that of Rast, the solidification point is determined instead of the melting point originally used.

clean test tube ; the internal diameter should be about 2 mm. Cut off pieces about 3 cm. long and seal them at one end by drawing the softened glass away sideways with a second capillary in the micro-flame of a Bunsen burner base, from which the tube has been screwed off. In this way tubes with very thin walls are obtained, which is important for a rapid exchange of heat. Introduce a column of the mixture about 1·5 mm. high into the melting tube, ram it down with a thinner tube, draw out the melting tube in a micro-flame to a capillary about 1 cm. above the column and stick it on the thermometer ; the melting tube should be well submerged in the sulphuric acid.

Now heat in a melting point apparatus (Fig. 33) till a clear melt is formed, allow to cool and thus determine the solidification point approximately. In order to obtain the accurate value, heat again, this time very carefully in the micro-flame, the tip of which should be about 4 cm. below the apparatus, until the contents of the tube, with the exception of a few quite small crystals adhering to the bottom, melt to a clear liquid. The temperature so observed lies usually 2° above the former solidification point. Now regulate the temperature by turning down the flame so that the temperature falls at about 1° per minute. In this way it is possible to ascertain very distinctly through a lens when the surviving crystals begin to grow. At this point read the temperature. As a check repeat the operation ; with careful working almost the same solidification point should again be found. It is desirable to surround the flame with a shield, a cylinder of glass or of paper 8 cm. in diameter, which should reach to the melting point apparatus.

The solidification point of the *camphor* employed is determined beforehand in the same way. Use a quite pure preparation. If such is not available make sure of a homogeneous supply by melting together separate pieces.

If the solidification temperature of camphor (177°) is depressed by Δ, and a = amount of substance, b = weight of camphor, then the molecular weight $M = \frac{40.a.1000}{b.\Delta}$.

The limits of error are ± 5 per cent.

To compounds which decompose at the temperature of melting or react with camphor this method cannot, of course, be applied.

C. PREPARATIVE PART

ON THE PREVENTION OF ACCIDENTS

Whoever sets about preparative work carelessly and thoughtlessly may easily come to harm. But even the careful are not secure against all danger. The serious accidents which again and again occur in chemical laboratories make it imperative that every member of the laboratory should fully and seriously consider his duty towards his fellows.

It is most important to protect the eyes. Strong *goggles* with stout glass must be worn in all *vacuum* and *pressure* work, and hence when distilling in a vacuum, evacuating a new desiccator for the first time, or working with sealed tubes, pressure bottles, or autoclaves. Goggles must also be worn when carrying out *fusions with alkali* and during all operations in which *caustic* or *inflammable* substances may spurt, and therefore, in particular, during all work with *metallic potassium* and *sodium* (cf. the special directions on p. 108). Finally, work with *explosive substances* should *never* be done *without wearing goggles*, and the behaviour in the flame of unknown substances should always first be tested with small amounts on a metallic spatula after the preparation itself has been put in a place of safety.

In order to guard the eyes against unforeseeable explosions, which can never be absolutely excluded, every worker in the laboratory should wear plain glasses, while not dispensing with goggles in the cases mentioned.

Care must invariably be taken when working with *ether* and other *volatile, readily inflammable liquids* that *no flame is burning in the neighbourhood.* If a *fire* occurs, *everything which may ignite must immediately be removed.* The fire should then be extinguished with moist *towels* or by *pouring on carbon tetrachloride,* but not water. The best extinguisher is a small portable CO_2 cylinder which should be kept in every laboratory; the so-called "Selbstretter", introduced for the use of the army in the field, is well adapted to this purpose. A larger fire may be tackled by pouring on sand, but here also a large carbon dioxide cylinder is usually preferable.

Parts *injured with acids or caustic alkalis* should first be washed thoroughly with *water* and then with *bicarbonate* solution or *dilute acetic acid* respectively. *Slight burns* should be washed with *alcohol* and then covered with *linseed oil* or an ointment.

Cotton wool, bandages, and *plaster* must always be at hand. When *serious accidents* occur the nearest *doctor* should at once be called.

When a *caustic* or irritant *organic* substance attacks the skin, washing with water is usually without effect. Immediate removal by washing with *copious amounts* of a suitable solvent such as alcohol or benzene is the procedure indicated. It must be borne in mind that the organic solvent itself facilitates the penetration of the harmful substance into the skin and therefore the formation of concentrated solutions on the skin must be avoided.

Especial care is required when working with the following much used substances: *hydrocyanic acid, phosgene, dimethyl sulphate,* the lower *acid chlorides, chlorine, bromine, nitric oxide* and *nitrogen peroxide, carbon monoxide, sodium,* and *potassium.* Large scale operations with these should be carried out in a special room; otherwise always in a *good fume chamber.*

Undiluted halogen compounds of the aliphatic series, such as *ethyl bromide, chloroform, bromoform,* and the like, should not be brought into contact with *metallic sodium* or *potassium*; thus they must not be dried with these metals since very violent explosions may occur as a result of detonation (Staudinger).

EQUIPMENT REQUIRED BY THE BEGINNER

I. Apparatus

Beakers, 100, 500, and 1000 c.c., one of each.
Burette.[1]
Calcium chloride tubes, straight, 3.
Cardboard for weighing (playing cards).

[1] The student should be able to carry out at once, at any time, the usual volumetric determinations. As a rule the readiness to carry out titrations in the organic laboratory is not so great as is urgently desirable.

Condensers, Liebig type, one about 60 cm. long ; a shorter one, 10–12 cm. long ; one of the Dimroth or coil type.
Conical flasks, 50, 100, 250, 500 c.c., two of each.
Copper wire for halogen test.
Cork-borers.
Corks, various sizes.
Crystallising basins, glass, 3, 5, 7 cm., one of each.
Distilling flasks, ordinary and Claisen types, 25, 50, 100 cc., one of each.
Files, round and triangular, one of each.
Filter flasks, 100, 500, 1000 c.c.
Filter paper, a few sheets.
Filter jars, 500 and 1000 c.c., one of each.
Filter plates, porcelain, 1, 3, 5 cm.
Filter tubes, long and short, 3 of each.
Funnels, 2 of smallest size, one each of other sizes up to 12 cm. Cylindrical *Büchner funnel*, about 8 cm. *Dropping funnel*, 25 c.c. capacity with *short* stem and 100 c.c. capacity with long stem.
Glass buttons, for filtration, 2.
Glass rods, 20 of various thicknesses and lengths, rounded at both ends but not thickened.
Glass tubes, drawn out to form pipettes (dropping tube, p. 12).
Glass tubing, straight and bent.
Labels.
Measuring cylinders, 10 c.c., 100 c.c.
Melting-point tubes, thin (to be made by the student).
Metal spatula.
Mortar.
Pipettes, 5, 10, 20 c.c., and at least 6 small ones. Also one of 2 c.c. capacity graduated in tenths of a c.c.
Porcelain basins, 15, 20, 25 cm. in diameter.
Porcelain spatulae, 3.
Porous pot. Pieces of porous plate about 3 mm. across.
Round-bottomed flasks, 50, 100, 250, 500 c.c., two of each.
Scales (apothecaries, capacity 50–100 g. Protect knives from rust by smearing with vaseline). *Set of weights for the balance*, 0·02–50·0 g.
Scissors.
Separating funnels, 500, 1000 c.c.

Slides (microscope), 3.
String.
Test papers. Litmus, blue and red, turmeric, Congo. Potassium iodide-starch paper.
Test tube holder.
Test tubes, at least 50 *of ordinary size,* 20 *of small size. Half of the test tubes must always be kept clean and dry.*
Thermometers. One standardised for melting point determination and two others, including a short one, for ordinary use.
Tongs.
Towel.
Vacuum desiccators, two large ones (16 and 18 cm. diameter).
Wash bottles, 2.
Watch glasses, chiefly small.

II. Solvents

Acetic acid, glacial, 500 c.c.
Acetone, 100 c.c.
Alcohol, 96 per cent, one litre.
Alcohol, absolute, 500 c.c.
Benzene (over sodium wire), 500 c.c.
Chloroform, 100 c.c.
Ether, absolute, over sodium, 500 c.c.
Ethyl acetate, 250 c.c.
Methyl alcohol, 250 c.c.
Petrol ether, low-boiling, 250 c.c. (over sodium wire).
Petrol ether, high-boiling, 250 c.c. (over sodium wire).
Xylene (over sodium wire).

III. Reagents, Drying Agents

Calcium chloride.
Decolorising charcoal.
Glycerol (bottle with a cork stopper carrying a glass rod).
Hydrochloric acid, concentrated.
Nitric acid, concentrated (d. 1·4) and *fuming* (d. 1·5).
Potassium hydroxide, commercial and pure.
Silver nitrate solution, 5 per cent.
Sodium hydroxide solution, about 10 N (=40 per cent).
Sodium metal.

Sodium sulphate, anhydrous.
Standard solutions: 0·1 *N*—HCl, 0·1 *N*—NaOH, 0·1 *N*-iodine solution, 0·1 *N*-thiosulphate solution.
Sulphuric acid, pure, concentrated.

IV. Books

Laboratory Notebook.[1]
Abstracts from the literature.

[1] From the outset the student should accustom himself to keep a notebook in which all records of experiments and all observations are entered. *Never rely on the memory in scientific work.*

CHAPTER I

THE REPLACEMENT OF HYDROXYL AND HYDROGEN BY HALOGEN

ALCOHOLS AND OLEFINES

1. ETHYL BROMIDE FROM ETHYL ALCOHOL

ALCOHOL (110 c.c. ; 90 g. of 95 per cent) is rapidly poured with continuous shaking but without cooling into a round-bottomed flask (capacity about 1 l.) containing 200 g. (110 c.c.) of concentrated sulphuric acid ; the warm mixture is then cooled to room temperature, 75 g. of ice-water are cautiously added with continued cooling, and finally 100 g. of finely powdered potassium bromide are dropped in. The reaction mixture is then heated on a small sand bath over the largest possible flame so that a fairly rapid distillation takes place (Fig. 42). Since the ethyl bromide has a low boiling point the longest condenser available is used along with an adapter or else a coil condenser (Fig. 24) and a really rapid current of water is passed through. Before the distillation is begun sufficient water is poured into the receiver to submerge the end of the adapter and a few pieces of ice are thrown in. When oily drops, which sink in the water, cease to pass over, the reaction is at an end. If the distillate tends to be sucked back into the condenser during the distillation, the

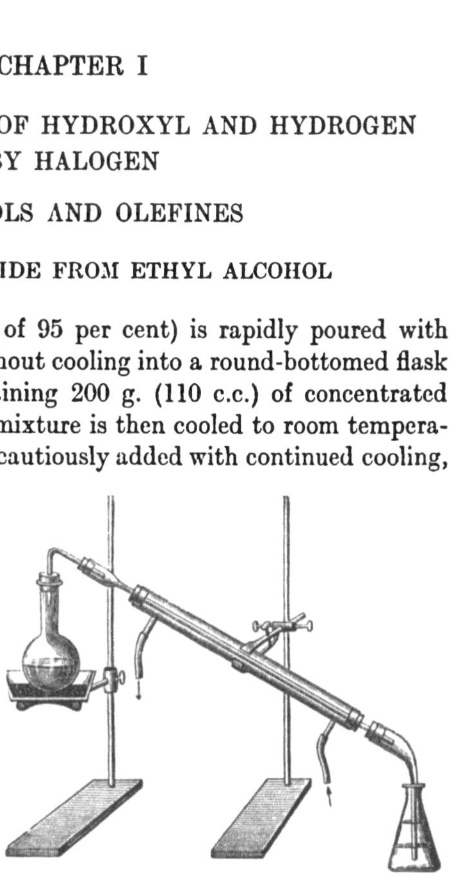

FIG. 42.

receiver should be lowered so that the adapter dips only slightly into the liquid. This can also be brought about by turning the adapter sideways. When distillation is over the contents of the receiver are transferred to a suitable separating funnel, the ethyl bromide which forms the lower layer is run into a conical flask (capacity 250 c.c.) and then the ethyl ether, which is also produced during the reaction, is extracted from the ethyl bromide with concentrated sulphuric acid. Since heat is liberated during this process and material would be lost by evaporation, the crude ethyl bromide is cooled in a freezing mixture and the sulphuric acid is added drop by drop from a funnel with shaking until a lower layer of the acid separates. The layers are again separated in a smaller funnel and the ethyl bromide, which has been dried by the sulphuric acid, is finally distilled into a receiver cooled in a freezing mixture. The distilling flask is heated in water in a porcelain pot or basin over a small burner. Between 35° and 40° the ethyl bromide passes over, the bulk at 38°–39°.

On account of the low boiling point care must be taken during the experiment that the preparation is never left for any length of time in an open vessel. Moreover, the finished preparation which is to be used for further work (cf. ethylbenzene) should *not* be kept in a thin-walled flask, least of all in summer, but should be stored in a thick-walled reagent bottle. Yield 70–80 g.

In this case, as in all others, a calculation should be made at the conclusion of the experiment of the percentage of the theoretical yield which has been obtained, keeping in mind the following considerations. According to the chemical equation one mole of alcohol (46) should be used for one mole of potassium bromide (119). Actually, however, in the case of organic reactions, which as a rule do not proceed quantitatively, one of the components is used in excess, in keeping with the law of mass action (pp. 131-134), and its choice is often determined by economic considerations. Thus, for example, 1 kg. of potassium bromide costs about 6s., and 1 kg. of duty-free alcohol, 1s. 2d. The price of a mole of KBr (119 × 6s.) is therefore to that of a mole of alcohol (95 per cent) (46 × 1s. 2d.) approximately as 14 : 1. From the economic standpoint it is therefore advisable to use the cheaper alcohol in excess in order that as much as possible of the dearer bromine compound may be converted into ethyl bromide. The proportions used in the above experiment have been adapted to this consideration. Theoretically 39 g. of alcohol are required for 100 g. of KBr, whilst actually 86 g.

(90 g. of 95 per cent) are used, *i.e.* more than double the theoretical quantity. Hence when the theoretically possible yield is calculated here, the amount of potassium bromide used must be taken as the basis of reckoning. If it is desired to convert an alcohol which is more valuable than KBr into its bromide, the inorganic substance would naturally be used in excess.

The preparation is used for making *ethylbenzene* (p. 107) and *diethyl ethylmalonate* (p. 245).

Methyl Bromide.—The simplest alkyl bromide is prepared in an essentially similar way (Bygdén, *J. pr. Chem.*, 1911, **83**, 421). Since it boils at 4·5° it is difficult to keep in stock, but its direct application in the Grignard reaction, in place of the dearer iodine compound, is very much to be recommended. Uses analogous to those of ethyl bromide.

2. ETHYL IODIDE FROM ETHYL ALCOHOL[1]

Absolute alcohol (50 c.c.) is poured into a small flask (capacity about 200 c.c.) containing 5 g. of red phosphorus and then, during the course of a quarter of an hour, 50 g. of finely powdered iodine are gradually added with frequent shaking and occasional cooling, by dipping the flask into cold water. An efficient water condenser is then attached to the flask, the mixture is left for two hours with frequent shaking and heated for two hours on the water bath under reflux. Then the ethyl iodide is distilled through a downward condenser, preferably with the flask dipping into the vigorously boiling water. If the last part of the material distils only with difficulty, the water bath is removed and the flask, after drying, is heated over a luminous flame which is kept in constant motion. The distillate, which is coloured brown by dissolved free iodine, is shaken in a separating funnel first with water (repeatedly) to remove the alcohol, then with a few drops of bisulphite solution to remove the iodine, and, finally, with an equal volume of sodium hydroxide solution. The colourless oil so obtained is drawn off from the funnel, dried with a little granulated calcium chloride, and distilled directly over a small flame. If the calcium chloride should float on the ethyl iodide, pour the latter into the distilling flask through a funnel plugged with asbestos or glass wool. Ethyl iodide boils at 72°. Yield about 50 g. What percentage of the theoretical is this?

[1] F. Beilstein, *Annalen*, 1863, **126**, 250.

Use in the synthesis of ethyl diethylmalonate and for Grignard reactions.

Methyl iodide is prepared in a completely analogous manner. Boiling point 44°. Used like ethyl iodide.

Remarks on 1 and 2.—These two reactions are special cases of a general change, namely, the substitution of an alcoholic hydroxyl group by a halogen atom. This substitution can be carried out in two ways; first, as in the preparation of ethyl bromide, **1**, by acting on alcohols with hydrogen halides, *e.g.*

$$C_2H_5.\boxed{OH + H}Br = H_2O + C_2H_5.Br,$$
$$(HCl,\ HI)$$

or, second, as in the preparation of ethyl iodide, **2**, by treating alcohols with halogen compounds of phosphorus, *e.g.*

$$3C_2H_5.OH + PI_3 = 3C_2H_5.I + PO_3H_3$$
$$(PCl_3,\ PBr_3).$$

The first reaction proceeds most easily with hydrogen iodide since in many cases mere saturation with the gaseous acid suffices to bring it about. Hydrogen bromide reacts with greater difficulty, and in its case it is frequently necessary to heat the alcohol saturated with this acid in a sealed tube. The preparation of ethyl bromide described above, in which the HBr is liberated from the potassium bromide by means of concentrated sulphuric acid, constitutes a very smooth example of this reaction.

Hydrogen chloride reacts most difficultly, and here it is necessary (as, for example, in the preparation of methyl and ethyl chloride), to use a dehydrating agent, preferably zinc chloride, or, as in the case of the higher alcohols, to heat under pressure in a sealed vessel.

Aromatic alcohols, *e.g.* benzyl alcohol, are thus esterified by concentrated halogen acids more easily than the aliphatic.

The reaction cannot be extended to phenols, however, but it can be applied to di- and polyhydric alcohols. Here the number of hydroxyl groups replaced depends on the experimental conditions, *e.g.* quantity of halogen hydride, temperature, and so on, *e.g.*

$$\begin{array}{l} CH_2.OH \\ | \\ CH_2.OH \end{array} + HBr = \begin{array}{l} CH_2.Br \\ | \\ CH_2.OH \end{array} + H_2O,$$

Ethylene glycol — Ethylene chlorohydrin

$$\begin{array}{l} CH_2.OH \\ | \\ CH.OH \\ | \\ CH_2.OH \end{array} + 2HCl = \begin{array}{l} CH_2.OH \\ | \\ CH.Cl \\ | \\ CH_2.Cl \end{array} + 2H_2O.$$

Glycerol — Dichlorohydrin

Hydrogen iodide not only esterifies polyhydric alcohols but also reduces them. Thus glycerol passes by way of 1:2:3-triiodopropane into isopropyl iodide:

$$CH_2OH.CHOH.CH_2OH + 3\,HI = CH_2I.CHI.CH_2I + 3\,H_2O$$
$$CH_2I.CHI.CH_2I + 2\,HI = CH_3.CHI.CH_3 + 2\,I_2\,.$$

Similarly the tetrahydric *erythritol* and the hexahydric *mannitol* are converted respectively into 2-iodobutane and 2-iodohexane. Formulate! Naturally the reaction also takes place with hydroxy-acids.

The second reaction proceeds much more energetically than the first, especially if preformed phosphorus halide is used. This is, however, not always necessary, at least not in the case of replacements by bromine and iodine; in many cases the procedure is rather to produce the halide only during the reaction, either by dropping bromine from a separating funnel into a mixture of alcohol and red phosphorus or, as above, by adding finely powdered iodine. Like the former, this reaction can also be applied to polyhydric and to substituted alcohols; indeed, it is possible to replace all the OH groups by halogens, and in particular also by chlorine.

In many cases the solid, more active and much less volatile *pentachloride* is used instead of phosphorus trichloride. Then it is necessary to use a whole mole of PCl_5 for each mole of alcohol since the reaction leads to the formation of the much more sluggish *phosphorus oxychloride, e.g.*

$$CH_3.CH_2OH + PCl_5 = CH_3.CH_2Cl + POCl_3 + HCl\,.$$

In recent years *thionyl chloride* has also been brought into use for the same reaction; it has the advantage that its decomposition products are gaseous and hence do not interfere with the working up of the reaction mixture.

$$\underset{\text{Amyl alcohol}}{C_4H_9.CH_2OH} + SOCl_2 = \underset{\text{Amyl chloride}}{C_4H_9.CH_2Cl} + SO_2 + HCl\,.$$

The more energetic action of the halide of phosphorus also shows itself in the fact that even the hydroxyl groups of phenols can be replaced by this reaction:

$$\underset{\text{Phenol}}{C_6H_5.OH} + \underset{(Br_5)}{PCl_5} = \underset{(Br)}{C_6H_5.Cl} + \underset{(Br_3)}{POCl_3} + \underset{(Br)}{HCl},$$

The yields thus obtained are much less satisfactory since the phosphorus oxychloride, acting on unchanged phenol, produces esters of phosphoric acid, *e.g.*

$$POCl_3 + 3\,C_6H_5.OH = PO(OC_6H_5)_3 + 3\,HCl\,.$$

The alkyl halides $C_n\,H_{2n+1}$ Cl (Br, I) are colourless and usually liquid; methyl chloride, methyl bromide, and also ethyl chloride, are gaseous at ordinary temperatures and the members of high molecular weight, such as cetyl iodide $C_{16}H_{33}I$, are semi-solid, paste-like masses.

In addition to " methyl iodide " the term " *iodomethane* " can be used. For purposes of classification the alkyl halides are to be regarded as esters. The halogen does not ionize and its mobility increases in the order chloride, bromide, iodide.

As esters the alkyl halides are hydrolysed by alkalis to alcohols and salts of halogen acids. They are converted by nascent hydrogen into hydrocarbons, by ammonia into amines, by alkoxides into ethers, by alkali hydrogen sulphides into mercaptans, by potassium cyanide into nitriles, and by sodium acetate into acetic esters. (Formulate these reactions.) The alkyl halides are practically insoluble in water but are, on the other hand, miscible with organic solvents. As a consequence of the great affinity of iodine for silver, the alkyl iodides are almost instantaneously decomposed by aqueous-alcoholic silver nitrate solution, and so yield silver iodide and alcohol. The important method of Zeisel for the quantitative determination of alkyl groups combined in the form of ethers, depends on this property (cf. p. 68).

Chlorine and bromine can be replaced by iodine by means of alkali iodide, and this is of importance in cases where direct treatment of alcohols with hydriodic acid gives a bad yield or none at all, *e.g.* in the preparation of ethylene iodohydrin :

$$CH_2OH.CH_2Cl + NaI = CH_2OH.CH_2I + NaCl\,.$$

Since the reaction requires heat, it cannot be carried out in an aqueous medium for fear of hydrolysis ; moreover, in most cases, the insolubility of the chloride in water precludes the use of this solvent. Following Finkelstein, acetone well dried with calcium chloride and anhydrous sodium iodide, which is rather soluble in acetone, are used.

The iodo-compounds soon turn brown (iodine) on keeping, especially if exposed to light. In order to decolorise them again they are shaken with finely divided silver.

Since halogen hydride can be removed from the alkyl halides, this group is also directly connected with the olefines:

$$H_3C.CH_2Br \xrightarrow{-HBr} H_2C = CH_2 + HBr\,.$$

The removal is most suitably accomplished by means of alcoholic potassium hydroxide solution.[1] In many cases tertiary bases are also used, such as pyridine, quinoline, or dimethylaniline.

Of the *synthetic reactions* of the alkyl halides that with potassium cyanide, which enabled H. Kolbe to synthesise acetic acid from a methane derivative, has already been mentioned (cf. the preparations

[1] This reagent is very much used and it is best to keep a stock. Dissolve 25 g. of potassium hydroxide sticks in 100 c.c. of *methyl* alcohol by warming or by leaving over night in the cold ; potassium hydroxide in *ethyl* alcohol soon resinifies. Remove carbonate by filtration and determine the KOH content by titration.

on pp. 126 and 244). Of the simpler syntheses that of Wurtz may be mentioned here. Metallic sodium removes the halogen from two molecules and the two radicles combine. Thus, in the simplest case, ethane is formed from methyl bromide:

$$\underset{Na}{H_3C.Br} \quad \underset{Na}{Br.CH_3} \longrightarrow H_3C.CH_3 + 2NaBr.$$

The mechanism of this reaction is discussed on p. 109. It is applied preparatively to triphenylchloromethane, p. 344.

The alkyl halides have also acquired extraordinary importance as starting materials for the Grignard reaction, discussed on pp. 328-332.

3. BENZYL CHLORIDE FROM TOLUENE [1]

The use of cork or rubber connections should be avoided when working with chlorine, bromine, or halogen acids. For the present preparation the flask [2] shown in Fig. 42 (with inlet tube) is used. A *short* thermometer, the lower part of which rests in a support consisting of a glass tube 3–4 cm. long and constricted in the middle, is slipped into the flask, held horizontally. The end of the support which rests on the bottom of the flask is rounded in a flame so as to prevent scratching. Next toluene (100 g.) is poured into the flask and heated to boiling in the air bath. A rapid current of chlorine from a cylinder is now passed in through a wash-bottle with sulphuric acid and the ground-in inlet tube, until the temperature of the *vigorously* boiling liquid has reached 156°. In order that escaping chlorine may be rendered harmless the upper end of the condenser is connected by means of a tube, which should *not* dip into the liquid, to a vessel containing caustic alkali solution. Experience has shown that the time during which chlorine has to be passed in depends on the illumination [3]; in bright sunlight the reaction is over in a few hours, whereas on dull days it occupies half a working day. As far as possible, therefore, advantage is taken of the light.

Next the contents of the flask are directly distilled. After a fraction consisting of unchanged toluene has passed over, the bulk of the distillate is collected over a range of 7° (at 12 mm. between 63° and 70°). Pure benzyl chloride boils at 64°/12 mm. Yield 65–70 per cent of the theoretical.

The preparation obtained by vacuum distillation is purer and

[1] Cannizzaro, *Ann. Chim.*, 1855, [3] 45, 468; Beilstein and Geitner, *Annalen*, 1866, 139, 332; Schramm, *Ber.*, 1885, 18, 608.

[2] It should be obtainable on loan from the demonstrator.

[3] G. Book and J. Eggert, *Z. Elektrochem.*, 1923, 29, 521; *Ber.*, 1926, 59, 1192; F. Bergel, *Ber.*, 1926, 59, 153.

keeps better than that distilled under atmospheric pressure, since in the latter case elimination of HCl always occurs.

Uses: for *benzyl cyanide* (p. 126), ethyl benzylmalonate (p. 246), the Grignard reaction.

The theoretically simplest method of replacing hydrogen united to carbon by halogen consists in acting on saturated hydrocarbons with free halogen. The process, like the reaction between chlorine and hydrogen, is catalytically accelerated by light and when applied to methane and chlorine leads to the conversion of this hydrocarbon into mono-, di-, tri-, and tetrachloromethane. The higher paraffins are also chlorinated in this way, but, as a preparative method, the process is inconvenient and has the disadvantage that various reaction products are produced simultaneously and are difficult to separate. As a general rule, the chlorine first attacks that carbon atom which is poorest in hydrogen. In the aliphatic series the alcohols, which are more easily obtainable in the pure state than the hydrocarbons, constitute the sole starting materials for the preparation of the halogen compounds (reactions 1 and 2). Substitution by chlorine is much simpler in the case of toluene and the homologous methylbenzenes (xylenes, etc.). Here we have two quite distinct processes.

1. In the cold substitution only occurs when typical halogen-carriers such as iron filings or iodine are used. From toluene the *o*- and *p*-derivatives are produced together.

2. In the absence of such a carrier the benzene ring is not attacked in the slightest, even at the boiling point. The rate of substitution in the methyl group (side chain), which is unmeasurably small in the cold, increases to an extent sufficient for preparative purposes, in accordance with the general law that each rise of 10° in the temperature produces a two to threefold increase in the velocity of the reaction. This reaction is sensitive to light, like all direct replacement of hydrogen by chlorine. The statement that addition of phosphorus pentachloride also accelerates the reaction is incorrect. The reaction between toluene and chlorine presents a very fine example of the specific influence of various catalysts on a reacting system.

For preparative purposes it is of great importance that the entrance of the second chlorine atom into the side chain proceeds at a much lower rate than the first phase of the reaction. Hence almost all the chlorine is used up by the toluene present before any significant further chlorination of the benzyl chloride occurs. The proximity of the benzene ring makes the chlorine less firmly attached to the side chain, *i.e.* more mobile than in the true paraffins. In explanation it is suggested that the benzene ring uses up more of the combining energy of the methane carbon atom than does alkyl or hydrogen, and that therefore less is available for the chlorine. (Thiele and Werner's theory of the distribution of affinity.) From this example we learn that although the connecting lines in our formulae give formal expression to the quadrivalency of carbon, they tell us nothing about the energy relationships of the individual bonds. It is only through a command of the systematic principles that the chemist can acquire the insight and understanding which enable him to read more into the diagrammatic formulae than is indicated by the uniform links between the individual atoms.

Benzyl chloride undergoes all the transformations of the alkyl halides. Hydrolysis with hot aqueous alkalis yields the corresponding alcohol, *benzyl alcohol* $C_6H_5.CH_2OH$, a colourless liquid which boils at 206°. (Chap. V. 4, p. 212 top.)

When benzyl chloride is treated with ammonia under suitable conditions *benzylamine*, $C_6H_5.CH_2NH_2$, is obtained. It is a rather strong, liquid base which exhibits all the characteristics of the aliphatic amines and differs completely from the ring-substituted aminotoluenes (toluidines) which are isomeric with it.

In general it may be said that all changes in the methyl group of toluene and of compounds of analogous constitution take place in the same way as in purely aliphatic alkyl groups.

Continued chlorination of toluene introduces a second and finally a third chlorine atom into the side chain.

Benzylidene chloride, $C_6H_5.CHCl_2$, a colourless liquid which, like benzyl chloride, provokes tears, is the technical starting material for the preparation of benzaldehyde. Cf. Chap. V. 3, p. 200.

Benzotrichloride (phenylchloroform), $C_6H_5.CCl_3$.

The effect of the benzene ring on the combining power of the adjacent carbon atom here shows itself with special clarity. Whereas chloroform is rather resistant towards alkalis, benzotrichloride is hydrolysed by them with extraordinary ease; all three chlorine atoms are removed and benzoic acid is produced. It would be wrong to think, however, that in this reaction all the chlorine is removed at the same time in accordance with the equation:

$$C_6H_5.C\begin{matrix}\diagup Cl & NaOH\\ -Cl & NaOH\\ \diagdown Cl & NaOH\end{matrix} \rightarrow C_6H_5.C\begin{matrix}\diagup OH\\ -OH\\ \diagdown OH\end{matrix} + 3NaCl \rightarrow C_6H_5.C\begin{matrix}\diagup OH\\ \\ \diagdown\!\!\diagdown O\end{matrix} + H_2O.$$

All chemical reactions proceed in stages and usually by the interaction of two molecules (reactions of the second order or bimolecular reactions). Hence our reaction will take place in stages and should be formulated as follows:

$$C_6H_5.C\begin{matrix}\diagup Cl\\ -Cl\\ \diagdown Cl\end{matrix} + NaOH \longrightarrow \underset{I}{C_6H_5.C\begin{matrix}\diagup OH\\ -Cl\\ \diagdown Cl\end{matrix}} + NaCl \xrightarrow{+NaOH} \underset{II}{C_6H_5.C\begin{matrix}\diagup\!\!\diagup O\\ \\ \diagdown Cl\end{matrix}}$$

$$+ NaCl + H_2O \xrightarrow{+NaOH} C_6H_5.C\begin{matrix}\diagup\!\!\diagup O\\ \\ \diagdown OH\end{matrix} + NaCl.$$

The intermediate products are hydrolysed much more rapidly than benzotrichloride and therefore escape detection.

As regards intermediate product I it should also be remarked that compounds of this type, containing hydroxyl and halogen attached to the same carbon atom, are not capable of existence, but immediately undergo the change

$$>C<^{OH}_{Cl} \longrightarrow >C{=}O + HCl.$$

Experiment.—Boil a few drops of benzyl chloride with (halogen-free) alcoholic potassium hydroxide solution for a few minutes in a test tube on the water bath. Then dilute with water, acidify with nitric acid, remove undissolved material by extraction with ether, and add a few drops of silver nitrate solution.

No bromine ion is liberated when the analogous experiment is performed with bromobenzene (next preparation). Distinction between halogen in aliphatic and aromatic combination.

Analysis of the Benzyl Chloride.—The quantitative determination of halogen in substances containing halogen in aliphatic combination is not carried out in a sealed tube by the Carius method (cf. p. 65), but by hydrolysis with standard alcoholic potassium hydroxide solution. Since this method is very often used, a check on the purity of the present preparation may be combined with practice in this method of analysis.

In a small round-bottomed flask which has been frequently used and well steamed, an accurately weighed amount of benzyl chloride (about 1 g.) is boiled for one hour under reflux with one and a half times the calculated amount of approximately normal alcoholic sodium hydroxide solution; the solution is then diluted with two volumes of water, and the excess of alkali is titrated with $0{\cdot}5\,N$-hydrochloric acid after addition of phenolphthalein.

This method can naturally only be used when no *other* acids are formed. In that case the halogen is titrated with thiocyanate by Volhard's method.

4. BROMOBENZENE

A round-bottomed flask (capacity 500 c.c.) has a side tube, to which a condenser is attached by means of a ground joint, and a dropping funnel in the upright neck, also provided with a ground joint (Fig. 43; corks and rubber stoppers are so seriously attacked by bromine that clean working without ground joints is made very difficult). The upper end of the condenser is connected by means of a waxed cork to a large Péligot tube (Fig. 44), or to a conical flask, in which the hydrogen bromide produced is absorbed (in the latter case the delivery tube does not dip into the water).

Benzene (90 c.c; 1 mole) and iron filings (2 g.) are poured into the flask and 53 c.c. (160 g.) of bromine are added little by little from the funnel with shaking. The beginning of the reaction and the accompanying evolution of hydrogen bromide are awaited, and the addition of bromine is then so regulated that the reaction proceeds briskly without becoming violent. If, towards the end, the substitution becomes too sluggish, the flask is heated for a short time on the water bath until all the bromine is consumed. The reaction mixture is then transferred to a larger round-bottomed flask and distilled with steam. As soon as crystals of *p*-dibromobenzene appear in the condenser the receiver is changed and this by-product is then driven over completely.

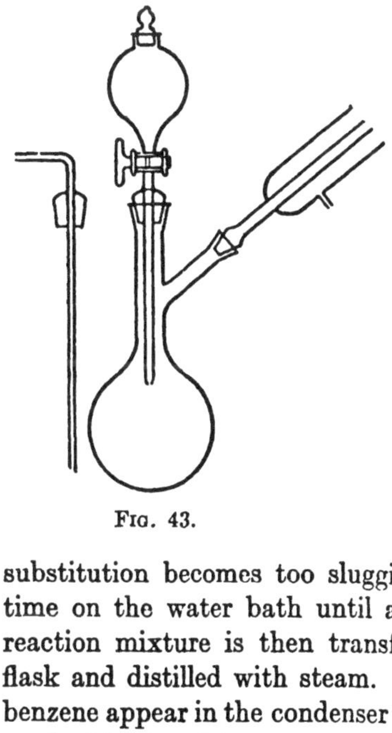

FIG. 43.

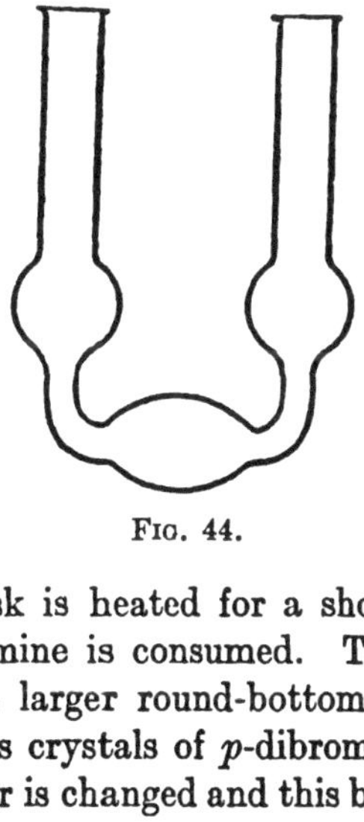

FIG. 44.

The monobromobenzene, which came over first, is allowed to settle in a separating funnel, drawn off, dried for one hour with calcium chloride, and then distilled. When the bulk of the fraction which passes over between 140° and 170° is redistilled it yields a distillate which mainly boils between 152° and 158° and consists of fairly pure bromobenzene. Yield 70–80 g. The pure compound boils at 155°. Use in the preparation of ethylbenzene (p. 107) and in the Grignard reaction (p. 328).

p-**Dibromobenzene.**—The residue which remains in the flask at the first distillation is poured while still hot into a small porcelain basin, and after it has solidified it is freed from oily matter on a porous plate; the product from the steam distillation is dried in the same way.

In this operation and in all similar ones the substance should not be pressed down on the plate with the spatula but laid on with gentle pressure, so that the capillary action of the plate exerts its full effect. In the case of very sticky substances the material is transferred after a few hours with the spatula to an unused part of the plate.

After drying, the *p*-dibromobenzene is recrystallised from a little alcohol from which it separates in magnificent colourless prisms. Melting point 89°.

Hydrobromic Acid as By-Product.—During the reaction 80 g. of hydrogen bromide are formed; about 200 c.c. of water are required for absorption. Therefore, if an insufficiently large receiver is used, its contents must be renewed as soon as fumes begin to appear. The hydrobromic acid is purified by distillation from a flask having a condenser slipped over the side tube (Fig. 19). After some water has passed over, the boiling point rises to 126°, and at this temperature 48 per cent acid passes over. This material is very useful for all kinds of work in the laboratory. Thus, for example, the potassium bromide required for the preparation of the alkyl bromides can be prepared from it by transferring to a capacious vessel and adding the calculated amount of potassium carbonate until the neutral point is reached.

A useful rule for operations of this kind: Set aside a small part of the less accessible substance—here the hydrobromic acid—so that, if the neutral point is accidentally passed, no difficulty is experienced.

Pure bromobenzene loses no bromine ions when boiled with potassium hydroxide solution. Check this.

The halogen is very firmly fixed to the benzene nucleus; the aryl halides do not give the reactions characteristic of the alkyl halides. Only by hydrogen activated by a catalyst or by vigorous nascent hydrogen (sodium in alcohol) can the halogen be replaced, although the aryl halides also react with magnesium (Chap. IX. 1, p. 328), and the halogen is likewise removed in the Fittig synthesis described in one of the subsequent sections.

If it is desired to compare bromobenzene with a halide of the aliphatic series, the saturated ethyl bromide must naturally not be chosen for comparison, but rather substances of the type of *vinyl bromide,*

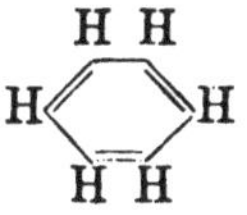

$$\underset{Br}{HC}=CH_2,$$

i.e. substances which have the halogen attached to a *doubly* linked carbon atom. Then it is found that the halogen is also very firmly bound in halogenated olefines of this type, so that *no* fundamental difference exists between them and the halogen substitution products of benzene.

The reactivity of aromatically bound halogen is increased by nitro-groups in the *ortho-* and *para*-positions; likewise the chlorine in *o*-chlorobenzoic acid is rather loosely held.

How is the *course of halogen substitution in the benzene nucleus* to be explained? It is not at all probable that direct replacement of hydrogen occurs, such as we must assume in the formation of benzyl chloride and in the reaction between methane and chlorine, since the hydrogen attached to the doubly bound carbon atom of olefines exhibits no special reactivity. However, various facts which will be considered later (p. 155) indicate that benzene reacts with halogen in fundamentally the same way as does ethylene. The behaviour of ethylene towards bromine is the subject of the next preparation.

In the case of either substance the bromine is doubtless first added on at the double bond. Whilst the reaction takes place easily with the reactive double bond of the olefines, carriers such as iron, iron halide, and aluminium bromide are required for the sluggish double bond of the benzene ring:

$$H_2C=CH_2 \xrightarrow{Br_2} BrCH_2-CH_2Br;$$

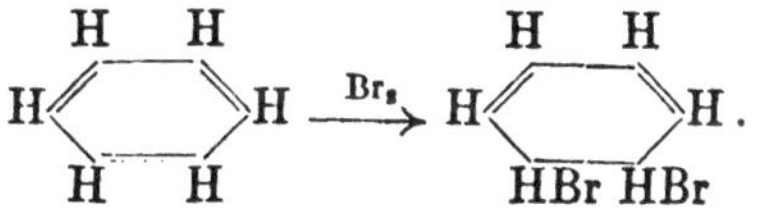

The ethylene addition product is saturated, that of benzene, on the other hand, is more unsaturated than is benzene itself, since the symmetrical cancelling of residual valency (Thiele) is upset and the "aromatic" character is destroyed. In order to re-establish this character it is only necessary for hydrogen bromide to be eliminated, which takes place with liberation of energy. The elimination occurs with extraordinary speed, even before the other double bonds, which have become reactive, have time to take up bromine.

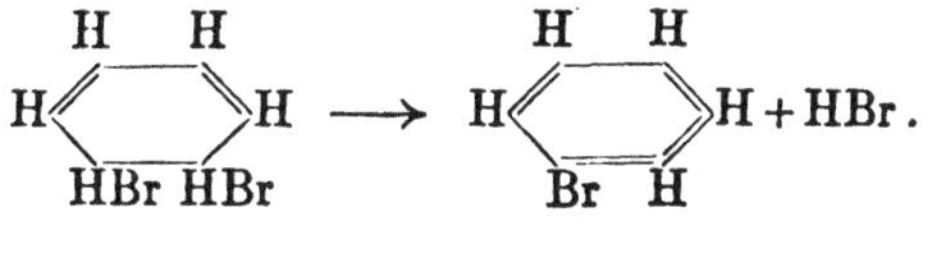

In direct sunlight three molecules of chlorine or bromine add themselves to the three double bonds of benzene to form *hexachlorocyclohexane* or *hexabromocyclohexane* (benzene hexachloride):

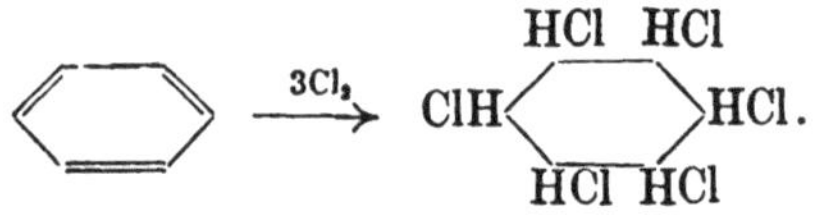

Whereas in *benzene* and in its derivatives the six substituents lie in the same plane, namely, that of the ring, they are distributed in *cyclohexane* in two planes parallel to that of the ring. Hence there results a special type of spatial isomerism when two hydrogen atoms united to different carbon atoms are replaced. The isomerism is caused by the position of the two substituents, for they may lie in the same plane (*cis*-form), or one in each plane (*trans*-form). The phenomenon is closely related to the *cis-trans* isomerism of the ethylenes of which the best-known example is that of maleic and fumaric acids.

Thus two stereo-isomeric forms of 1:4-dihydroxycyclohexane (quinitol) are known:

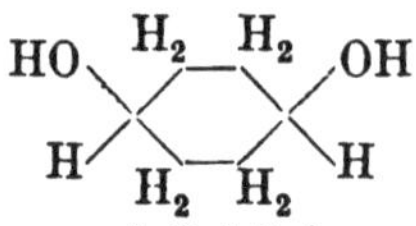

cis-Quinitol

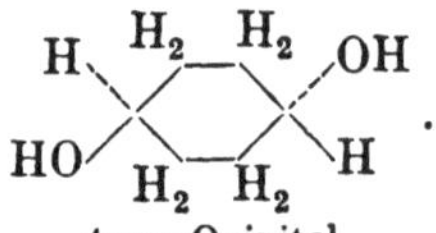

trans-Quinitol

The isomerism of the two known benzene hexachlorides is also to be attributed to a spatial difference of this kind.

5. ETHYLENE FROM ETHYL ALCOHOL.[1] ETHYLENE DIBROMIDE

A freshly prepared and preferably still warm mixture of 25 g. (30 c.c.) of ordinary alcohol and 150 g. (90 c.c.) of concentrated sulphuric acid is heated *not too strongly* (to 160°) with 60 g. of fine sea sand[2] or an equal weight of anhydrous aluminium sulphate in a large round-bottomed flask (capacity about 3 l.) on an asbestos wire gauze or on a sand bath. The flask is stoppered with a *very tightly* fitting two-holed cork. Through one hole is pushed a thermometer which dips into the liquid and through the other a T-tube (internal diameter 0·6–0·8 cm.). A dropping funnel with a long stem is inserted through the upper constricted opening of this tube and fixed with a small piece of rubber tubing, while the end of the side tube is also constricted, and is connected with the receivers (Fig. 45). Before the cork is pushed in, the stem of the funnel, which is con-

[1] Erlenmeyer and Bunte, *Annalen*, 1873, 168, 64; 1878, 192, 244.
[2] Quartz accelerates the elimination of water (Senderens).

stricted at the end by drawing out, is filled by suction from a mixture of 190 c.c. (150 g.) of alcohol and 170 c.c. (300 g.) of concentrated sulphuric acid.

As soon as vigorous evolution of ethylene has set in the alcohol-sulphuric acid mixture is dropped in at such a rate that no copious frothing occurs and a regular current of ethylene is evolved. The temperature is kept continually under control (small flame !).

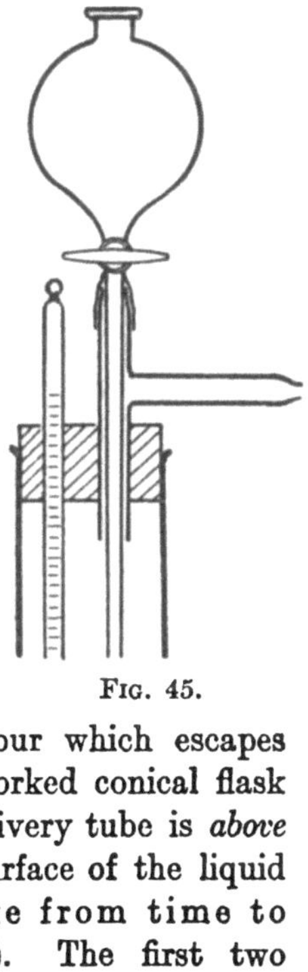

FIG. 45.

In order to remove alcohol and ether the gas is passed through a wash-bottle containing concentrated sulphuric acid[1], and sulphurous acid is removed in a three-necked wash-bottle[2] containing 4 *N*-sodium hydroxide and provided with a safety tube (Fig. 46). The gas then passes into two moderately wide wash-bottles each containing 25 c.c. of bromine ; in order to reduce loss by evaporation the bromine is covered with a layer of water 1 cm. deep and the two bottles are kept cool in a vessel of cold water. If the pressure permits, the last receiver in the train is a Péligot tube (Fig. 44, p. 91) containing 2 *N*-sodium hydroxide solution ; if not, the bromine vapour which escapes is absorbed in sodium hydroxide solution in a corked conical flask (notch the side of the cork !) in which the delivery tube is *above* the surface of the liquid (shake from time to time !). The first two wash-bottles are conveniently closed with rubber stoppers. As soon as the bromine has been decolorised or, at least, as soon as bromine vapour is no longer visible over the brownish-red reaction product (in normal circumstances after two to three hours) the flask is

FIG. 46.

[1] Since ethylene combines with hot sulphuric acid to form ethylsulphuric acid, it may be necessary to cool this wash-bottle.

[2] Notice that during the evolution of the gas the sodium hydroxide solution must rise in the middle safety tube to about 20–30 cm. above the inner level. Why ?

disconnected from the receivers. Then the crude ethylene dibromide is shaken in a separating funnel with water and sodium hydroxide solution until decolorised and washed repeatedly with water. After drying with calcium chloride it is completely purified by distillation. Boiling point 130°. Yield 125–150 g. Used in the preparation of glycol (p. 102) and also as a solvent.

The use of *highly concentrated phosphoric acid* obviates the very troublesome frothing which occasionally occurs and is to be attributed to the oxidising action of the sulphuric acid. Only by cautious heating can the frothing otherwise be kept within bounds. Commercial syrupy phosphoric acid (150 g.) is dehydrated by heating slowly to 220° in a porcelain basin with constant stirring. The ethylene is produced in a small flask set up according to the directions given; the alcohol is added drop by drop from the funnel to the acid, which is poured in while cold and then heated to 210°–220°. The funnel is filled with alcohol beforehand. In order to absorb alcohol vapour it is sufficient to introduce into the train *one* wash-bottle, containing a saturated aqueous solution of calcium chloride and cooled in ice. In this case the amount of alcohol required is considerably smaller.

Calculate how much alcohol is required to yield the amount of ethylene necessary to decolorise the bromine.

What volume does this amount of ethylene occupy ?

If the ethylene produced with sulphuric acid is analysed (method ?) it is found that very much carbon monoxide is present. The phosphoric acid method is better suited for the production of the pure gas, but the best way is to remove with zinc dust and glacial acetic acid the bromine from the ethylene dibromide already prepared. The dibromide is dropped into a suspension of an excess (not too great) of zinc dust in alcohol and glacial acetic acid (2·5 moles) and the ethylene is collected over water in a gas-holder.

The intramolecular elimination of water from alcohols, which constitutes the usual method for the preparation of olefines, does not proceed so simply, when concentrated acids are used, as the equation

$$CH_3—CH_2OH \longrightarrow CH_2{=}CH_2 + H_2O$$

would suggest. Already on gentle warming alcohol is esterified by concentrated sulphuric acid and yields ethylsulphuric acid. It is from the decomposition of the latter that the ethylene arises.

$$CH_3.CH_2OH \xrightarrow{H_2SO_4} CH_3.CH_2.O.SO_3H \longrightarrow CH_2{:}CH_2 + H_2SO_4 .$$

It will be remembered that the ethylsulphuric acid first formed is decomposed when hot (130°) by *excess* of alcohol and in this way ether is prepared.

$$CH_3.CH_2.O.SO_3H + HO.CH_2.CH_3 \longrightarrow CH_3.CH_2.O.CH_2.CH_3 + H_2SO_4 .$$

Ethyl ether is also formed as a by-product during the preparation of ethylene.

Ethylene, the "*oil-forming gas*", was prepared already in 1795 from spirits of wine and oil of vitriol by the five Dutch chemists, Deiman, Troostwyk, Bondt, Louwerenburgh, and Crells.

Technically, ethylene is obtained from alcohol by catalytic removal of water with *alumina* heated to 200°–300° (Senderens). The alcohol vapour is led over the hot material.[1] For the preparative application of such reactions aluminium phosphate is as suitable as alumina.

Instead of the acid sulphuric ester of the alcohol, as in our example, the ester of another acid, *e.g.* benzoic, is often submitted to thermal decomposition, and so the charring action of the sulphuric acid is avoided.

Potassium bisulphate and anhydrous boric or oxalic acid are also used (acrolein from glycerol, pyruvic acid from tartaric acid).

The chemical characteristics of the olefines depend on their double bond, which can take part in a great variety of additive reactions.

The following substances are added :

1. **Halogens,** chlorine and bromine with especial ease, yielding dihalides.

2. **Hydrogen Halides** to give alkyl halides. For preparative purposes hydrobromic acid dissolved in glacial acetic acid is generally used, and since the reaction proceeds slowly, the components are heated in a sealed tube.

3. **Sulphuric Acid** (cf. supra) and other acids, *e.g.* acetic acid (technical application in the terpene group).

4. **Nitric Acid.**—In the presence of concentrated sulphuric acid ethylene yields the nitric ester of nitroethyl alcohol.

[1] Directions suitable for laboratory experiments are given by W. Kesting, *Z. angew. Chem.*, 1925, 38, 362.

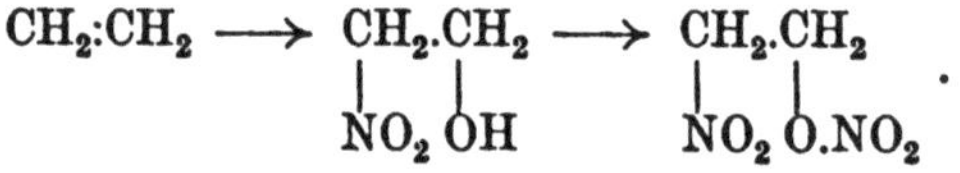

5. **Hypochlorous Acid** according to the equation:

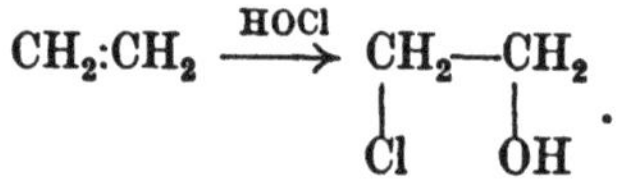

Thus ethylene chlorhydrin is obtained by passing ethylene and carbon dioxide together into bleaching powder solution.

6. **Nitrogen Peroxide,** to form dinitroethanes:

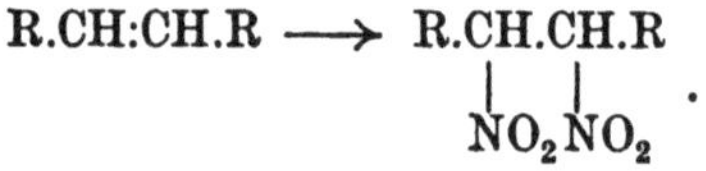

With nitrogen trioxide N_2O_3 is added and the bimolecular pseudonitrosites are formed.

7. **Ozone** (Harries, Staudinger):

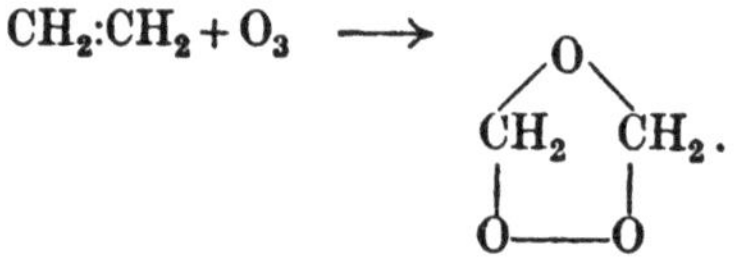

Since the ozonides are decomposed on heating with water, in accordance with the equation

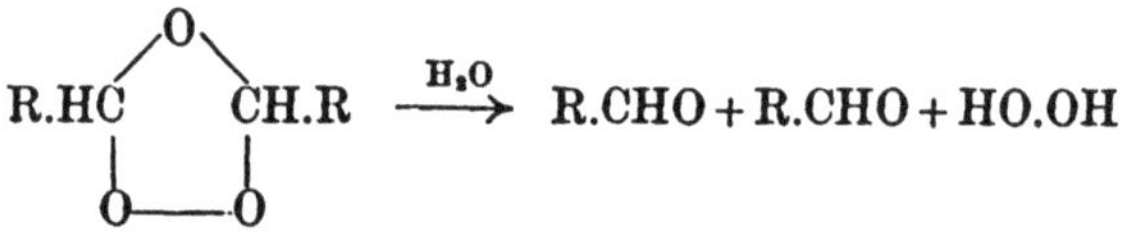

they provide a means of synthesising *aldehydes* (or ketones). The hydrolysis takes place at the ether linkage and *dihydroxyalkyl peroxides* RH(OH)C.O—O.C(OH)HR are formed as intermediate compounds (see also p. 196). These latter further decompose into aldehyde (or ketone) and hydrogen peroxide (Rieche). Benzene adds three molecules of ozone, and the triozonide (ozobenzene) $C_6H_6O_9$ is decomposed by water into three molecules of glyoxal.

8. **Hydrogen.**—The olefines cannot be hydrogenated with nascent hydrogen from any of the usual reducing agents. The reduction only succeeds catalytically with hydrogen in the presence of finely divided metals, such as nickel (Sabatier), palladium (Paal, Skita), or platinum (Fokin, Willstätter). Cf. the preparations on pp. 367-374.

9. **Perbenzoic Acid** (Prileschaiev's reaction). Alkylene oxides are formed.

$$R.CH{:}CH.R + C_6H_5.\overset{\overset{O.OH}{|}}{C}{:}O \longrightarrow \underset{\diagdown O \diagup}{R.CH.CH.R} + C_6H_5.COOH\,.$$

10. **Hydroxyl.**—At low temperature the olefines are converted into their glycols by permanganate :

$$R.CH{:}CH.R \longrightarrow R.CHOH.CHOH.R\,.$$

This oxidising agent, however, readily breaks the double bond, and the carbon atoms which it joined are oxidised further.

If they are also united to hydrogen, carboxylic acids are formed, otherwise ketones.

$$R.CH{:}C\begin{matrix}\diagup CH_3\\ \diagdown CH_3\end{matrix} \longrightarrow R.COOH + OC\begin{matrix}\diagup CH_3\\ \diagdown CH_3\end{matrix}\,.$$

The reaction with permanganate constitutes a valuable and much-used test for unsaturation in an organic compound. The substance is dissolved in cold alcohol, a few drops of sodium carbonate solution are added, and then a drop of dilute permanganate solution. Rapid disappearance of the red colour indicates the presence of a double bond. The " Baeyer test " can also be carried out in pure glacial acetic acid, which is stable towards permanganate. Another method of detecting double bonds is by the decolorisation of bromine. As a rule, chloroform is used as solvent.

In regard to the rate at which the above-mentioned addition reactions proceed, the olefines exhibit very great differences, depending on the nature of the molecule. When a double bond is shown in a formula it does not necessarily follow that it is also capable of all the above transformations. Thus, for example, it is quite impossible to add bromine to *tetraphenylethylene* $(C_6H_5)_2C{:}C(C_6H_5)_2$. The affinity of the double bond, therefore, differs from case to case.

If two double bonds are adjacent they can react as a closed system in additive reactions. Thus *butadiene* adds on bromine partly in the manner shown in the following equation :

$$\overset{1}{C}H_2{=}\overset{2}{C}H.\overset{3}{C}H{=}\overset{4}{C}H_2 \xrightarrow{Br_2} BrCH_2.CH{=}CH.CH_2Br\,.$$

Its dicarboxylic acid, *muconic acid*, yields the $\beta\gamma$-unsaturated dihydromuconic acid on reduction :

$$HOOC.CH{=}CH.CH{=}CH.COOH \longrightarrow HOOC.CH_2.CH{=}CH.CH_2.COOH.$$

In both cases the two original double bonds disappear and a new one appears between them; addition has taken place in the 1:4-positions.

The principle of 1:4-addition has found a particularly interesting and, from the standpoint of preparative work, important application in the "*diene synthesis*" discovered by Diels.[1] In this synthesis butadiene and numerous butadiene derivatives (isoprene, cyclopentadiene) add themselves to the single carbon bond yielding derivatives of tetrahydrobenzene. Thus from butadiene and acrolein *tetrahydrobenzaldehyde* is formed :

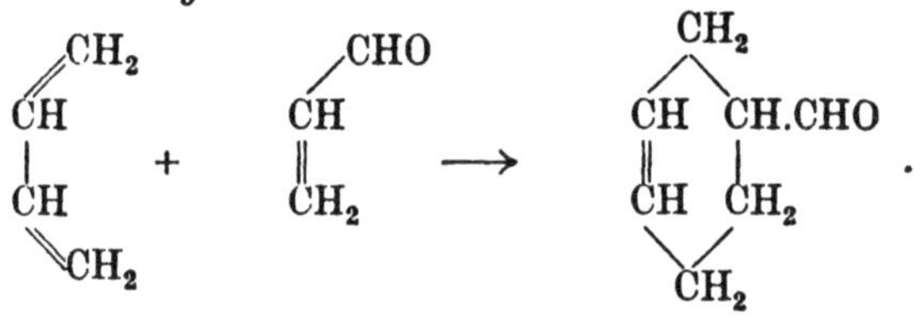

With cyclic olefines bicyclic compounds are obtained, *e.g.* from butadiene and cyclohexene *octahydronaphthalene* :

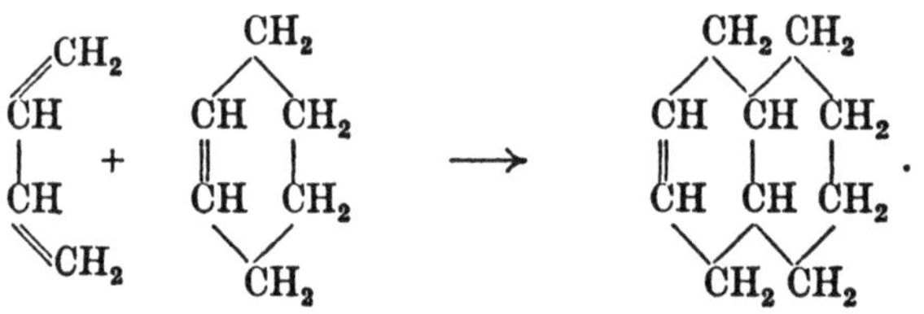

When the "diene" used is *cyclopentadiene*, endo-cyclic ring systems result such as are also produced by vegetable cells in camphor and other terpenes.

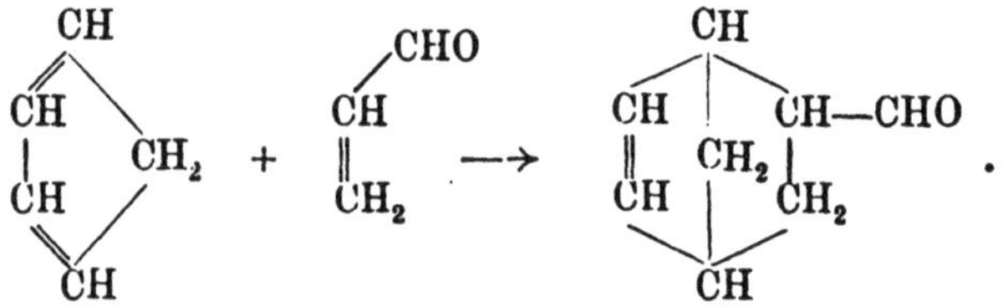

According to Thiele, the explanation of these reactions is that the fields of force, which surround the carbon atoms separated by

[1] *Annalen*, 1928, 460, 98; see also *Z. angew. Chem.*, 1929, 42, 911.

double bonds, partially cancel each other between C_2 and C_3 because of the proximity of these atoms in space, so that a higher chemical potential exists at C_1 and C_4 than at C_2 and C_3. Consequently the places at which addition occurs preferentially are C_1 and C_4.

Applied to benzene this conception provides for a structure which is much more saturated than that possible in olefines with several double bonds. In benzene the points of attack, which are always present in the open chain, are lacking :

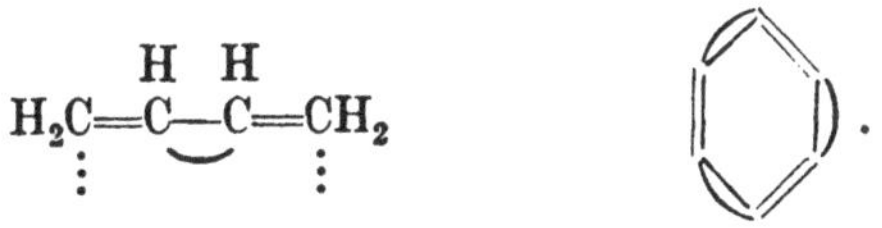

If, with Thiele, we indicate the inactivation of adjacent carbon atoms by a bracket, as shown above, we see that in benzene all the " partial valencies " cancel out.

This picture of the benzene structure, which thus results from Thiele's conceptions, seems still to-day the most suitable for deducing and understanding the "aromatic" character of benzene in relation to the nature of the olefines.

We believe, however, that the concept of " inactivation " of the partial valencies should be replaced by that of their enfeeblement. For, in the first place, by no means all substances which add themselves to systems of adjacent (" conjugated ") double bonds do so in the 1:4-positions, and secondly, benzene actually does exhibit the typical behaviour of a substance with three double bonds, since it directly adds halogen, hydrogen catalytically activated (to give cyclohexane), and ozone, albeit more slowly than an olefine does. It is just this adoption of Thiele's hypothesis in an attenuated degree which seems to render the lower velocity of the addition reactions of benzene comprehensible. Unexpectedly, the higher ring homologue of benzene, the hydrocarbon *cyclo-octatetraene* (Willstätter and Waser), is by no means the chemical analogue of benzene. It is yellow, and exhibits the great reactivity of an olefine with four double bonds :

CH.CH:CH
CH   CH .
CH:CH.CH

The higher conjugated unsaturated systems will be mentioned later in connection with the polyenes and the *carotinoids* (p. 224).

6. GLYCOL (ETHYLENE GLYCOL) FROM ETHYLENE DIBROMIDE[1]

Glycoldiacetate.—A mixture of 63 g. (0·33 mole) of ethylene dibromide, 20 g. of glacial acetic acid, and 60 g. of freshly fused and finely powdered potassium acetate (cf. p. 116) is placed in a short-necked, round-bottomed flask (capacity 0·5 l.) provided with a reflux condenser and is vigorously boiled for two hours on a sand bath or wire gauze over a large flame. The flask is then connected by means of a bent tube with a downward condenser and the reaction product is distilled over by means of a large luminous flame, which is kept in constant motion and is made progressively less luminous towards the end of the distillation. The distillate is then mixed with a further 60 g. of ethylene dibromide and 80 g. of potassium acetate; the mixture is vigorously boiled for two to three hours on a sand bath as described above and is again distilled. By means of a Hempel column of about 10 cm. length the distillate is fractionated and the following fractions are collected separately: 1. Up to 140°, 2. 140°–175°, 3. 175° to completion. Fractions 2 and 3 are then re-distilled separately, when pure glycoldiacetate passes over between 180° and 190° (the bulk at 186°). Yield about 70 g.

If it is desired to improve the yield, the portions which pass over below 180° are heated once again for three hours with an equal weight of potassium acetate and are then treated as above described. In this way the yield is increased by another 15 g.

Glycol.—In order to obtain free glycol from the ester, the latter is "transesterified" by boiling with a solution of hydrogen chloride in absolute methyl alcohol. This solution is prepared by passing hydrogen chloride into cooled absolute methyl alcohol, while excluding moisture, until the concentration is about 3 per cent, as indicated by the increase in weight on a balance sensitive to 0·1 g.; if too much hydrogen chloride has been passed in, more alcohol is added.

Glycoldiacetate (49 g. = 0·33 mole) is boiled in a 200 c.c. round-bottomed flask with 60 c.c. of the hydrogen chloride solution for half an hour and then the methyl acetate and part of the methyl alcohol are removed by distillation (slowly at first) through a downward condenser, the rest directly *in vacuo* at about 50°. In order to separate small quantities of unchanged ester from the glycol

[1] Henry, *Bull. Soc. Chim.*, 1897, [iii.], 17, 207; *Chem. Zentr.*, 1907, I, 1314.

remaining in the flask, the latter is closed by a rubber stopper, and the contents are shaken with 50 c.c. of absolute ether, in which glycol is insoluble. Adherent ether is then removed on a boiling water bath and the hot glycol is poured into a small distilling flask, fitted with an air condenser, and distilled. The main portion passes over at 195°. Yield 17–18 g. (80–90 per cent of the theoretical).

Ethylene dibromide can also be converted into glycol by direct hydrolysis with dilute alkali carbonate solution; since, however, the reaction proceeds very slowly (in the heterogeneous system), and since, moreover, large amounts of water have to be evaporated, the indirect method here adopted is to be preferred; it has the further advantage of teaching two new reactions. This method—much used for converting an alkyl halide into the corresponding alcohol—consists in first transforming it into the acetic ester by means of potassium acetate (often also by silver acetate); this ester is then hydrolysed, usually in the ordinary way by means of aqueous alkali or mineral acid. In the present case, where water-soluble glycol is the end product, we do not, however, wish to forego the advantage of working in organic solvents, so that we remove the acyl group from the ester by esterifying it with another alcohol. The acid indeed distributes itself between the two alcohols, glycol and methyl alcohol, but mass action favours the latter, which is present in large excess. Hydrolysis of this type may be called *trans-esterification*. Further details about the formation and hydrolysis of esters will be found on p. 131 *et seq.*

The reactions of the simplest dihydric alcohols, the 1:2-glycols, may be illustrated by means of the initial member of the series.

When heated with sulphuric acid glycol loses water and yields *acetaldehyde.*

Concentrated hydrochloric acid forms *ethylene chlorhydrin*; the second OH-group is replaced by chlorine with much greater difficulty:

$$CH_2OH.CH_2OH \xrightarrow{HCl} CH_2OH.CH_2Cl + H_2O.$$

On a large scale ethylene chlorhydrin is prepared by the addition of hypochlorous acid to ethylene, by passing CO_2 and ethylene simultaneously into a solution of bleaching powder. Strong potassium hydroxide converts chloroethyl alcohol into *ethylene oxide*, by removal of HCl:

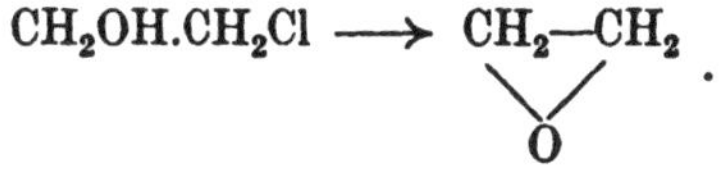

Special mention should be made of the ready conversion of the chlorhydrin, by means of *trimethylamine*, to *choline*, a substance of great physiological importance. Choline chloride is obtained very easily when equimolecular amounts of the two components are made to interact for some time under the influence of heat; a concentrated solution of the amine in absolute alcohol is used.

7. ISOAMYL ETHER[1]

A mixture of 500 g. of commercial amyl alcohol (b.p. 128°–132°) and 50 g. of concentrated sulphuric acid is heated to gentle boiling in a distilling flask having the side tube high up in the neck. Water and amyl alcohol distil over slowly and the temperature of the boiling mixture, as indicated on a thermometer which dips into the liquid, rises to 140° in the course of about eight to nine hours. Some time before this temperature is reached the amyl alcohol which has distilled is separated from the water and returned to the distilling flask. The contents of the flask are cooled to about 100° and distilled with steam; the distillate separates into two layers. Using a column or, still better, a Widmer spiral (see p. 18) the oily layer is fractionally distilled. The crude amyl ether passes over in a yield of 200–230 g. at 168°–172°. For complete purification it is boiled in an oil bath for two hours under a reflux condenser with finely powdered sodamide (1·5 g. per 100 g. of ether) and is then separated from the sodamide by distillation. The distillate is shaken with dilute hydrochloric acid, dried over calcium chloride and then carefully distilled over sodium.

8. CHLOROACETIC ACID FROM ACETIC ACID AND CHLORINE[2]

Dry chlorine is led into a mixture of 150 g. of glacial acetic acid and 12 g. of red phosphorus contained in a flask (Fig. 43) which is provided with a delivery tube and a reflux condenser and is heated on a vigorously boiling water bath. The flask should stand in a well-lit place, preferably in direct sunlight, since the rate of chlorination greatly depends on the illumination. As soon as a small portion of the mixture solidifies when cooled with ice-water and rubbed with a glass rod, the reaction is complete. In summer one day suffices, but on dull winter days the passing in of chlorine must be continued for a second day. To isolate the monochloroacetic acid the product of the reaction is fractionally distilled from a flask fitted with an air condenser; the fraction which passes over at 150°–200°

[1] G. Schroeter and W. Sondag, *Ber.*, 1908, **41**, 1924.
[2] R. Hoffmann, *Annalen*, 1857, **102**, 1; Russanow, *Ber.*, 1892, **25**, Ref. 334.

is separately collected in a beaker. The beaker is cooled with ice-water while its contents are stirred with a glass rod and the portion which solidifies, consisting of pure monochloroacetic acid, is rapidly filtered at the pump; the loose crystals are firmly compressed with a spatula or pestle. The filtration must not be continued too long or the chloroacetic acid will gradually be liquefied by the heat of the air. The filtrate is redistilled, the portion passing over between 170° and 200° being separately collected. By repetition of the process just described (cooling and filtering) a second portion of monochloroacetic acid is obtained. It is combined with the main portion and the whole is completely purified by a further distillation; b.p. 186°, m.p. 63°, yield, variable: 80–125 g. Used for the preparation of *nitromethane* (p. 147), *diethyl malonate* (p. 244), *glycine* (p. 264), *phenylglycine* (p. 360).

Since monochloroacetic acid, especially when warm, attacks the skin severely, care is indicated.

The chlorination proceeds considerably more rapidly and also in the absence of light, if 1·5 g. of iodine, 7 g. of phosphorus pentachloride, and 3 g. of red phosphorus are added to the 150 g. of glacial acetic acid mentioned above.[1] When the reaction is over, the hot mixture is decanted from the phosphorus, diluted with 40 c.c. of glacial acetic acid, and cooled. The monochloroacetic acid which crystallises is rapidly filtered and washed with a little glacial acetic acid. In this way a pale red product results, which is freed from adherent iodine by standing in a desiccator over potassium hydroxide.

The introduction of chlorine or bromine into a saturated chain is facilitated by the presence of a C:O-group. Thus, aldehydes and ketones are halogenated with great ease, the halogen entering exclusively in the α-position. It has been shown that in this process the enol-form which is present in minute amounts takes up bromine additively.[2]

$$CH_3.C(OH){:}CH_2 \longrightarrow CH_3.\underset{Br}{COH}.CH_2Br \longrightarrow CH_3.CO.CH_2Br\ .$$

Enol-form of acetone — Intermediate product

The intermediate product (which cannot be isolated) is then immediately converted into bromoacetone with elimination of HBr (Lapworth).

[1] H. Brückner, *Z. angew. Chem.*, 1927, 40, 973; 1928, 41, 226.

[2] The proof is based on the fact that the reaction between acetone and bromine is recognised as being *unimolecular* and not, as would be expected, bimolecular. In the determination of the velocity of reaction, therefore, the (slow) conversion of the ketone into the enol is measured, whilst the addition of the bromine occurs with immeasurable rapidity.

The "loosening" influence of the *carboxyl group* on adjacent hydrogen is much smaller. Hence, in carboxylic acids, substitution by halogen proceeds with much more difficulty but can be accelerated by illumination and by catalysts (carriers). The point of entry of the halogen is here also always at the carbon atom adjacent to the carboxyl group, the α-carbon atom.

Iodine is suitable as a carrier in chlorinations since it combines with chlorine to form the reactive iodine monochloride, viz.:

$$CH_3.COOH + ClI \longrightarrow CH_2Cl.COOH + HI.$$

Since the hydrogen iodide so formed is immediately reconverted by chlorine into iodine, iodine monochloride being then re-formed, we have here a good example of catalysis by a carrier, readily interpreted from the chemical standpoint.

The action of phosphorus is quite different and much more complicated. The phosphorus halide which is first formed reacts with the acid to form the acid chloride and this, with a second molecule of the acid, forms the anhydride:

$$(a)\quad CH_3.COCl + HOOC.CH_3 \longrightarrow CH_3.CO.O.CO.CH_3 + HCl.$$

Now the anhydride is much more easily chlorinated (by substitution) than the acid and the intermediate product thus formed is ultimately hydrolysed by the hydrogen chloride produced in the process:

$$(b)\quad \begin{matrix} CH_2Cl.CO \\ H_3C.CO \end{matrix}\!\!>\!O + HCl \longrightarrow \begin{matrix} CH_2Cl.COOH \\ CH_3.COCl \end{matrix}.$$

The regenerated acetyl chloride can again take part in the reaction according to equation *a*.

Although in the present case the amount of phosphorus used is small, equivalent amounts are often employed, particularly for the introduction of bromine, the acid bromide being produced and then substituted in the α-position. As a result, the reaction product is the bromide of the α-brominated acid which must be treated with water to convert it into the latter. Esters of the acid are often prepared from the acid bromide by the action of alcohols. (Hell-Volhard-Zelinsky process.)

The α-halogenated carboxylic acids, of which the simplest is chloroacetic acid, are widely used in synthesis. Their conversion into hydroxy-acids (by hydrolytic elimination of the halogen) and into aminoacids (p. 264) may be mentioned here:

$$ClCH_2.COOH + HOH \longrightarrow HOCH_2.COOH + HCl$$

$$ClCH_2.COOH + 2NH_3 \longrightarrow H_2N.CH_2.COOH + NH_4Cl.$$

The introduction of iodine is carried out according to the methods given on pp. 83, 84. β-Halogenated carboxylic acids are obtained by the addition of hydrogen halides to αβ-unsaturated acids:

$$\underset{\text{Acrylic acid}}{CH_2{:}CH.COOH} \xrightarrow{HBr} \underset{\beta\text{-Bromopropionic acid}}{CH_2Br.CH_2.COOH}.$$

9. ETHYLBENZENE FROM BROMOBENZENE AND ETHYL BROMIDE [1] (FITTIG'S REACTION)

100 c.c. of dry, alcohol-free ether [2] are poured into a dry round-bottomed flask (capacity 0·5 l.) fitted with a long reflux condenser and containing 27 g. of sodium cut as thinly as possible into slices with a sodium knife. The flask, which stands on a cork ring in an empty water bath, is left for some time, and when no further evolution of hydrogen is to be observed a mixture of bromobenzene (60 g.) and ethyl bromide (60 g.) is added. When boiling (which, in some cases, requires to be initiated by gentle warming) sets in, the flask is cooled with water and, after the beginning of the reaction has been observed, is left without further attention until the following day; no water is passed through the condenser during the night. Next day, after the ether has been removed by distillation from the water bath, the free ethylbenzene is distilled from the sodium bromide and excess of sodium.[3] For this purpose the flask is clamped as high as possible in an oblique position; a *large luminous flame* is used and *kept in constant motion*; an air condenser is employed and may be wrapped in a few strips of moist filter paper. The crude product is distilled twice through a column. The boiling point of pure ethylbenzene is 135°. Yield about 25 g.

[1] Tollens and Fittig, *Annalen*, 1864, 131, 303.

[2] In order to prepare absolute ether the commercial product (1–2 l.) is first dried for one to two weeks over calcium chloride (weight about 20 per cent of the weight of the ether taken) and poured rapidly through a folded filter into a dry flask into which sodium wire is driven from a press. Until the evolution of hydrogen ceases, a calcium chloride tube, drawn out to a fine capillary so as to reduce the evaporation of the ether, is fitted to the flask. For most purposes this absolute ether is used without further treatment.

When evaporating large amounts of impure ether which has been exposed to the air for a long time, the possibility of the occurrence of a violent explosion at the end of the evaporation, due to the presence of *peroxides*, must be taken into account. Such ether has a pungent odour, and liberates iodine from an acid solution of potassium iodide. The peroxides can be destroyed by shaking with a weakly acid solution of ferrous sulphate.

[3] According to Schlubach and Goes, the blue colour of this mixture is to be attributed to a colloidal solid solution of sodium in sodium bromide, analogous to blue rock salt.

The mixture of sodium bromide and sodium remaining in the flask must be handled with *extreme care*. Either the excess of metal is destroyed after cooling by pouring alcohol on to it (100 c.c. poured on in one lot while the flask stands in the fume chamber) or the mixture is shaken out of the flask in the open and water is sprayed on from *a great distance* or from under cover. Finally, the material adhering to the inside of the flask is removed with less alcohol than in the former method.

Many a serious accident has occurred in laboratories as a result of working with metallic sodium. When handling it, therefore, every precaution must be taken, and no residues should be thrown into sinks or waste buckets. It should never be allowed to lie exposed, but should be immediately returned to the stock bottle or destroyed as described above. For its dissolution sodium requires 15 to 20 times its weight of alcohol. The use of the boiling water bath or of the steam bath is to be avoided when carrying out a reaction with metallic sodium or potassium. An oil or sand bath should always be used, even when distilling dried ether from sodium wire. Students should pay special attention to the soundness of apparatus when working with sodium and potassium, and always bear in mind the consequences which may, in certain circumstances, result from the use of a leaky condenser or of a cracked flask. **Protective goggles should invariably be worn.**

Fittig's reaction is analogous to the Wurtz reaction of aliphatic hydrocarbons, *e.g.*

$$\underset{\text{Ethyl iodide}}{2\,C_2H_5I} + 2\,Na \longrightarrow \underset{\text{Butane}}{C_2H_5.C_2H_5} + 2\,NaI\,,$$

$$C_6H_5Br + C_2H_5Br + 2\,Na \longrightarrow \underset{\text{Ethylbenzene}}{C_6H_5.C_2H_5} + 2\,NaBr\,.$$

It can be applied generally; homologous bromobenzenes such as dibromobenzene on the one hand and all possible alkyl bromides on the other come within its scope. The reaction takes place also, although less readily, between two molecules of aryl bromide:

$$2\,C_6H_5Br + 2\,Na \longrightarrow C_6H_5.C_6H_5 + 2\,NaBr\,.$$

Diphenyl, however, is prepared by thermal dehydrogenation of benzene. (A current of benzene vapour is passed through a red-hot iron tube).

The Fittig reaction does not proceed so simply as the above equations indicate. The work of Acree[1] and Schlubach[2] has shown that

[1] *Amer. Chem. J.*, 1903, 29, 588.

[2] *Ber.*, 1922, 55, 2889; see also Schlenk, *Ber.*, 1917, 50, 262.

in the first phase of the reaction *sodium compounds of the hydrocarbons* are formed. These then react with the second molecule of the organic bromide and sodium bromide is eliminated:

$$1.\ R.Br + 2Na \longrightarrow RNa + NaBr,$$
$$2.\ RNa + BrR' \longrightarrow R{-}R' + NaBr.$$

Since both RBr and R'Br react at first with sodium and since both sodium compounds can react with RBr and R'Br in principle, three reactions, leading to the products R—R', R—R, and R'—R', are possible in Fittig's synthesis.

Bromobenzene reacts more rapidly with sodium than does ethyl bromide, but sodium phenyl reacts more rapidly with ethyl bromide than with bromobenzene: hence the ready production of ethylbenzene in the example given.

As regards the chemical properties of the homologues of benzene, all that need be mentioned here is that on oxidation with permanganate or chromic acid they are converted into the corresponding benzene carboxylic acids—toluene and ethylbenzene into benzoic acid, *p*-xylene into terephthalic acid.

CHAPTER II

CARBOXYLIC ACIDS AND THEIR SIMPLE DERIVATIVES

1. ACID CHLORIDES

(a) Acetyl Chloride[1]

Phosphorus Trichloride (72 g.) is run from a dropping funnel into a distilling flask containing 90 g. (1·5 mole) of anhydrous glacial acetic acid and cooled in water. The flask is fitted with a downward condenser, and next warmed in a porcelain basin of moderate size containing water at 40°–50°, until the vigorous evolution of hydrogen chloride slackens and the originally homogeneous liquid has separated into two layers. The acetyl chloride is then distilled from the lower layer of phosphorous acid by means of a vigorously boiling water bath. A small filter flask attached to the lower end of the condenser by a cork serves as a receiver and its contents are protected against atmospheric moisture by a calcium chloride tube connected by means of a rubber tube. The substance is purified by repeated distillation with the use of a thermometer. The fraction boiling between 48° and 53° is collected separately. (Boiling point of pure acetyl chloride 51°). Yield 70–80 g. Use in the preparation of acetic anhydride, see p. 115, and of acetophenone, p. 336.

The product is tested for the presence of phosphorus by decomposing a few drops with a little warm water and evaporating the solution in a porcelain capsule. The residue is twice evaporated with concentrated nitric acid or with bromine water and is finally tested for phosphoric acid in the usual way. If phosphorus is detected, the material should again be distilled with a few drops of glacial acetic acid.

(b) Benzoyl Chloride

Into a round-bottomed flask (capacity 0·25 l.) fitted if possible with a ground-in condenser and containing 40 g. (0·33 mole) of dry

[1] Béchamp, *Compt. rend.*, 1855, **40**, 946; 1856, **42**, 226.

benzoic acid, 100 c.c. of thionyl chloride are poured. The flask is now heated on the water bath to 80° in the fume chamber. When the vigorous evolution of gas (HCl and SO_2) has ceased (half an hour), the cooled mixture is poured into a distillation flask and the excess of thionyl chloride is as far as possible distilled through a downward condenser from a vigorously boiling water bath. (The distillate can be used again for the preparation of further amounts of benzoyl chloride.)

The benzoyl chloride is now distilled from a wire gauze or heated with a faintly luminous flame. A long air condenser with a receiver similar to that used for acetyl chloride is employed. The first portion of the distillate, which consists chiefly of thionyl chloride, is fairly large (this material also may be used again). Then the bulk passes over at 194°–199°. Pure benzoyl chloride boils at 194°. Yield 40–42 g.

Vacuum distillation, which yields a purer product, is also to be recommended here. Benzoyl chloride is a widely used laboratory product.

The hydroxyl of a COOH-group may be replaced by chlorine by means of reactions, in part similar to that described above for the replacement of alcoholic hydroxyl groups by halogen.

In practice acid chlorides are almost always prepared by the action of PCl_3, PCl_5, or $SOCl_2$ (in exceptional cases $POCl_3$) on the acids themselves or also, in many cases, on their alkali salts. The choice of the chloride to be used depends (1) on the ease with which the acid concerned reacts, and (2) on the boiling point of the acid chloride. If, for example, as in the case of acetic acid and its homologues, PCl_3 reacts readily with the formation of acid chlorides, then this chloride is used in preference to the more energetic PCl_5. In the reaction, the mechanism of which is discussed below, *phosphorous acid* is formed according to the equation:

$$3\,CH_3.C\begin{matrix}\nearrow O\\ \searrow OH\end{matrix} + PCl_3 = 3\,CH_3.C\begin{matrix}\nearrow O\\ \searrow Cl\end{matrix} + PO_3H_3.$$

The evolution of hydrochloric acid which is observed during the experiment is due to a reaction between the phosphorous acid and acetyl chloride or phosphorus trichloride.

Benzoic acid reacts less readily with PCl_3 but energetically with PCl_5. Since the separation of the excess of the chloride (or oxychloride) of phosphorus is much less simple than removal of excess of thionyl chloride, the latter, which is now cheap and easily obtained, is to be preferred (H. Meyer).

The course of the reaction may be regarded as proceeding in such a way that a mixed anhydride is first formed with elimination of HCl and that the anhydride then breaks down into acid chloride and SO_2.

$$\begin{array}{l}R.C{:}O\\ \quad\dot{}\\ \;OH\end{array} + SOCl_2 \xrightarrow[-HCl]{} \begin{array}{l}R.C{:}O\\ \quad\dot{}\\ \;O.SOCl\end{array} \longrightarrow \begin{array}{l}R.C{:}O\\ \quad\dot{}\\ \;Cl\end{array} + SO_2\,.$$

The course of the reaction is similar when the chlorides of phosphorus (or phosgene) are used.

In cases where the reaction proceeds violently, chloroform or benzene is used as a diluent; this holds for the reaction with alcohols also. As a rule phosphorus oxychloride is only used with salts of carboxylic acids. It reacts with these as follows:

$$2\ CH_3.CO.ONa + POCl_3 = 2\ CH_3.CO.Cl + NaPO_3 + NaCl\,.$$

This reaction can be considered as having the advantage that the chlorine of the PCl_5 is more completely utilised than when the free acid is treated.

Lower members of the series of acid chlorides are colourless liquids; the higher members are colourless crystalline solids. At ordinary pressure most of them boil without decomposition. Vacuum distillation is only indicated with those of high molecular weight. The boiling points of the acid chlorides are lower than those of the acids, for the replacement of hydroxyl by chlorine usually causes a lowering of the boiling point—

$CH_3.CO.Cl$	Boiling point 51°	$C_6H_5.CO.Cl$	Boiling point 198°
$CH_3.CO.OH$	„ „ 118°	$C_6H_5.CO.OH$	„ „ 250°.

The acid chlorides have a powerful pungent odour and fume when exposed to the air. Water decomposes them with the formation of acids and hydrogen chloride. Often this decomposition takes place with exceptional ease since the chlorine atom is much more loosely attached to an acid radicle than to an alkyl. Whereas the conversion of an alkyl halide into an alcohol requires prolonged boiling with water, often with addition of sodium or potassium hydroxide, or of a carbonate or acetate, the analogous decomposition of an acid chloride is much more easy. Amongst the lower members the reaction takes place, even in the cold, with exceptional violence and almost immediately, whilst with the higher members, as, for example, benzoyl chloride, heating is required to induce the decomposition. Sulphonic acid chlorides (see benzenesulphonyl chloride, p. 182) even resist for a long time the action of boiling water. Alkalis naturally act much more energetically than water. Alcohols and phenols react with acid chlorides to form esters.

Experiment (*a*).—Gradually pour 0·5 c.c. of *acetyl chloride* into a test tube containing 2 c.c. of water. If the water is very cold, drops of the chloride may be seen to sink in the water without mixing. When the tube is shaken, heat is developed and a vigorous reaction occurs.

Experiment (*b*).—The same should be done with *benzoyl chloride*. Even on prolonged shaking no perceptible change takes place; the mixture must be boiled for some time in order to cause complete decomposition. Treat benzoyl chloride in the same way with $2N$-alkali.

Experiment (*c*).—An equal volume of acetyl chloride is added drop by drop to 1 c.c. of alcohol in a test tube which is cooled in water. The mixture is then treated, while similarly cooled, with an equal volume of water and is carefully made slightly alkaline with sodium hydroxide. If a mobile layer of fragrant ethyl acetate fails to appear, finely powdered common salt is added until no more dissolves. This treatment causes the separation of the ethyl acetate.

Benzoyl chloride is mixed in the same way with excess of alcohol, and the velocity of the reaction and the odour are used as indications of the rate of reaction.

Acid chlorides are also used in order to determine whether or no an unidentified substance contains alcoholic or phenolic hydroxyl groups. If a substance reacts with an acid chloride, such a hydroxyl group is present, since all groups in which oxygen is combined in other ways, *e.g.* in ether linkage, are indifferent to this treatment. The reaction can be considerably facilitated by the addition of alkali or of alkali carbonate.

Finally, the action of acid chlorides on alcohols and phenols is also used to separate them from solutions or to characterise them. For this purpose benzoyl chloride is usually employed. Methyl alcohol, for example, gives with *p*-nitrobenzoyl chloride the beautifully crystalline methyl ester and small amounts of the alcohol can thus be separated from aqueous solution.

Acid chlorides react with salts of carboxylic acids to give acid anhydrides (see next preparation).

It must further be mentioned that the acylation of alcohols, phenols, and amines with acid chlorides (and also anhydrides) is now frequently carried out in *pyridine* solution instead of according to the older Schotten-Baumann method (action of acid chloride in aqueous-alkaline suspension). The hydrogen chloride is fixed by the pyridine.

Acid chlorides react also very easily with ammonia and with primary and secondary organic bases:

$$CH_3.CO.Cl + 2\,NH_3 = CH_3.CO.NH_2 + NH_4Cl$$

$$CH_3.CO.Cl + 2\,\underset{\text{Aniline}}{C_6H_5.NH_2} = \underset{\text{Acetanilide}}{C_6H_5.NH.CO.CH_3} + C_6H_5.NH_3Cl\,.$$

Experiments.—(*a*) Acetyl chloride is added drop by drop to aniline. Accompanied by strong hissing, a vigorous reaction occurs which ceases as soon as an approximately equal volume of the chloride has been added. The liquid is cooled in water while five volumes of water are added. A copious precipitate of *acetanilide* is thrown down ; the amount can be increased by rubbing the walls of the vessel with a glass rod. The precipitate is filtered off and crystallised from a little hot water. Melting point 115°.

(*b*) In the same way aniline is treated with benzoyl chloride.

This reaction is also used to characterise organic bases and to identify through a melting-point determination small amounts, particularly of the liquid bases, by converting them into their usually crystalline acyl derivatives. In order to cause the whole of the base to react—one mole is fixed by the hydrochloric acid liberated—alkali or carbonate is added when aqueous solutions or suspensions are used and dry potassium carbonate or pyridine when anhydrous solvents are employed. Since tertiary bases do not react with acid (acyl) chlorides, no further replaceable hydrogen atom being present, it is possible by the use of an acid chloride to determine also whether a base is, on the one hand, primary or secondary, or, on the other hand, tertiary.

The important application of acid chlorides in the Friedel-Crafts reaction (p. 334) may also be mentioned here.

Hydrogen peroxide can also be acylated by the Schotten-Baumann process. In this way acyl peroxides are obtained.

Preparation of Benzoyl Peroxide.[1]—Hydrogen peroxide (50 c.c. of about 10 per cent aqueous solution) kept well cooled in ice and continually shaken (preferably in a glass-stoppered bottle) is treated alternately with 4 *N*-sodium hydroxide solution and benzoyl chloride, added each a few drops at a time ; the solution is maintained faintly alkaline throughout. After about 30 c.c. of alkali and 15 g. of benzoyl chloride have been used up, the hydrogen peroxide has been decomposed and the benzoyl peroxide has separated in crystalline flocks, while the odour of the chloride has almost completely disappeared. The peroxide is filtered with suction, washed with water, and dried. Yield 10–12 g. Crystallised from a little alcohol, with which it should be boiled for a short time only, the substance forms beautiful colourless prisms. Melting point 106°–108° dec. Heat a *small* quantity rapidly in a dry test tube over a naked flame. An especially pure product is obtained when a concentrated chloro-

[1] von Pechmann and Vanino, *Ber.*, 1894, 27, 1510.

form solution of the peroxide is run into twice as much methyl alcohol.

The simplest method for preparing alkylene oxides involves the use of benzoyl peroxide (Prileschaiev). In absolute ether or, still better, in benzene solution, it is broken by sodium ethoxide into the sodium salt of perbenzoic acid and ethyl benzoate.[1]

$$C_6H_5.\underset{\overset{\|}{O}}{C}.O.O.\underset{\overset{\|}{O}}{C}.C_6H_5 \longrightarrow C_6H_5.\underset{\overset{\|}{O}}{C}.O.ONa + H_5C_2O.\underset{\overset{\|}{O}}{C}.C_6H_5.$$

The unstable per-acid, which like all per-acids is much weaker than the corresponding carboxylic acid, is taken up in chloroform after acidification of the solution of the sodium salt. The chloroform solution serves in the reaction with unsaturated substances which has already been formulated on p. 99. Ethylene itself does not give this reaction.

2. ACETIC ANHYDRIDE[2]

For the preparation of acetic anhydride the same apparatus as for acetyl chloride is employed.

Acetyl chloride (54 g. = 0·75 mole) is allowed to run drop by drop from a tap funnel on to 80 g. of finely powdered anhydrous sodium acetate prepared in the manner described below. When about half of the chloride has been added, the experiment is interrupted for a short time in order to stir the pasty mass of material with a bent glass rod, the lower end of which has been flattened. The rest of the acetyl chloride is then run in at such a rate that none passes over unchanged. The anhydride is now distilled from the residual salt by means of a *luminous* flame kept constantly in motion. Complete conversion of the last traces of unchanged acetyl chloride to acetic anhydride is attained by adding 3 g. of finely powdered anhydrous sodium acetate to the distillate which is finally fractionally distilled. Boiling point of acetic anhydride 138°. Yield 55–60 g. Use for *acetylation*, in *Perkin's synthesis* (Chap. V. 8, p. 223), preparation of *acetophenone* (Chap. IX. 3 *b*, p. 336).

The product should be tested for chlorine by boiling a sample with water and adding dilute nitric acid and a few drops of silver nitrate.

In a similar way fine crystals of *benzoic anhydride* (m.p. 42°) can be prepared. This is also obtained when benzoic acid is boiled with

[1] Baeyer and Villiger, *Ber.*, 1900, 33, 1575.
[2] C. Gerhardt, *Ann. Chim.*, 1853, [iii.], 37, 313.

an excess of acetic anhydride (transformation of one anhydride into another).

Preparation of Anhydrous Sodium Acetate.—The hydrated salt ($3\,H_2O$) is heated over a naked flame in a shallow iron or nickel dish. After the water of crystallisation has evaporated the material solidifies. By careful heating the anhydrous salt is now also fused. After resolidification the still warm substance is powdered and immediately transferred to an air-tight container. Commercial anhydrous acetate should also be fused before use.

Acetyl chloride acts on sodium acetate in the manner expressed in the following equation :

$$CH_3.CO.Cl + CH_3.CO.ONa = \begin{array}{l} CH_3.C{=}O \\ \qquad\quad \rangle O + NaCl\,. \\ CH_3.C{=}O \end{array}$$

Mixed anhydrides which contain two different acid radicles can be prepared in this way from the chloride and salt of two different acids.

Since, as has been shown above in the case of acetyl chloride, an acid chloride can be obtained from the alkali salt of an acid and $POCl_3$, it is not necessary when preparing an anhydride first to isolate the chloride. It may be preferable to make the latter act at once on an excess of the salt so that from $POCl_3$ and the salt an anhydride can be obtained directly in one operation (technical method). Write out the equation for this reaction.

The lower members of the series of acid anhydrides are colourless liquids, the higher ones crystalline solids. They have a pungent odour, are insoluble in water, but dissolve in indifferent organic solvents. Their specific gravities are greater than that of water and their boiling points higher than those of the corresponding acids :

Acetic acid 118°, acetic anhydride 138°.

As a rule, their melting points are lower than those of the corresponding acids. The lower members can be distilled at ordinary pressure without decomposition ; distillation of the higher members must be carried out *in vacuo*.

Towards water, alcohols, phenols, and bases the anhydrides behave, chemically, exactly as do the chlorides, except that the former react more slowly than the latter.

Experiment.—To 3 c.c. of water 0·5 c.c. of acetic anhydride is added. It sinks to the bottom and does not dissolve even on prolonged shaking. If, however, the mixture of water and anhydride is warmed for a short time, the anhydride is hydrated ; it goes into solution more rapidly if dilute alkali hydroxide solution is used instead of water.

Acetic anhydride is very often used to introduce the acetyl group into an alcoholic or phenolic hydroxyl group or into a derivative of ammonia $HNRR_1$. The reaction is accelerated to an extraordinary extent by the presence of a drop of concentrated sulphuric acid.

Experiment.—Acetic anhydride is added to alcohol, aqueous ammonia, aniline, and phenol. A drop of concentrated sulphuric acid is added to the phenol mixture.

By thermal decomposition on the surface of a glowing platinum wire, acetic anhydride loses water and is converted into ketene, the unimolecular anhydride of acetic acid (Wilsmore):

$$\begin{array}{l} H_3C.CO \\ \quad \dot{O} \\ H_3C.\dot{C}O \end{array} \; - H_2O \longrightarrow 2\,H_2C{:}CO\,.$$

In practice ketene is prepared by the thermal decomposition of acetone (Schmidlin):

$$CH_3.CO.CH_3 \longrightarrow CH_2{:}CO + CH_4\,.$$

When the close relationship between the two classes of compounds is examined more closely, the analogy of the anhydrides to the acid chlorides becomes more intelligible. In both, the hydroxyl of the carboxyl group is replaced by the anionic portion of an acid: in the chloride by Cl, in the anhydride by acetoxyl $O.CO.CH_3$.

The anhydrides of the organic acids can also be regarded as *diacyloxides* (acyl = acid radicle, *e.g.* $CH_3.CO$ = acetyl) and can be thus assimilated to the ethers, or *dialkyl oxides*. The ethers are amongst the most indifferent of all the compounds of organic chemistry. Whence, then, comes the great reactivity of the similarly constituted anhydrides? The weak point in the anhydride molecule is to be found, not at the oxygen bridge, but at the double linkage ${}^{O}{>}C{=}O$. Additions, *e.g.* of water and ammonia, take place here:

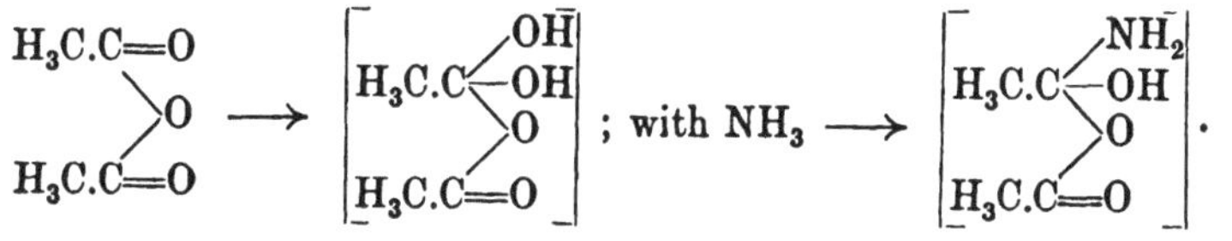

The intermediate products, which are represented within brackets, are exceptionally labile since they contain OH and the negative acetoxyl group attached to the same carbon atom (cf. p. 90); they therefore decompose into two molecules of acid or, in the case of ammonia, into acetic acid and acetamide. The reaction with alcohols is to be formulated in the same way. It will be observed that, when an acyl group is introduced (into an alcohol, amine, etc.) by means of an anhydride, one of the two acid radicles in the molecule is always converted into the acid and is consequently not used in the acylation. The great reactivity of the acid chlorides has the same cause as that discussed in connection with the anhydrides.

3. ACETAMIDE [1]

Ammonium acetate (80 g.)—from ammonium carbonate and glacial acetic acid [2]—and glacial acetic acid (60 c.c.) in a round-bottomed flask fitted with a column of two or three bulbs are heated on a wire gauze for one hour until the boiling point is almost reached and are then kept gently boiling for five to six hours longer. During this second period care is taken that the thermometer fixed in the upper part of the column scarcely registers over 103°; the acetic acid and the water formed in the reaction distil over slowly at the top of the column and can be condensed by means of a small condenser jacket slipped over the side tube. When about 80 c.c. of liquid have collected in the measuring cylinder used as a receiver the reaction is complete. The contents of the flask, after being allowed to cool somewhat, are transferred while still warm to an ordinary distilling flask and after a small preliminary fraction has passed over, the bulk is collected at 195°–220°. If the product does not solidify completely when cooled and stirred, the liquid portion is removed as completely as possible by filtration on a Büchner funnel and the residue is dried on a porous plate in an ordinary desiccator. A further quantity of acetamide can be obtained from the filtrate by distillation. The pure compound boils at 223°. A small sample may be recrystallised from benzene. Melting point 80°. Yield 40–45 g. Use for preparation of *acetonitrile* (Chap. II. 5, p. 130); of *methylamine* (Chap. II. 8, p. 143).

Amides are prepared, quite generally, from the acids by subjecting the ammonium salts to dry distillation, or still better, by heating them for a long time at a high temperature.

[1] In principle according to François, *Chem. Zentr.*, 1906, I, 1089. Hitch and Gilbert, *J. Amer. Chem. Soc.*, 1913, 35, 1780; W. A. Noyes and Goebel, *ibid.* 1922, 44, 2294.

[2] Finely powdered ammonium carbonate is added to 60 c.c. of glacial acetic acid until a sample of the solution, after dilution with water, reacts alkaline. It must be remembered that in this process 0·5 mole of water is formed per mole of ammonium acetate.

Formerly acetamide was usually prepared by heating ammonium acetate to 200° in a sealed tube. By this method, however, the reaction cannot proceed to completion because the water set free in the reaction partly hydrolyses the amide again:

$$CH_3.\underset{\underset{O}{\|}}{C}—ONH_4 \rightleftarrows CH_3.\underset{\underset{O}{\|}}{C}—NH_2 + H_2O.$$

When, as in the process given, the water formed is removed from the reaction mixture by distillation, the back reaction is restricted and the yield increased. At the same time the excess of acetic acid counteracts the dissociation of the salt which takes place according to the equation:

$$CH_3.\underset{\underset{O}{\|}}{C}—ONH_4 \longrightarrow CH_3.COOH + NH_3.$$

See, in this connection, the discussion of the law of mass action on p. 131 *et seq.*

A good method for the preparation of acetamide is to pass ammonia gas into an ethereal solution of acetic anhydride, evaporate the ether and extract the residual mixture of ammonium acetate and acetamide with benzene in an extraction apparatus (Fig. 28). The salt remains undissolved. Amides can also be prepared by the action of ammonia on acid chlorides and esters.

In addition, they are formed in the partial hydration of nitriles by the action of strong mineral acids. An example of this reaction is given on p. 130, middle.

Experiment.—Finely powdered ammonium carbonate (10 g.) is thoroughly mixed in a porcelain basin with 5 g. of benzoyl chloride, by stirring with a pestle; the mixture is then warmed on the water bath till the odour of the chloride has disappeared. Water is now added and the mixture is filtered with suction, washed on the funnel with water, and crystallised from the same liquid. Melting point of *benzamide* 128°.

Apart from the first member of the series, formamide, $HCO.NH_2$, which is a liquid, the amides are colourless crystalline solids. The lower members are readily soluble in water and the higher ones can generally be recrystallised from hot water. The boiling points are much higher than those of the corresponding acids:

Acetic acid, boiling point	118°.	Propionic acid, boiling point	141°.
Acetamide, " "	223°.	Propionamide, " "	213°.

Owing to the acyl group to which it is bound, the amino-group loses its basic character almost entirely. Salts of the amides with strong acids are indeed known, but such salts are immediately and completely decomposed into their constituents by water. Only urea, the diamide of carbonic acid, forms stable salts, the existence of which is made possible by the second NH_2-group.

The compounds of the amides with bivalent mercury are characteristic. In them the metal is united to the nitrogen and is not in ionic combination, as in a salt. They are formed by the action of mercuric oxide, *e.g.*

$$2\,CH_3.CO.NH_2 + HgO \longrightarrow (CH_3.CO.NH)_2Hg + H_2O\,.$$

Experiment.—A small quantity of acetamide is dissolved in water, mixed with a little yellow oxide of mercury, and warmed. The oxide goes into solution and the compound formulated above is formed.

The dehydration which leads to the formation of nitriles and the action of hypohalides on amides are dealt with in the following preparations. The amino-group of the amides, as distinguished from that of the amines, is more or less easily removed by hydrolytic agents, acids being re-formed. For the cause of this difference in behaviour compare the explanation given on p. 117.

Experiment.—A little acetamide is heated in a test tube with 2 *N*-sodium hydroxide solution. An intense odour of ammonia develops and sodium acetate is formed in the solution. The presence of acetic acid is proved by making the solution just acid to Congo red paper with hydrochloric acid, shaking the tube while closing its mouth with the thumb and then heating the solution to the boiling point (porous pot to prevent bumping). Litmus paper held over the mouth of the tube turns red. (General test for volatile acids.)

The reaction between amides and PCl_5, which leads by way of the *amide chlorides* to the *imide chlorides*, need merely be mentioned here.

4. UREA AND SEMICARBAZIDE

(*a*) Potassium Cyanate [1] by Oxidative Fusion [2]

Potassium ferrocyanide (200 g.) is completely dehydrated by careful heating in a porcelain basin or on an iron plate; the crystals must be completely disintegrated and a sample may not produce a

[1] Since there is only one cyanic acid, we do not consider it correct to drop this name and, as is often done, to call the substance *iso*-cyanic acid.

[2] C. A. Bell, *Chem. News*, 1875, 32, 99; Gattermann, *Ber.*, 1890, 23, 1223; H. Erdmann, *Ber.*, 1893, 26, 2442.

film of moisture on the walls of a tube in which it is heated. In the same way 150 g. of potassium dichromate are freed from adherent moisture. After being powdered separately, the two salts are thoroughly mixed in a mortar and the mixture, in portions of 4 to 5 g., is dropped into an iron basin or on to a large iron plate which is strongly heated (but not, however, to red heat) by a powerful burner (Teclu or triple burner). The temperature should be high enough to cause vigorous glowing of each portion, but the spongy black mass thus formed must on no account be allowed to melt. After the very rapid completion of its oxidation, each portion is scraped aside by means of a broad metallic spatula or is entirely removed from the plate. The whole amount can be treated in this way in one to one and a half hours. The various portions of the material, collected in a round-bottomed flask, are covered with hot 80 per cent alcohol (800 c.c.), which is then kept boiling for three minutes by heating the flask on a vigorously boiling water bath. The clear solution which is formed is decanted from the black deposit into a conical flask, which is immediately placed on ice and shaken well so that its contents are cooled as quickly as possible. After the flask has been allowed to stand for a short time, the mother liquor from the deposited cyanate crystals is poured off into the round-bottomed flask containing the melt and the extraction of the latter is repeated until all the salt has been extracted. (Five to six times. A sample in a test tube should no longer deposit crystals on cooling.) The salt is now filtered as dry as possible at the pump, washed twice with alcohol, and then three times with ether, and finally thoroughly dried in a desiccator. Average yield 80 g.

(*b*) Urea

Potassium cyanate (40 g.) and ammonium sulphate (40 g.) are dissolved in 500 c.c. of water in a porcelain basin, and the solution is evaporated to dryness on the water bath. The residue is exhaustively extracted in a round-bottomed flask with boiling absolute alcohol, and the alcoholic solution is concentrated until, on cooling and inoculating, crystallisation sets in. Melting point 132°. The remainder of the urea in the mother liquor is isolated as nitrate after evaporation of the alcohol.

To prepare the *nitrate* a few grammes of urea are dissolved in water and concentrated nitric acid is added drop by drop. The salt separates in fine crystals. Urea nitrate is rather soluble in water, and this fact must be kept in mind when washing it.

(c) Semicarbazide [1]

Hydrazine sulphate (52 g.) and 21 g. of anhydrous sodium carbonate are dissolved in 200 c.c. of boiling water. The solution is cooled to 50°, a solution of 35 g. of potassium cyanate in 100 c.c. of water is added, and the whole is allowed to stand over night. After small amounts of hydrazodicarbonamide, produced in the reaction $H_2N.CO.NH.NH_2 + O{=}C{=}NH \longrightarrow H_2N.CO.NH.NH.CO.NH_2$, have been removed by filtration, 60 c.c. of acetone are added to the solution, which is again allowed to stand, with frequent shaking, for twenty-four hours. The *acetone semicarbazone* which has then crystallised out is filtered dry at the pump, washed with a little water, and dried on a porous plate or *in vacuo*.

The mother liquor is evaporated to dryness on the water bath, and the powdered residue is extracted with alcohol in an extraction apparatus. Semicarbazone crystallises out in the flask of the apparatus. If a sample of the main portion of the material leaves a considerable amount of ash when ignited on platinum foil, this portion also should be extracted in the same way.

To decompose the semicarbazone it is gently warmed with concentrated hydrochloric acid (8 c.c. for each 10 g. of material) until dissolution is just complete. On cooling the solution semicarbazide hydrochloride sets to a thick crystalline mass which is filtered dry at the pump and washed, first with a little cold hydrochloric acid (1 : 1), then twice with 3 to 5 c.c. portions of ice-cold alcohol. The salt is dried in a desiccator. Yield 22–25 g.

In order to obtain the *free semicarbazide*, 5·5 g. of the hydrochloride are ground in a small mortar with 4·5 g. of anhydrous sodium acetate (see p. 116). The paste produced by the formation of free acetic acid is transferred with a spatula to a 100 c.c. conical flask; the last portions are washed in with alcohol and the contents of the flask are then boiled on the water bath with (altogether) 50 c.c. of alcohol. During the boiling the flask is frequently shaken. The hot mixture is freed from precipitated sodium chloride by immediate filtration through a Büchner funnel provided with a hardened filter paper. On cooling, the free semicarbazide slowly crystallises from

[1] Thiele and Stange, *Ber.*, 1894, 27, 31; H. Biltz, *Annalen*, 1905, 339, 250.

the clear filtrate in large prisms which have the appearance of urea crystals. Sparingly soluble in alcohol, very soluble in water.

Experiments.—Being a primary hydrazide (of carbamic acid), semicarbazide reduces ammoniacal silver solutions and Fehling's solution. It reacts readily with aldehydes and ketones with the elimination of water and formation of *semicarbazones*, which, since they are more easily hydrolysed than are phenylhydrazones and oximes, are to be preferred to the latter for purposes of separation and purification of carbonyl compounds. Shake an aqueous solution of the hydrochloride (prepared as described above) with a few drops of benzaldehyde, isolate the semicarbazone and purify it by recrystallisation from alcohol. Melting point 214° dec. Benzaldehyde semicarbazone is decomposed into its constituents by gentle warming with concentrated hydrochloric acid.

The ketones and aldehydes, the preparation of which is described later, should be characterised in the same way by their semicarbazones.

(*d*) Oxidation of Cyanide in Solution [1]

The copper hydroxide obtained by treating 40 g. of crystallised copper sulphate with KOH is added to a solution of 20 g. of potassium cyanide and 20 g. of potassium hydroxide in 40 c.c. of water. The copper hydroxide must be thoroughly washed until sulphate ions are absent. A solution of 32 g. of potassium permanganate in 200 c.c. of water, prepared by heating and rapidly supercooled to 50°, is added to the mixture, which is then heated for two hours at 60° with frequent shaking. The excess of permanganate is then decomposed in the cooled solution by slow addition of hydrogen peroxide, and the clear cyanide-free (tested!) solution is separated by filtration through a Büchner funnel from manganese dioxide which is washed with a little water. The filtrate is concentrated *in vacuo* to a volume of 40–50 c.c. When the concentrated solution is cooled in an ice-salt freezing mixture, the *potassium cyanate* which has been formed crystallises out after a short time and is filtered dry at the pump. It is recrystallised from 80 per cent alcohol as described under 4 (*a*). The yield amounts to 15–18 g.

In this oxidation method the powerful catalytic effect of the

[1] J. Volhard, *Annalen*, 1890, 259, 378; F. Ullmann and Uzbachian, *Ber.*, 1903, 36, 1806; Marckwald, *Ber.*, 1923, 56, 1325. The directions given here are based on the work of Gall and Lehmann, *Ber.*, 1928, 61, 675.

copper is noteworthy. A considerable acceleration of the action of the permanganate is achieved.

If it is desired to prepare *urea* without isolating the potassium cyanate, 55 g. of ammonium chloride are added to the filtrate from the manganese dioxide, the solution is evaporated on the water bath and the urea is isolated as described under (*b*).

Wöhler's synthesis of urea by which a product of the living cell was first prepared artificially a century ago is the prototype of many addition reactions which take place with the reactive molecules of cyanic acid and its esters, as well as with the series of analogous thio-compounds. In these reactions NH_3 is added to the C=N double bond :

$$O{:}\dot{C}{:}NH \xrightarrow{NH_3} O{:}C\left\langle\begin{matrix}NH_2\\NH_2\end{matrix}\right. .$$

It is immaterial for this explanation whether it is assumed that the salt itself is converted or that dissociation first occurs.

Reaction with amines gives substituted ureas—with hydrazine for example, *semicarbazide* :

$$O{:}C{:}NH + H_2N.NH_2 \longrightarrow O{:}C\left\langle\begin{matrix}NH_2\\NH.NH_2\end{matrix}\right. .$$

Reactions of the same kind as those described above, but involving the use of compounds related to cyanic acid, will suggest themselves.

Experiment.—A few c.c. of the cyanate solution are acidified with dilute hydrochloric acid. Carbon dioxide is evolved and the pungent odour of the free cyanic acid, an odour very similar to that of SO_2, is observed.

The decomposition of the free cyanic acid in aqueous solution proceeds in a way which is analogous to that given above in the case of the addition of ammonia or amines. Water is added on and the carbamic acid thus formed decomposes into NH_3 and CO_2 :

$$O{:}C{:}NH \xrightarrow{H_2O} O{:}C\left\langle\begin{matrix}NH_2\\OH\end{matrix}\right. \longrightarrow CO_2 + NH_3 .$$

Both these changes occur together in the decomposition of *phenyl cyanate* (for preparation see p. 144), CO_2 and diphenylurea being formed:

$$O{:}C{:}N.C_6H_5 + H_2O \longrightarrow CO_2 + NH_2.C_6H_5$$

$$O{:}C{:}N.C_6H_5 + NH_2.C_6H_5 \longrightarrow O{:}C(NH.C_6H_5)_2 .$$

The esters of carbamic acid, the *urethanes*, which are formed by combination of alcohols with compounds of the cyanic acid series, are stable substances. The reaction by which they are formed is, likewise, capable of undergoing many variations. It may be recalled that a second method of synthesising them consists in acting on the esters of chloroformic acid with ammonia and amines.

(*e*) Urea (and Uric Acid) from Urine [1]

Urine (2 l.) in a porcelain basin is evaporated to a syrup on the water bath. The flame is extinguished and the hot syrup is stirred with 500 c.c. of alcohol. After some time the clear extract is decanted and the residue is again warmed and once more digested in the same way with 500 c.c. of alcohol. If necessary, the combined extracts are filtered, most of the alcohol they contain is removed by distillation, and the aqueous-alcoholic residue, after transference to a smaller porcelain basin, is evaporated to dryness on the water bath. The dry residue is well cooled and is kept in an efficient freezing mixture while two volumes of colourless concentrated nitric acid are slowly added with thorough stirring. After the product has stood for twelve hours, the paste of urea nitrate is filtered dry at the pump, washed with a little ice-cold nitric acid (1 : 1), again filtered with suction till no more liquid drains off, and suspended in 100–150 c.c. of warm water. To this suspension barium carbonate is added gradually in small portions until the liquid is neutral (addition of excess of the carbonate is to be avoided). When this stage is reached the liquid is boiled with a little animal charcoal (a few portions from the point of a knife) filtered with suction while hot, and, after the material on the funnel has been washed once with hot water, the filtrate is evaporated to dryness. Urea, which is readily soluble in hot alcohol, is completely extracted from the residue with this solvent and is crystallised after concentration of the alcoholic solution. Yield 20–25 g.

The amount of urea excreted daily by an adult is 25 to 30 g. (in an average of 1·5 l. of urine).

Further Experiments.—A solution of urea is made alkaline with sodium hydroxide solution and is then shaken with a few drops of bromine. Nitrogen is evolved. Compare, in this connection, the Hofmann reaction (p. 143).

Nitrite solution is added to an acidified urea solution. Use of urea for the removal of nitrous acid, *e.g.* in the preparation of ethyl nitrate (p. 137).

[1] Salkowski, *Prakt. d. physiol. u. path. Chemie*, p. 161, Berlin, 1900.

Urea is but slowly hydrolysed. A solution is boiled with barium hydroxide solution. How can hydrolysis be detected ?

Uric Acid.—The residue from which the urea was first extracted with alcohol is freed from the solvent by heating on the water bath and 50 c.c. of concentrated hydrochloric acid are added to it. When the resulting mixture has been allowed to stand for a day or two, 0·3–0·5 g. of uric acid will be found to have separated and can be purified by dissolving in a little hot potassium hydroxide solution, boiling this with animal charcoal and adding an excess of hydrochloric acid to the filtrate from the charcoal.

Murexide Reaction.—A few centigrams of uric acid are evaporated to dryness in a small porcelain basin on the water bath with some drops of slightly diluted nitric acid. When ammonia is added to the residue an intense purple colour is produced.

Uric acid is a normal product of metabolism. The chemistry of the purines, the uric acid syntheses of Baeyer and Fischer, Behrend and Roosen, and W. Traube, the chemistry of adenine, guanine, caffeine, and their relationships to uric acid, should be studied.

5. NITRILES

(a) ACETONITRILE [1]

Dry *acetamide* (12 g. = 0·2 mole) is well mixed with phosphorus pentoxide (20 g.) in a small dry flask fitted with a short downward condenser and the mixture is *cautiously* heated with a moderate-sized *luminous* flame. A reaction, accompanied by frothing and bubbling, takes place. After a few minutes the acetonitrile is distilled by stronger heating and is collected in a test tube. To the distillate half its volume of water is added and then solid potassium carbonate is dissolved in the lower aqueous layer of liquid until saturation point is reached. The layers are now separated in a funnel having a short delivery tube, and the acetonitrile is redistilled from a small distilling flask containing a little phosphorus pentoxide, which ensures complete dehydration. Boiling point 82°. Yield about 6 g.

(b) BENZYL CYANIDE

In a round-bottomed flask (capacity 0·5 l.) fitted with a double neck attachment (Anschütz attachment), reflux condenser, and

[1] Dumas, *Annalen*, 1847, 64, 332; Buckton and W. Hofmann, *Annalen*, 1856, 100, 131.

dropping funnel, 30 g. of sodium cyanide are dissolved in 35 c.c. of hot water and 50 c.c. of alcohol are added. Pure benzyl chloride (63 g. = 0·5 mole) is then run in from the funnel in the course of ten minutes. Then the contents of the flask are boiled for three hours longer, cooled, and filtered through a small Büchner funnel. The alcohol is distilled off from the filtrate in the filter flask (a capillary is used during the distillation and the temperature of the bath is kept at 40°–50°); the benzyl cyanide, separated from the solution of sodium chloride by means of a small separating funnel, is distilled *in vacuo* from a Claisen flask after drying for a short time over a small stick of calcium chloride. Boiling point 105°–109°/12 mm. The boiling point of the completely pure substance at 760 mm. is 232°. Yield about 45 g. By redistillation of the first and last portions of the distillate this yield can be increased.

Use for the preparation of *phenylacetic* acid (p. 129) and of *phenylnitromethane* (Chap. VI. 8, p. 246).

When an amide is heated with a dehydrating agent (P_2O_5, P_2S_5, PCl_5) it loses water and is converted into a nitrile, *e.g.*

$$CH_3.CO.NH_2 \longrightarrow CH_3.C{\vdots}N + H_2O\,.$$

Since, as has been described above, amides can be obtained by removal of water from the ammonium salts of acids, nitriles can be prepared directly from ammonium salts in one operation by heating them with a powerful dehydrating agent, *e.g.* ammonium acetate with P_2O_5:

$$CH_3.COONH_4 \;=\; CH_3.CN + 2H_2O\,.$$

Nitriles can also be obtained by the method of Kolbe: alkyl iodides (bromides or chlorides) are heated with alkali cyanide (example, benzyl cyanide), or salts of alkylsulphuric acids are subjected to dry distillation with potassium cyanide:

$$KO_3S.OC_2H_5 + CNK \longrightarrow K_2SO_4 + C_2H_5.CN\,.$$

The preparation of aromatic nitriles from diazo-compounds will be discussed later (p. 281).

The lower nitriles are colourless liquids, the higher are crystalline solids. The solubility in water decreases progressively with increasing molecular weight.

Acetonitrile possesses in a high degree the power of ionising electrolytes, *i.e.* salts, acids and bases dissolved in it conduct the electric current and do so much better than do similar solutions in alcohol, ether, or chloroform (Walden). The reactivity of the nitriles is due to the triple linkage between carbon and nitrogen, a linkage which permits a series of addition reactions to take place. Thus, on heating with water to 180° (in a sealed tube) or at a lower temperature in the presence of acids or alkalis, a molecule of water is added and the corresponding *amide* is re-formed :

$$R.C{:}N + H_2O \longrightarrow \underset{\dot{H}O\,H}{R.C{:}N} \longrightarrow \underset{\ddot{O}}{R.C.NH_2}.$$

The reaction is analogous to the conversion of acetylene into acetaldehyde :

$$HC{:}CH + H_2O \longrightarrow \underset{\dot{H}O\,H}{HC{:}CH} \longrightarrow \underset{\ddot{O}}{HC.CH_3}.$$

In both cases the intermediate product, the "enol form", is unstable.

Energetic hydrolysis by heating with slightly diluted sulphuric acid or with strong alkalis naturally converts the amide into a carboxylic acid and ammonia, so that, with such reagents, a nitrile is in practice directly converted into an acid. The method of carrying out this reaction is given on p. 129.

If nascent hydrogen (*e.g.* from zinc and sulphuric acid or from sodium and alcohol) is allowed to act on nitriles, four hydrogen atoms are added and *primary amines* are formed (reaction of Mendius) :

$$CH_3.CN + 4H \longrightarrow \underset{\text{Ethylamine}}{CH_3.CH_2.NH_2}.$$

Other less important but general reaction may be indicated by the following equations :

$$CH_3.CN + H_2S \longrightarrow \underset{\text{Thioacetamide}}{CH_3.CS.NH_2},$$

$$CH_3.CN + NH_2.OH \longrightarrow \underset{\text{Oxime of acetamide}}{CH_3.C\begin{matrix}\nearrow N.OH \\ \searrow NH_2\end{matrix}},$$

$$CH_3.CN + HCl = \underset{\text{Imidochloride}}{CH_3.C\begin{matrix}\nearrow NH \\ \searrow Cl\end{matrix}}.$$

$$CH_3.CN + CH_3.CH_2OH + HCl = \underset{\text{Hydrochloride of iminoether}}{CH_3.C\begin{matrix}\nearrow OC_2H_5 \\ \searrow NH.HCl\end{matrix}}.$$

In many of its reactions *hydrocyanic acid* behaves as the nitrile of formic acid H.CN. Many facts, in particular its great chemical and pharmacological similarity to the *isonitriles* >C=NR suggest another constitution, namely, that of *carbimide* >C=NH with bivalent carbon. The addition reactions of the nitriles (see above), which reactions are also characteristic of hydrocyanic acid, can equally well be explained on the basis of this second structural formula. In the nitrile form it is at the triple bond between carbon and nitrogen that addition takes place. In the " methylene form " this occurs at the two free valencies of the bivalent carbon atom, *e.g.*:

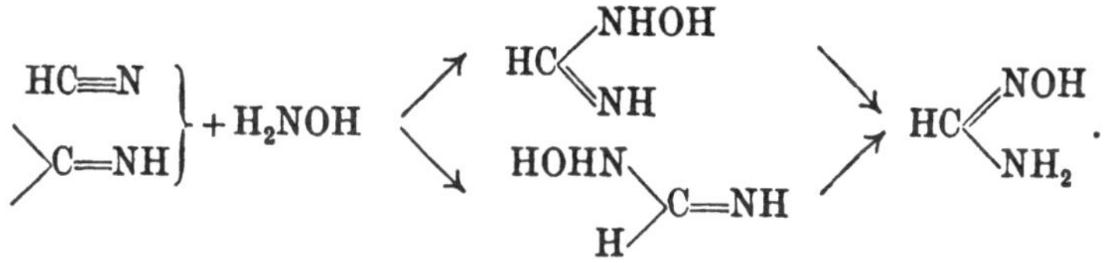

The hydrochloride of formiminoether is of practical importance. It is produced in fine colourless crystals by passing dry hydrogen chloride into a solution of equimolecular amounts of anhydrous hydrogen cyanide and ethyl alcohol in absolute ether. The salt is slowly decomposed by standing for a long time in the cold with alcohol, when ethyl orthoformate and ammonium chloride are formed:

$$HC\begin{cases}OC_2H_5\\NH.HCl\end{cases} + 2\,HOCH_2.CH_3 = HC\begin{cases}OC_2H_5\\OC_2H_5\\OC_2H_5\end{cases} + NH_4Cl.$$

This synthesis of ethyl orthoformate is more elegant and smooth than that from chloroform and sodium ethoxide. Ethyl orthoformate is used for preparing *acetals* of ketones, *e.g.*

$$CH_3.CO.CH_2.COOC_2H_5 + HC(OC_2H_5)_3 = CH_3.C(OC_2H_5)_2.CH_2.COOC_2H_5 + HC(:O)OC_2H_5.$$

6. HYDROLYSIS OF A NITRILE TO THE ACID PHENYLACETIC ACID [1]

Benzyl cyanide (40 g. = 0·33 mole) is heated in a round-bottomed flask (capacity 0·5 l.), with a mixture of 50 c.c. of concentrated sulphuric acid and 30 c.c. of water. The flask is provided with an upright air condenser, and is placed in a conical (Babo) air bath. The heating is continued until the appearance of small bubbles of vapour indicates that a reaction, which rapidly becomes violent, has set in; the liquid boils up, and white fumes are emitted. It is allowed to cool and then two volumes of water are added. After some time the phenylacetic acid which has crystallised out is filtered off with suction. If a sample of the material does not form a clear solution with sodium carbonate in water (presence of phenylacetamide), the

[1] Staedel, *Ber.*, 1886, 19, 1951.

whole of the crude material is shaken with sodium carbonate solution and the mixture is filtered. From the clear filtrate phenylacetic acid is reprecipitated with sulphuric acid, and can be recrystallised directly from a rather large volume of hot water or, after drying, from petrol ether. Because of its low melting point (76°) it often separates at first as an oil, but it can also be conveniently purified by distillation *in vacuo* from a sausage flask.[1]

The yield amounts to 34–38 g., and can be slightly increased by extracting the first mother liquor, from the sulphuric acid treatment, with ether.

Benzyl cyanide is always lost by evaporation in the rather violent hydrolysis described above; this can be avoided by boiling the benzyl cyanide (40 g.) under a reflux condenser for forty-five minutes with a mixture of water, concentrated sulphuric acid, and glacial acetic acid (40 c.c. of each). After cooling, the product is poured into water.

Under milder conditions (3 g. of benzyl cyanide dissolved in 8 c.c. of concentrated sulphuric acid and after standing for six hours, pouring into 50 c.c. of water) the hydrolysis leads, for the most part, only to the amide. How is any phenylacetic acid, formed at the same time, to be separated? *Phenylacetamide* melts at 155°.

There are various methods of hydrolysing nitriles; the operation usually requires the use of powerful reagents, strong acids or strong alkalis. Sometimes indirect methods are adopted. The ready addition of hydrogen sulphide to the nitrile, forming a thioamide (p. 128), may, for example, be utilised, since the thioamides are readily hydrolysed. Esters of the required acid are also easily obtained by passing hydrogen chloride into the hot alcoholic solution of the nitrile. Imino-esters are formed, and their NH-group is hydrolytically replaced by O with extraordinary ease. Compare the example on pp. 244, 245.

7. ESTERS

(a) Ethyl Acetate from Acetic Acid and Alcohol[2]

A half-litre flask is fitted with a two-holed cork through which are inserted a dropping funnel and a tube connected with a long downward condenser. Alcohol and concentrated sulphuric acid (50 c.c. of each) are poured into the flask and the mixture is heated

[1] Adams and Thal, *Organic Syntheses*, II, 1922, p. 64.
[2] *Bull. Soc. chim.*, 1880, 33, 350.

in an oil bath until the temperature reaches 140°. As soon as this is the case a mixture of alcohol and glacial acetic acid (400 c.c. of each) is gradually run in from the dropping funnel at the same rate as that at which the ethyl acetate formed distils. In order to free the distillate from acetic acid which has been carried over, it is shaken in an open flask with moderately concentrated sodium carbonate solution until the upper layer no longer reddens blue litmus paper. This layer is then separated in a tap funnel and, after filtration through a dry folded filter, is *freed from alcohol* by being shaken with a solution of 100 g. of calcium chloride in 100 g. of water. The two liquids are separated again in a funnel and the upper one is dried with granulated calcium chloride and finally distilled on the water bath. Boiling point 78°. Yield 80–85 per cent of the theoretical. Use in the preparation of *ethyl acetoacetate* and of *acetylacetone*, see pp. 242, 243.

Ethyl benzoate is prepared in a similar way, by boiling benzoic acid (30 g.) in 100 c.c. of absolute alcohol for four hours under reflux condenser with 3 c.c. of concentrated sulphuric acid. Most of the alcohol is then removed by distillation, 300 c.c. of water are added, and the liquid is extracted with ether. By shaking with sodium carbonate solution, acid is removed from the ethereal solution, which is next dried over sodium sulphate over night. When the ether has been evaporated, the residual ethyl benzoate is distilled. Boiling point 212°. Yield 30 g.

The formation of an ester from acid and alcohol corresponds in a formal way to salt formation from acid and metallic hydroxide.

$$NO_3H + NaOH = NO_3Na + H_2O$$
$$CH_3.COOH + C_2H_5OH = CH_3.COOC_2H_5 + H_2O.$$

As far as the velocity and the extent of the conversion are concerned, the two processes are, however, altogether different. Whereas an acid is practically instantaneously and completely converted into a salt by an equivalent amount of a sufficiently strong base (neutralisation), a process on which, indeed, alkalimetry and acidimetry depend, it is not possible to obtain from equimolecular amounts of acid and alcohol the theoretical (calculated) amount of ester. A certain maximal quantity of ester is formed, but always falls short of the theoretical, and it is impossible, even by indefinitely extending the duration of the reaction, to make the unchanged acid and alcohol produce ester in excess of that maximum. If, for example, equimolecular amounts of acetic acid and alcohol are allowed to interact in a closed system, only two-thirds of each enter into reaction, and it is impossible to induce the remaining third of acetic acid to react with that of alcohol. The maximum yield of ester therefore amounts to only two-thirds, or 66·7 per cent, of the theoretical quantity. The quantitative difference in the course of the two reactions mentioned above depends on the fact that esterification is a so-called "reversible reaction", *i.e.* one in which the reaction products represented on the right-hand side of the equation (ester and water) also interact in the opposite direction:

$$CH_3.COOH + HO.C_2H_5 \rightleftarrows CH_3.COOC_2H_5 + H_2O.$$

Strictly speaking, all chemical reactions are reversible, but when the extent of the reverse reaction is immeasurably small and, consequently, the transformation as a whole proceeds practically in one direction only, the reversibility is neglected. The synthesis of water from its elements is a reaction of this kind. But we should nevertheless bear in mind that already at +2000° the opposite reaction (decomposition) takes place to such an extent that we must use the equation:

$$2H_2 + O_2 \rightleftarrows 2H_2O.$$

The conditions of the reaction are thus controlled by the temperature, and indeed in such a way that endothermic reactions (*i.e.* those which involve the absorption of energy) are, in general, favoured by a rise of the temperature, whilst on the other hand exothermic reactions are favoured by lowering it.

After this brief characterisation of reversibility, we may use the example of esterification to consider next the question how the limitation of the reaction is to be explained. To the extent that acid and alcohol interact, and their reaction products, ester and water, are formed, the reverse reaction (ester + water = acid + alcohol) also gains in extent. A point is eventually reached at which just as many molecules of acid and alcohol react to form ester as molecules of ester and water are decomposed to form acid and alcohol. The two reactions balance each other, and it would seem as if the reacting system had come to a state of rest. But this apparent rest is simulated by the fact that, in unit time, equal numbers of ester molecules are formed and decomposed. A *state of equilibrium* has been attained, and, as the above considerations indicate, this state would also have been reached had the reaction proceeded at the outset from the opposite side between equimolecular amounts of ester and water. In the latter case the hydrolysis of the ester would likewise have been balanced sooner or later, according to the conditions prevailing, by the opposing esterification—in this case when 33·3 per cent of the ester had been decomposed. The equilibrium is therefore the same, no matter from which side it is approached; on this depends its exact experimental investigation, both here and in many other reactions.

As has already been mentioned (p. 3), the velocity of a bimolecular reaction is proportional to the product of the reacting masses. Hence, if their concentrations are expressed in grammolecular weights, the following simple relations hold:

$$v = C_{Ac} \,.\, C_{Al} \,.\, k\,; \qquad v' = C_E \,.\, C_W \,.\, k'.$$

In these equations v is the velocity of esterification, v' is that of hydrolysis. C_{Ac}, C_{Al}, C_E and C_W are the concentrations of the four substances involved, and k and k' are the velocity constants of the two reactions. The state of equilibrium is attained when on both sides of the chemical equation equal numbers of molecules interact, *i.e.* when the velocity is the same in both directions:

$$v = v'$$

Then also

$$C_{Ac} \,.\, C_{Al} \,.\, k = C_E \,.\, C_W \,.\, k',$$

or

$$\frac{C_{Ac} \,.\, C_{Al}}{C_E \,.\, C_W} = \frac{k'}{k} = K.$$

The last equation expresses the fact that at the point of equilibrium the products of each pair of the concentrations are inversely proportional to the velocity constants. The quotient $\frac{k'}{k}$, usually designated by K, is the equilibrium constant.

If this important constant is known it is readily possible, as will next be shown, to calculate in practice the extent of the transformation in an equilibrium reaction when the reactants are no longer present in equimolecular amounts, *i.e.* are not in equal molar concentrations.

The calculation of K causes no difficulty in the present example, after what has been said about the position of the equilibrium. The resultant mixture contained one third of a mole each of acid and alcohol and two-thirds of a mole each of ester and water. Therefore:

$$K = \frac{\frac{1}{3} \,.\, \frac{1}{3}}{\frac{2}{3} \,.\, \frac{2}{3}} = \frac{1}{4}.$$

It will now be possible to discover how and to what extent the formation of the ester can be affected if acetic acid and alcohol are not used in equimolecular proportions but, for example, in the ratio 1:2 moles. If the amount (in moles) of ester present at equilibrium be x, then $C_E = x$; and since just as many molecules of water as of ester are formed, C_W is also equal to x. The concentration of the acid is then $1 - x$, and that of the alcohol $2 - x$, and, therefore:

$$\frac{(1-x) \,.\, (2-x)}{x \,.\, x} = \frac{1}{4}.$$

From this equation it follows that $x = 0{\cdot}85$, *i.e.* by increasing the concentration of the alcohol (or the acetic acid) the position of the equilibrium can be so shifted that the yield of ester is increased to 85 per cent. Advantage is very frequently taken of this practical possibility of altering the position of the equilibrium.

In this connection the following problems should be solved:

How much ester will be formed when 3 moles of alcohol react with 1 mole of acetic acid? How much when 30 g. of acetic acid and 50 g. of alcohol are used? In what proportions by weight must acetic acid and alcohol be caused to interact in order to convert 75 per cent of the former into ester?

In all reactions which proceed entirely in one direction the amount of end product which could theoretically be expected can be calculated from the stoicheiometrical ratios. But in reactions in which an equilibrium is set up, it follows from the above considerations that we must know the equilibrium constant, which must be deduced by analysis. In the example chosen this evidently causes no difficulties; it is only necessary to determine by titration the concentration of the acetic acid in the equilibrium mixture.

The Rôle of the Sulphuric Acid.—If acetic acid and alcohol alone are heated together no noticeable reaction takes place even after a long time. The reaction is only started by added sulphuric acid, which may be replaced by gaseous hydrogen chloride. In all probability these mineral acids form an unstable addition compound (more likely with the acetic acid) and this compound reacts with alcohol to form the ester more rapidly than does the organic acid itself. The hydrolysis is also accelerated to the same extent as is the esterification, so that exactly the same equilibrium is attained as in the absence of sulphuric acid, for instance, by heating the components to a high temperature in a sealed tube. *The influence of the catalyst (the sulphuric acid) consists, therefore, only in an increase in the velocity of the reaction; the position of the equilibrium is left unchanged. This important law holds for all reactions which are catalytically accelerated.*

It has been shown that the catalytic action of acids is proportional to their strength, of which their degree of ionisation is an expression. Conversely, the strength of an acid can be determined by measuring the rate of hydrolysis of an ester (usually methyl acetate) in aqueous solution in the presence of the acid concerned.

An understanding of the theoretical principles which have been but briefly discussed in this section is indispensable to anyone who does not wish to practise organic chemistry merely as a culinary art.

Other Methods for the preparation of Esters.—Esters are extremely easily formed from silver salts by the action of alkyl iodides:

$$R.COOAg + I.R' \longrightarrow R.COOR' + AgI.$$

The same object is attained by treating alkali salts with a dialkyl sulphate. Usually in this case only one alkyl group takes part in the reaction in accordance with the equation:

$$R.COONa + (CH_3)_2SO_4 \longrightarrow R.COOCH_3 + CH_3.SO_4Na.$$

The alkyl group of the salt of the alkylsulphuric acid can also be made available for esterification if the temperature is raised sufficiently. The formation of esters from acid chlorides or anhydrides need only be recalled here. This method also has practical importance.

For the esterification of difficultly accessible acids the elegant diazomethane method (p. 263) is most appropriate ; it usually proceeds very smoothly.

The lower members of the series of esters are colourless liquids having pleasant fruity odours. The higher members, as well as most esters of aromatic acids, are crystalline substances. The boiling points of esters containing alkyl groups of low molecular weight (CH_3, C_2H_5, C_3H_7) are lower than those of the corresponding acids :

$CH_3.COOCH_3$	Boiling point		57°,
$CH_3.COOC_2H_5$	„	„	78°,
$CH_3.COOH$	„	„	118°.

It is worthy of note that the melting points of the *methyl* esters are generally higher than those of the corresponding *ethyl* esters ; thus, to take a well-known example, dimethyl oxalate (melting point 54°) is a solid whilst the diethyl ester is a liquid.

The esters are often prepared as an end in themselves, and are used technically, as solvents, perfumes, essences for fruit drinks, etc. But their most important rôle is connected with the transformation of the carboxyl group. Thus the alkyloxy group can be replaced by —NRR′ as a result of the action of ammonia and of numerous ammonia derivatives containing at least one hydrogen atom directly attached to the nitrogen (primary and secondary amines, hydroxylamine, hydrazine). In this way *amides* and *hydrazides* of carboxylic acids are prepared. Special attention is called to the hydrazides: they are a stage in the Curtius degradation (pp. 143, 144). Attention may also be directed here to the large subject of ester condensations.

Vigorous reduction with metallic sodium (and a little alcohol) converts esters into the corresponding primary alcohols (Bouveault) :

$$R.COOR' \longrightarrow R.CH_2OH + R'.OH.$$

Finally, esters are very often prepared in order to purify acids, since the majority of esters—often in contrast to acids—can conveniently be distilled, especially *in vacuo*. From the pure esters the pure acids are then obtained by hydrolysis.

The hydrolysis of the esters is carried out by prolonged heating with aqueous mineral acids or with solutions of alkali hydroxides. See saponification of fats on p. 139. Alcoholic potassium hydroxide solution is an especially rapid hydrolytic agent.

(b) Isoamyl Nitrite[1]

A mixture of amyl alcohol (44 g. = 0·5 mole) and technical sodium nitrite solution (37 g. in 70 c.c. of water) in a filtration jar is cooled to 0° in an ice-salt mixture. During the cooling the mixture is mechanically stirred, and this stirring is continued while concentrated hydrochloric acid (44 c.c., d. 1·18) is slowly dropped in from a tap funnel (Fig. 47); the temperature is not allowed to rise above +5°. The reaction mixture is now shaken in a separating funnel with about 200 c.c. of water, the aqueous layer is run off, and the remaining liquid is washed first with dilute sodium carbonate solution and then several times with water. After separation of the two layers, the reaction product is clarified and dried over a little calcium chloride in a small conical flask. The dried liquid is then distilled at 50–60 mm. pressure (cf. p. 23) into a well-cooled receiver. The bulk of the liquid passes over at about 30° as a yellow oil. Yield 75 per cent of the theoretical.

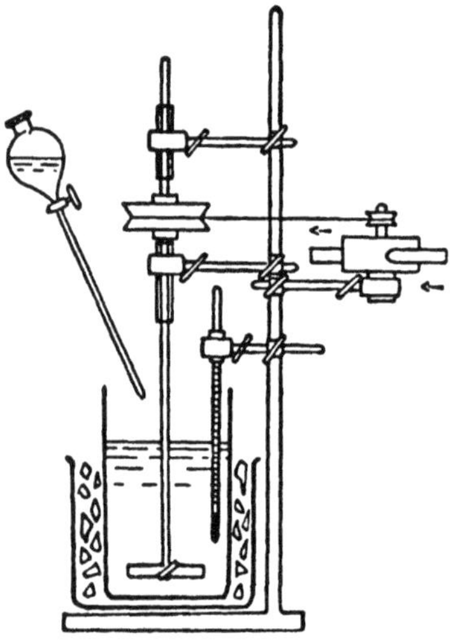

Fig. 47.

The esters of nitrous acid are characterised by their high velocities of formation and hydrolysis. They are almost instantaneously decomposed by mineral acids and in the method of preparation given this has been taken into account. The slightest excess of hydrochloric acid must be avoided. Advantage is taken of this property of the alkyl nitrites in all cases where it is desired to liberate nitrous acid in organic solvents (in which metallic nitrites are insoluble). Examples: addition of N_2O_3 to olefines, preparation of solid diazonium salts (p. 273), production of isonitroso-derivatives from ketones by the action of HNO_2. This synthesis is often also carried out in the manner of the acetoacetic ester synthesis, with ketone, alkyl nitrite, and sodium ethylate; the sodium salt of the isonitrosoketone is formed (cf. in this connection p. 250):

[1] Witt, *Ber.*, 1886, 19, 915.

$$\begin{array}{l} R{-}CH_2 \\ \quad | \\ R{-}CO \end{array} + \begin{array}{l} RO.N{:}O \\ \\ RONa \end{array} \longrightarrow \begin{array}{l} R{-}C{:}NONa \\ \quad | \\ R{-}C{:}O \end{array} + 2ROH.$$

The elegant synthesis of sodium azide from hydrazine and alkyl nitrite is carried out in the same way (Stollé) :

$$H_2N{-}NH_2 + RO{-}N{=}O + RONa \longrightarrow \begin{array}{c} N \\ \| \\ N \end{array}\!\!\!> N.Na + 2ROH + H_2O.$$

Ethyl nitrite is often used in preference to isoamyl nitrite because the removal of the amyl alcohol produced from the latter may be troublesome on account of its high boiling point (136°).

Ethyl Nitrite.—To the mixture of the above sodium nitrite solution with 60 c.c. of spirit, contained in a distilling flask and cooled in ice, 42 c.c. of concentrated hydrochloric acid are gradually added drop by drop. During the addition of the acid the flask is shaken. The flask is then attached to an efficient condenser, through which ice-water may be run with advantage ; the receiver (a filter flask) should stand in a freezing mixture. After addition of the acid the ethyl nitrite is distilled by heating the flask first at 25°, finally at 40°, in a basin of warm water. After being dried for a short time over potassium carbonate the product is sufficiently pure for most purposes and, because of its great volatility (boiling point 17°), is best used immediately.

Experiment.—A few drops of amyl or ethyl nitrite are shaken with dilute potassium iodide solution. No brown colour should be produced. Addition of a drop of dilute hydrochloric acid leads, in a few seconds, to a copious liberation of iodine.

(c) Ethyl Nitrate [1]

Concentrated nitric acid (d. 1·4, 250 c.c.) is boiled with 30 g. of urea nitrate. After the solution has been cooled one half is poured into a tubulated retort containing 30 g. of urea nitrate and 150 c.c. of alcohol. A moderate-sized condenser is attached to the retort, which is then carefully heated on a sand bath. After about one third of the material has been distilled the remainder of the boiled urea nitrate-nitric acid solution, mixed with 100 c.c. of alcohol, is allowed to run slowly into the retort from a dropping funnel inserted through the tubulure. The operation must be carried out without

[1] Lossen, *Annalen*, 1868, supplementary vol. 6, 220.

interruption: the mixture of alcohol and nitric acid must not be allowed to stand for any length of time. When all the liquid in the funnel has been added and that in the retort has been reduced by distillation to 50–100 c.c., the ethyl nitrate which has passed over is freed from alcohol by being shaken in a separating funnel twice with water, then once with dilute sodium carbonate solution, and finally once more with water (ethyl nitrate is heavier than water !). After being dried over calcium chloride it is distilled from a water bath in which the distilling flask should be immersed. Boiling point of ethyl nitrate 86°. (Goggles should be worn in this experiment.)

The use of ethyl nitrate in the preparation of phenylnitromethane is described later (Chap. VI. 8, p. 246).

When ethyl nitrate is rapidly heated, for example in a flame, it decomposes explosively; it belongs to the same class of substances as nitroglycerine. Precautions should therefore be taken. Under the conditions employed ethyl alcohol is *not oxidised* by *pure* nitric acid but merely esterified. Oxidation sets in immediately if traces of nitrous acid are present. Since the nitric oxide which is thus produced from the nitrous acid, is at once oxidised to NO_2 by the nitric acid, the oxidation, which is initially slight, becomes progressively greater, its velocity increases continuously as a result of the heat produced in the reaction, and it finally proceeds violently and almost explosively. Reactions of this type, in which the velocity is progressively increased by their own intermediate products, are said to be "autocatalytic".

The first product of the *oxidation* of alcohol is *acetaldehyde* and an important end-product is *fulminic acid*, which latter can, however, only be isolated if silver or mercury ions are present. With these ions it forms salts—fulminates—which are stable towards nitric acid; in them, it must be presumed, the linkage with the metal is complex and non-ionogenic, as in mercuric cyanide. The formation of fulminic acid takes place because the carbonyl group of the aldehyde confers reactivity on the adjacent methyl group which then forms a point of attack for the nitrous acid. The various stages in the process are indicated by the following formulae:

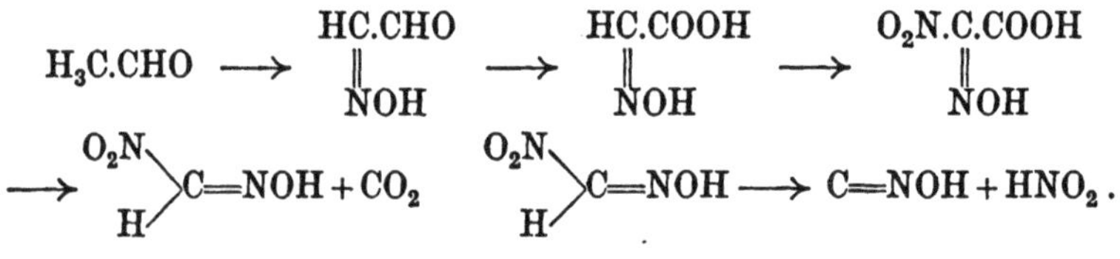

The nitrous acid here acts on the alcohol in the same way as do the halogens in the production of chloroform and iodoform.

(*d*) Hydrolysis of a Fat or Vegetable Oil

Of any fat or oil 600 g. (about 0·66 mole) are hydrolysed with 600 c.c. of approximately $5N$-sodium hydroxide solution.

The fat is poured into a warm mixture of 100 c.c. of the sodium hydroxide solution and 100 c.c. of water. After one hour another 150 c.c. of hydroxide solution are added, and again one hour later 200 c.c. of the solution and 200 c.c. of water. The mixture must be frequently shaken and should be warmed to gentle boiling only. After a further period of four hours the remainder of the hydroxide solution is added; if necessary, water is first added to replace that lost by evaporation. Again after one hour half a litre of water is added and boiling is continued until a thick homogeneous mass is produced (about two to three hours). Then 3–4 litres of hot water are poured in with vigorous stirring. A thick transparent gum is thus formed. Finally the soap is salted out at 100° by the addition of 200 g. of common salt and the mixture is left to stand over night.

The whole operation is best carried out in a large enamelled pot because of the voluminous froth produced.

Next morning, when the material has cooled, the solid cake of soap is lifted out and washed to remove adherent alkali. It may be cut into small pieces by means of a thin wire and then left to dry for several weeks.

The sodium salts of the higher fatty acids are sparingly soluble in cold water but more readily in hot. Dissolve a small piece of soap in the minimum amount of boiling water in a beaker and allow the solution to cool: a stiff jelly is formed.

To purify the soap 20–30 g. are dissolved in boiling water, salted out from the hot solution, and again allowed to solidify. In this way the small amount of alkali in the crude product is removed. The soap, however, remains alkaline to litmus and turmeric papers. The hydrolysis of the *quite pure* soap is, however, not sufficiently extensive, the concentration of OH-ions not sufficiently large, for phenolphthalein to be coloured.

Preparation of the Free Fatty Acids.—About 150 g. of the crude moist soap in a litre of water is heated almost to boiling. With vigorous stirring $2N$-sulphuric acid is added, until the solution is distinctly acid to Congo paper and the mixture of fatty acids has separated as a supernatant oily layer. If the original fat was solid this layer solidifies after standing in the cold for some time. The cake is lifted out, melted once again with a little water in a small beaker on the water bath, and, after it has again solidified, the acids are distilled *in vacuo*. Boiling point 220°–225°/12 mm.

If an oil is hydrolysed the soap obtained is less solid and the acids crystallise only partially. (Why ?) In this case they are dissolved in ether and worked up in the usual way.

Glycerol.—The glycerol remains in the brown liquor from which the soap has been taken. This liquor is first accurately neutralised (to Congo paper) with hydrochloric acid, shaken with animal charcoal in order to remove precipitated fatty acids,[1] filtered through a folded filter paper, and then concentrated *in vacuo* in the apparatus shown on p. 30. When after some time sodium chloride separates the capillary tube sometimes fails to act, and in order to save time the concentration is continued on the water bath. The highly concentrated solution is filtered at the pump from the sodium chloride, which is washed with a little alcohol, and the filtrate is again concentrated in a Claisen flask till almost quite free from water. The residue is digested with 150 c.c. of alcohol and filtered through a small Büchner funnel; the material on the filter is then washed with 50 c.c. of alcohol. The alcoholic solution thus obtained is concentrated as far as possible on the water bath and is then washed into a Claisen flask with a little alcohol. When the contents of the flask are distilled *in vacuo* alcohol and water, and then glycerol pass over. The bulk of the glycerol is collected between 180° and 195°/13 mm. Yield about 35 g.

In order to obtain the glycerol quite pure and anhydrous the distillation must be repeated.

(e) Linolenic Acid from Linseed Oil[2]

Unbleached linseed oil (100 g.) is boiled in a flask connected with a downward condenser with 35 g. of potassium hydroxide dissolved in 150 c.c. of methyl alcohol. After most of the alcohol has distilled—a sample of the liquid in the flask must form a clear solution in water—200 c.c. of 20 per cent sulphuric acid are added to the hydrolysed material and the acids, which have been liberated, are taken up in ether. After the ethereal solution has been dried over fused sodium sulphate the solvent is evaporated. Almost the whole of the residue distils at 215°–230°/12 mm. Yield 70–85 g.

[1] Clarification with animal charcoal is often unnecessary.
[2] Erdmann and Bedford, *Ber.*, 1909, **42**, 1324. See also *Annalen*, 1928, **464**, 214.

Hexabromide of Linolenic Acid.—Into a solution of 75 g. of the distilled linolenic acid (some of which may have crystallised) in 150 c.c. of glacial acetic acid, cooled in ice and mechanically stirred, bromine is slowly dropped from a burette until a permanent orange colour indicates that an excess of halogen has been added. About 30 c.c. (90 g.) of bromine are required. The thick paste, composed of the hexabromide and of the precipitated saturated fatty acids, is allowed to stand for six hours at ordinary temperature and is then filtered on a medium-sized porcelain funnel. By this means most of the mother liquor is removed. The filtration, which requires several hours, is carried out with gentle suction only. The sticky mass is now transferred to a round-bottomed flask into which 200 c.c. of ethyl acetate are poured, and is heated until the liquid boils gently. Meanwhile the paste is continuously stirred with a glass rod (flattened at its lower end) until, after a few minutes, the gummy matter has dissolved and the hexabromide has settled to the bottom in crystalline form. After cooling, the mixture is filtered as dry as possible at the pump, and the material on the filter is washed with ethyl acetate until colourless. The hexabromide is now sufficiently pure for further treatment and melts at 178°. Yield 20–25 g.

Removal of Bromine from the Hexabromide.—The material (25 g.) is finely powdered and suspended in 250 c.c. of spirit. To the suspension, gently warmed on the water bath and vigorously shaken, 50 g. of zinc dust are added in one lot. Whenever the reaction becomes too vigorous it is moderated, but should not be completely suppressed, by cooling the container in ice-water, which is kept ready for the purpose. Finally the mixture is heated under a reflux condenser for an hour. The excess of zinc dust is then removed at the filter pump and is washed twice with alcohol. Most of the alcohol in the filtrate is distilled on the water bath, and the cooled residue, almost free from alcohol, is poured into a separating funnel containing 200 c.c. of 20 per cent sulphuric acid. The linolenic acid is obtained by repeated extraction with ether. Since, during the debromination, part of the acid is esterified, a solution of 5 g. of potassium hydroxide in 25 c.c. of methyl alcohol is added to the undried ethereal solution in order to hydrolyse the ester. Ether and alcohol are then removed by distillation on the water bath. The residue, which should be alkaline to phenolphthalein, is again treated with 20 per cent sulphuric acid and extracted with ether. When the ethereal solution has been dried over $CaCl_2$ and the solvent has been removed by distillation, almost pure, nearly colourless, linolenic acid remains. The yield hardly falls short of the theoretical. The acid can be completely purified by distillation in a good vacuum. Boiling point 197°/4 mm.

The linolenic acid thus prepared serves as material for catalytic hydrogenation. See p. 370.

Linseed oil is the most important of the so-called "drying" oils. These oils contain highly unsaturated acids, such as linolenic $C_{17}H_{29}.CO_2H$ and linolic acid $C_{17}H_{31}.CO_2H$, and are therefore able, by virtue of a system of conjugated double bonds, to combine directly with the oxygen of the air, forming solid peroxides and their transformation products. Oleic acid cannot act in the same way. Olive oil and sesame oil, for example, do not "dry". Linseed oil is used as a medium in oil painting and in the manufacture of varnishes.

The "iodine value" of a fat or oil is a quantitative measure of the number of carbon-carbon double bonds which it contains. This number is the amount of iodine in grammes which combines chemically with 100 g. of the fat or oil. Nowadays the number of double bonds in organic compounds is usually determined with perbenzoic acid (cf. p. 99).

Determination of the Iodine Value.—Pure iodine (2·5 g.) and mercuric chloride (3 g.) are each dissolved in 50 c.c. of pure spirit and the clear solutions are mixed. After twelve hours the iodine titre of the mixture is determined in a 10 c.c. portion by adding 10 c.c. of 10 per cent potassium iodide solution and titrating with 0·1 N-thiosulphate solution.

The fat to be examined (0·5–0·7 g.) is dissolved in 15 c.c. of chloroform in a dry conical flask (capacity 500 c.c.) and 25 c.c. of the standardised iodine solution are added. If, after a short time, the colour of the solution diminishes to a light brown, it is necessary to add a further 10 c.c. of the iodine solution. After four hours the colour of the solution should still be dark brown. Potassium iodide solution (20 c.c. of 10 per cent solution) is now added and the uncombined iodine still present titrated as above. The calculation is made in accordance with the definition of "iodine value". Lard, olive oil, or linseed oil should be examined.

To determine the *saponification value*[1] of a fat 0·5–1·0 g. is boiled under reflux condenser for half an hour with 10 c.c. of 0·5 N-alcoholic KOH solution and the excess of alkali is then titrated with N-HCl, and phenolphthalein as indicator.

The method is a general one and can be applied to esters to determine the equivalent of an acid combined in them. Ester equivalent = $a.1000/b$ where a is the weight of ester taken (in grammes) and b is the number of c.c. of N-alkali used up.

[1] The number of milligrammes of KOH which is used up by 1 g. of fat.

8. DESCENDING A HOMOLOGOUS SERIES. CONVERSION OF CARBOXYLIC ACIDS INTO THE NEXT LOWER AMINES

(a) The Hofmann Reaction. Methylamine from Acetamide[1]

A mixture of acetamide (30 g. = 0·5 mole) and bromine (80 g. = 26 c.c.) in a half-litre flask is kept well cooled with water while enough of a solution of 50 g. of potassium hydroxide in 350 c.c. of water is added to change the initially red-brown colour into a pale yellow; this requires most of the alkali. The solution is now run from a dropping funnel in an unbroken jet into a solution of 80 g. of potassium hydroxide in 150 c.c. of water, maintained at 70°–75° in a litre flask. The operation lasts for several minutes. Until the reaction mixture becomes colourless (one quarter to half an hour) the temperature is maintained at 70°–75°, and then the methylamine is distilled with steam. An adapter is fixed to the lower end of the condenser and dips 1 cm. below the surface of the liquid in the receiver (100 c.c. of approximately 5*N*-hydrochloric acid[2]). As soon as the liquid which forms in the condenser is no longer alkaline the distillation is discontinued and the contents of the receiver are evaporated to dryness in a porcelain basin on the water bath. The last traces of water are removed by allowing the basin to stand over night in a vacuum desiccator. The dried material is boiled with absolute alcohol, which dissolves the methylamine hydrochloride but not the ammonium chloride with which it is mixed. The clear filtrate obtained when the ammonium chloride is removed by filtration is concentrated to a small volume and the methylamine hydrochloride is allowed to crystallise out in the cold. The salt is filtered with suction, washed with a little alcohol, and dried in a desiccator. Yield 15–20 g.

The *isonitrile reaction* (p. 158) should be carried out with this salt, and its behaviour on warming with a little nitrite in faintly acid aqueous solution should be examined.

(b) The Curtius Reaction. Phenyl Cyanate

Benzhydrazide.[3]—24 g. (0·15 mole) of ethyl benzoate (p. 131) are heated for six hours under a small reflux condenser on the water bath

[1] *Ber.*, 1882, **15**, 762; 1884, **17**, 1406, 1920.
[2] 50 c.c. of conc. hydrochloric acid and 50 c.c. of water.
[3] T. Curtius, *J. pr. Chem.*, 1894, **50**, 295.

with 9 g. of hydrazine hydrate.[1] The solid crystalline cake which is formed on cooling is, after some time, filtered as dry as possible at the pump and washed with a little ice-cold methyl alcohol. If the yield is too small the filtrate is concentrated and heated again.

The crude product (16–18 g.) is sufficiently pure for further treatment. A sample can be recrystallised from hot water or from a little alcohol. Melting point 112°.

Benzoylazide.[2]—14 g. (0·1 mole) of the *dry* hydrazide are made into a clear solution with 200 c.c. of approximately *N*-hydrochloric acid in a filter jar (capacity 0·5 l.). The solution is cooled in ice and stirred, while 8 g. of sodium nitrite in 50 c.c. of water are added. An immediate reaction takes place and the azide separates in crystalline form. When a filtered sample of the solution is no longer made turbid by the addition of a drop of nitrite solution, the precipitate is filtered dry at the pump, washed well with water, and dried, first on porous plate and then in a vacuum desiccator over concentrated sulphuric acid and potassium hydroxide. Yield 14 g.

Phenyl Cyanate.[3]—The azide used for the preparation of the cyanic ester must be *absolutely dry.* Test for constant weight on a good hand balance.

Since benzoylazide explodes when rapidly heated or when in contact with concentrated sulphuric acid, it must be handled carefully. **Goggles** *must be worn until the distillation of the phenyl cyanate has been completed.*

The distillation of the end-product is carried out in the flask which is used for the decomposition; a Claisen flask of capacity 75–100 c.c. is suitable. A capillary tube and a thermometer are got ready before the decomposition of the azide is carried out. All the apparatus must be thoroughly dry.

The Claisen flask is fixed in an inclined position and a small condenser jacket is slipped over the side tube, which slopes upwards. Atmospheric moisture is excluded by fixing a calcium chloride tube to the upper end of the side tube. Benzoylazide (12 g.) and benzene (40 c.c.; dried over sodium) are placed in the flask, which is *slowly* heated to 60°–70° in a basin of water. The flask must not be in contact with the bottom of the basin. When the vigorous evolution of nitrogen which sets in has slackened, the temperature is raised to

[1] This can be obtained cheaply from the firm of Dr. F. Raschig, Ludwigshafen am Rhein.

[2] T. Curtius, *Ber.*, 1890, 23, 3029.

[3] G. Schroeter, *Ber.*, 1909, 42, 2339.

about 80°. The flask is then cooled and adjusted for vacuum distillation. The benzene is first distilled at *ordinary pressure* from the boiling water bath. Then the bath is cooled and the phenyl cyanate is distilled at 20–25 mm. pressure. Boiling point 60°/20 mm. Yield 7–8 g.

The distillate, which should be colourless, must be immediately transferred to a tight vessel (preferably to a sealed tube) after making the following two tests. A few drops are poured into a little water. The crystalline substance formed is *diphenylurea.* How is it produced ?

Phenylurethane.—Another portion is poured into alcohol and the solution is evaporated to dryness.

The portion of azide which was not used for conversion to cyanate (about 2 g.) is boiled for half an hour under reflux condenser with 5 c.c. of absolute alcohol.[1] On concentrating the solution, phenylurethane likewise separates. Melting point 52°.

The *decomposition of urethanes into amine,* CO_2, and alcohol is usually carried out with hydrochloric acid in sealed tubes. Decomposition by distillation with calcium hydroxide is more convenient, but gives a poorer yield.

The phenylurethane which has been obtained is mixed with three times its weight of slaked lime and cautiously distilled from a small retort. The *aniline* which passes over can, with a little skill, be redistilled from a small flask, but in any case it should be identified by conversion into *acetanilide* and by the *bleaching powder* reaction.

In the determination of constitution it is often necessary to remove carboxyl groups (such as are formed by oxidation, for example) and so to break down the molecule. The simplest method of doing this, namely, removal of carbon dioxide by distillation of a salt with soda-lime :

$$R.COONa + NaOH \longrightarrow RH + Na_2CO_3 ,$$

generally does not proceed smoothly and leads, moreover, to a hydrocarbon unsuitable for further reactions.

Hence the great preparative importance of the two related methods for the degradation of acids, due to Hofmann and to Curtius, and starting respectively from the amide and the hydrazide.

Both processes furnish the primary amine corresponding to the next lower homologue of the series, and both lead to this result by way of the same intermediate product, namely, the cyanic ester.

By the action of hypobromite on the —$CONH_2$-group one hydrogen

[1] T. Curtius, *Ber.*, 1894, 27, 779.

atom of the NH_2-group is replaced by bromine. The first product of the Hofmann reaction, the N-bromoamide, can be isolated in certain cases.

By the action of alkali the bromoamide loses HBr, and the transient radicle which is thus formed undergoes rearrangement to a cyanic ester which, under the experimental conditions of the reaction, is decomposed into a primary amine and CO_2:

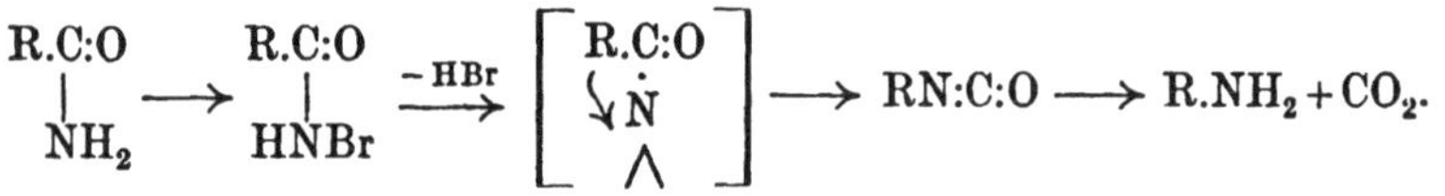

In this way *acetamide* furnishes *methylamine, benzamide* yields *aniline*, and *urea*, albeit in small amount, forms *hydrazine.*

Hydroxamic acids are transformed in a similar way by loss of water into cyanates and so into amines.

The reaction of Curtius, which is especially to be preferred in the case of the higher members on account of the favourable solubilities of the intermediate products, involves as its first stage the preparation of the hydrazide from an ester (or acid chloride). The hydrazide is then converted, usually very readily, by the action of nitrous acid into the azide. Azides easily undergo thermal decomposition, the two "azo" nitrogen atoms being eliminated as elementary nitrogen. In this way, however, the same radicle is formed as was invoked above to explain the Hofmann reaction :

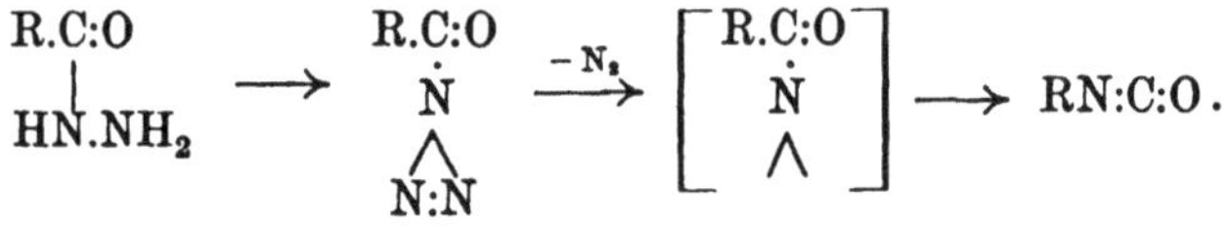

Curtius usually decomposed the azides in alcohol, which at once explains the formation of urethanes; these, by vigorous hydrolysis, decompose into primary amine, CO_2, and alcohol.

An important application of the Hofmann reaction occurs in the first technical synthesis of indigo and consists in the degradation of phthalimide to anthranilic acid. See p. 363.

CHAPTER III

NITRO-COMPOUNDS AND THEIR REDUCTION PRODUCTS

1. NITROMETHANE [1]

CHLOROACETIC acid (94 g.) dissolved in 200 c.c. of water in a wide beaker is accurately neutralised with 53 g. of anhydrous sodium carbonate, and 75 g. of sodium nitrite dissolved in 120 c.c. of water are then added. About 100 c.c. of the mixture thus obtained are transferred to a round-bottomed flask (capacity 750 c.c.) which is fitted with a dropping funnel and is connected to a downward condenser. The flask is now strongly heated in a conical air bath or on a wire gauze (raise the temperature slowly). A vigorous reaction sets in with evolution of CO_2 even before boiling begins. By allowing the remainder of the solution to run slowly into the boiling liquid the reaction is maintained, but is not allowed to become too vigorous. The nitromethane distils with the steam and forms the heavier layer in the receiver. As soon as drops of oil no longer pass over, the receiver is changed and a further 100 c.c. of water containing dissolved nitromethane are collected separately. The nitromethane is separated from the first distillate and the aqueous layer of the latter is combined with the second distillate. The combined solution is saturated with sodium chloride (35 g. for each 100 c.c.) and again subjected to distillation; about a quarter of the whole volume is collected. Then a clear distillate again passes over.

The nitromethane in this second distillate is separated from the water, combined with the material first obtained, thoroughly dried over calcium chloride, and then distilled. Boiling point 101°. Yield 20–24 g. (30–36 per cent of theoretical).

Nitromethane is the most easily accessible aliphatic nitro-compound; Kolbe's method of preparation is much less satisfactory when applied to higher members of the series. The course of the reaction is

[1] H. Kolbe, *J. pr. Chem.*, 1872, 5, 429; Steinkopf, *Ber.*, 1909, 42, 3438.

clear, and the reasons for the decomposition which takes place are similar to those used to explain the decomposition of malonic acid: the nitroacetic acid first formed decomposes into CH_3NO_2 and CO_2. Other nitroparaffins are usually obtained by the method discovered by V. Meyer—by the action of silver nitrite on alkyl iodides. The method of Konovalov—heating in a sealed tube to 120°–130° with very dilute nitric acid—is often successful in the case of saturated hydrocarbons, particularly those of the hydroaromatic series. *Phenylnitromethane* is discussed in Chap. VI. 8, pp. 246, 254. Recall the isomerism with the alkyl nitrites. How do the reactions of the two classes of compounds differ?

The primary and secondary nitroparaffins are neutral substances, but are transformed by alkalis into salts of isomeric *aci*-forms (Hantzsch):

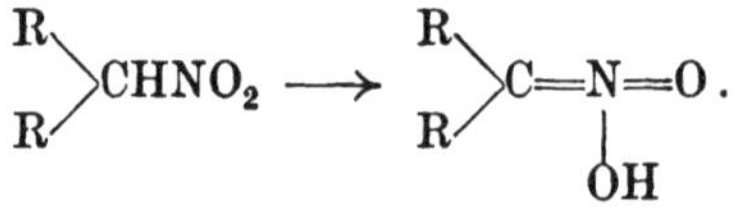

This change is more fully discussed in the section on tautomerism on p. 254.

Experiment.—Dissolve 1 c.c. of nitromethane in water and test the solution with litmus paper. Then add 10 c.c. of *N*-alkali and test again with red litmus paper or with turmeric paper. The fact that the solution is not alkaline proves that the alkali has been neutralised by the acid formed. A small sample of this solution gives with ferric chloride a blood-red colour, characteristic of *aci*-nitro-compounds. If to the solution which was treated with alkali 5 c.c. of *N*-hydrochloric acid are added and the solution is at once tested, blue litmus paper is turned red. Since an excess of alkali was used (10 c.c. to 5 c.c. of hydrochloric acid) the reaction can only be due to the acid liberated, *i.e.* the *aci-nitromethane.* The reconversion of this to neutral nitromethane proceeds so rapidly that after a few minutes the acid reaction of the solution has disappeared.

When the nitroparaffins are reduced with powerful agents, the corresponding amines are formed in a way similar to that described in the next section for nitrobenzene. But just as in the case of nitrobenzene, so also with nitroparaffins, the reaction can be stopped at the hydroxylamine stage by using zinc dust in a neutral medium.

Experiment.—To a few drops of nitromethane dissolved in a little water some pieces of granulated zinc and then some concentrated hydrochloric acid are added. A vigorous reaction takes place. When this has subsided the mixture is heated for a short time on the water bath and then the liquid is decanted. When a large excess of concentrated caustic alkali is added it is possible to recognise, by the odour produced and by the fact that turmeric paper is turned brown, the formation of a volatile amine. If it is desired to use the reaction for the preparation of *methylamine* the nitromethane must be added to the reducing mixture in small portions. Compare also p. 143.

N-**Methylhydroxylamine.**—An aqueous solution of nitromethane is mixed with about its own weight of ammonium chloride, the mixture is then cooled (temperature about 10°) and three parts of zinc dust are added in small portions with constant shaking. Zinc dust is now removed by filtration, and it is found that the filtrate reduces ammoniacal silver and Fehling's solutions. The preparation of this easily accessible alkylhydroxylamine as hydrochloride is described by Beckmann, *Annalen*, 1909, **365**, 204.

Almost without exception the numerous transformations which primary and secondary nitroparaffins undergo involve the *aci*-form, *i.e.* they take place under conditions in which the salt of the *aci*-form is produced. Qualitatively the nitroparaffins greatly resemble ketones in their mode of action, although the much greater reaction velocity of the nitro-compounds brings about a quantitative difference.

1. By the action of bromine *bromonitro-compounds* are formed, *e.g.*

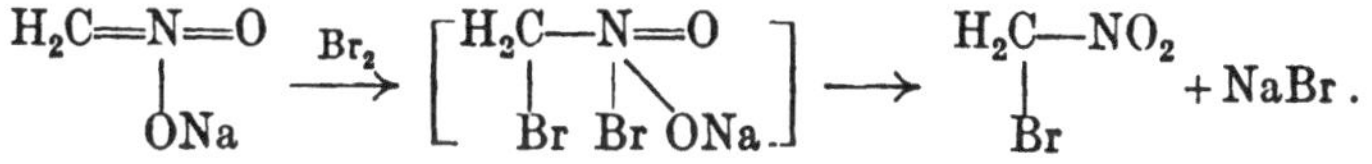

2. Nitrous acid converts *primary* nitroparaffins into *nitrolic acids* and the *secondary* ones into so-called *pseudonitroles*, which latter, being nitroso-compounds, are coloured green or blue:

(*a*) 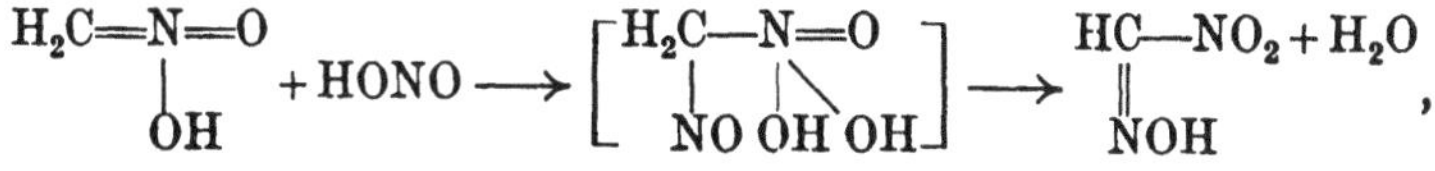

(*b*) 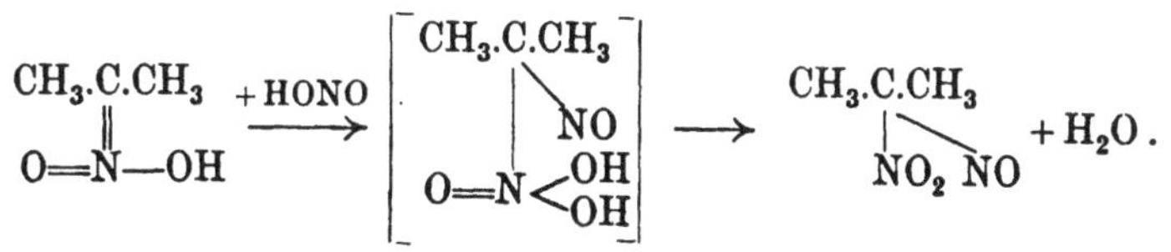

Experiment.—*Methylnitrolic acid.*[1]—Nitromethane (3·2 g.) is dissolved in 30 c.c. of ice-cold 2*N*-sodium hydroxide solution and a

[1] *Ber.*, 1909, **42**, 808.

concentrated solution of sodium nitrite (3·5 g.) is added. Without further cooling 4 *N*-sulphuric acid is run in from a dropping funnel until the solution, which is at first deep red, has just become yellow and does not yet turn potassium iodide-starch paper blue. The mixture is now extracted twice with ether and the aqueous layer is again cooled. Sulphuric acid is again dropped in until the evolution of nitrous acid becomes distinct and the solution is then again made so strongly alkaline with 5 *N*-sodium hydroxide solution, that a deep orange colour results. Once more the solution is acidified to such an extent that nitrous acid cannot yet be detected, and is again extracted twice with ether. The combined ether extracts are dried for two hours over calcium chloride in a vessel kept on ice. The ethereal solution is now transferred to a small round flask and the solvent removed by distillation with a capillary *in vacuo* on the water bath at 15°–20°. A residue consisting of about 1 g. of well-crystallised pale yellow methylnitrolic acid remains. The preparation decomposes in the course of a few hours. Test its behaviour towards alkalis.

Silver Fulminate.[1]—Freshly prepared methylnitrolic acid (0·5 g.) dissolved in 4 c.c. of water is heated in a wide test tube over a naked flame with 1 c.c. of 5 *N*-nitric acid (concentrated nitric acid of density 1·4 diluted with an equal volume of water) and 4 c.c. of 10 per cent silver nitrate solution. After a short time the reaction begins, a vigorous evolution of gas (NO) occurs, and crystalline silver fulminate is precipitated. Boiling is continued for a few minutes longer with constant shaking. The mixture is then cooled and the product filtered at the pump and washed with water. On a small piece of porous plate a small sample weighing about 10 mg. is dried without rubbing and its shattering power is tested in the flame and under a blow from a hammer. (*Wear goggles.*)

While still moist the bulk of the material is transferred to a test tube; even in the moist condition pressing with a metal spatula or other hard object is to be avoided. Then 2 c.c. of concentrated hydrochloric acid are poured into the test tube, when the odour of free fulminic acid can be perceived. This odour so closely resembles that of hydrocyanic acid as to make confusion possible. After half an hour the contents of the test tube are heated for a short time in the boiling water bath, 4 c.c. of water are added, silver chloride is removed by filtration, and the filtrate is evaporated to dryness on

[1] *Ber.*, 1907, 40, 419.

the water bath in a small glass dish. The *hydroxylamine hydrochloride* constituting the residue is recognised by its reducing effect on ammoniacal silver solution and on Fehling's solution.

Silver fulminate must invariably be destroyed immediately after it is prepared. This is most readily done with hydrochloric acid.

The nitrolic acids are colourless, but dissolve in alkalis with a deep red colour because in addition to the chromogenic nitroso-group the *aci*-nitro-group is formed. The following formula is ascribed to the red salt, *e.g.*

$$\begin{array}{l} HC{=}N{=}O \\ \;| \quad\; | \\ NO\;ONa \end{array} .$$

When methylnitrolic acid is heated in nitric acid solution it decomposes into nitrous and fulminic acids. The latter can be isolated as silver fulminate if silver nitrate is present.

$$\begin{array}{l} H \diagdown \\ \quad\;\; C{:}NOH \\ O_2N \diagup \end{array} \longrightarrow NO_2H + C{:}NOH .$$

The production of fulminates (silver and mercury fulminates) from ethyl alcohol and nitric acid takes place by way of methylnitrolic acid. This question was discussed on p. 138.

The mercuric salt of nitromethane decomposes directly into mercury fulminate and water (Nef).

$$(H_2C=NO_2)_2Hg \longrightarrow (C=NO)_2Hg + 2H_2O .$$

3. Like the ketones the primary nitro-compounds condense with aldehydes, water being eliminated. *Phenylnitroethylene* is conveniently prepared in this way.

$$C_6H_5.CHO + H_3C.NO_2 \longrightarrow C_6H_5.CH{:}CH.NO_2 .$$

Phenylnitroethylene.[1]—Nitromethane (3·2 g.) and benzaldehyde (5·3 g.) are dissolved in 20 c.c. of alcohol, and to the solution, which is well cooled in a freezing mixture and is vigorously shaken, 3·5 g. of potassium hydroxide dissolved in a mixture of water (5 c.c.) and methyl alcohol (10 c.c.) are gradually added. Shaking is continued until a sample of the resulting crystalline paste forms a clear solution in water. The potassium salt of phenylnitroethyl alcohol, C_5H_6.CH(OH).CH:NOOK, has been formed, and the corresponding free acid changes into phenylnitroethylene with elimination of water. The change takes place if the reaction product is dissolved in ice-

[1] Thiele and Haeckel, *Annalen*, 1902, 325, 7; Bouveault and Wahl, *Compt. rend.*, 1902, 135, 41.

water and 60 c.c. of ice-cold *N*-sulphuric acid are run in with stirring. The oil which is formed soon solidifies and is then filtered with suction, dried on porous plate for a short time, and recrystallised from little alcohol. About 5 g. of phenylnitroethylene, in the form of splendid yellow needles, are obtained. Melting point 58°.

4. All primary nitro-compounds couple with diazobenzene, but instead of the expected azo-compounds, phenylhydrazones of α-nitro-aldehydes are formed by rearrangement of the molecule :

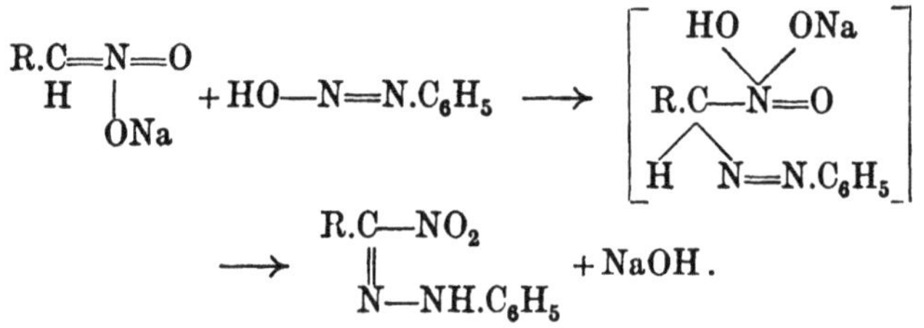

5. A very interesting reaction of nitromethane produced by strong alkali must also be mentioned here.

Two molecules condense with elimination of water to the so-called *methazonic acid*, which has the constitution of nitroacetaldoxine (I) (Meister).[1]

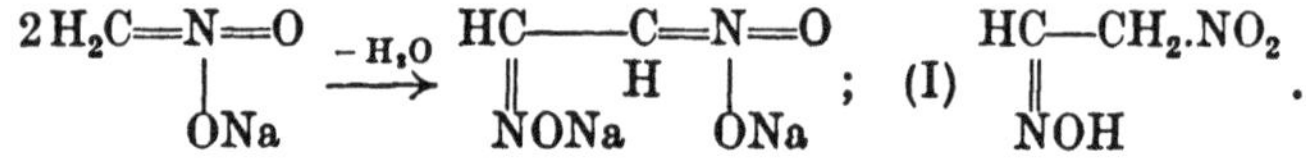

By the action of thionyl chloride Steinkopf prepared from this substance the long-sought *nitroacetonitrile*[2] $NO_2.CH_2.CN$, and by the hydrolysis of this nitrile, *nitroacetic acid*.[3]

2. NITRATION OF AN AROMATIC HYDROCARBON

Nitrobenzene and Dinitrobenzene

(a) Nitrobenzene

Concentrated nitric acid (d. 1·4 ; 100 c.c. = 140 g.) is poured gradually, with shaking, into a flask (capacity about 0·5 l.) containing concentrated sulphuric acid (125 c.c. = 230 g.). After the warm mixture has been cooled to room temperature by immersion in cold water, benzene (90 c.c. = 78 g. 1 mole) is gradually added with frequent shaking. If, during the addition, the temperature rises

[1] On the mechanism of this reaction, see *Annalen*, 1925, 444, 15.
[2] *Ber.*, 1908, 41, 1048. [3] *Ber.*, 1909, 42, 3925.

above 50°–60° the flask is dipped for a short time in ice-water before further quantities of benzene are added. Each time benzene is added the production of a transient intense brown colour is observed. After the flask has been warmed on the water bath at 60° under an air condenser for half an hour, the lower layer of liquid, which consists of sulphuric and nitric acids, is separated from the upper layer, which contains the nitrobenzene.[1] The latter is shaken in a separating funnel, first with water, then with dilute sodium hydroxide solution, and finally again with water. It must be borne in mind that the nitrobenzene now forms the lower layer. After washing and settling, the nitrobenzene is run into a dry flask and warmed with calcium chloride under an air condenser on the water bath until the original milkiness has disappeared. Finally, the substance is purified by distilling (not quite to dryness) from a flask with air condenser. Boiling point 206°–207°. Yield 100–105 g.

(*b*) Dinitrobenzene

To a mixture of 14 c.c. (25 g.) of concentrated sulphuric acid and 10 c.c. (15 g.) of *fuming* nitric acid in an open flask which is heated on the water bath for half an hour and frequently shaken, 10 g. of nitrobenzene are gradually added (fume chamber). The reaction mixture is cooled somewhat and then poured with stirring into cold water. The dinitrobenzene solidifies, is filtered with suction, washed with water, pressed on porous plate, and crystallised from alcohol. Melting point 90°. Yield 10–12 g.

The property of yielding nitro-derivatives by the action of nitric acid is a characteristic of aromatic substances. According to the conditions under which the nitration is carried out one or more nitro-groups can be introduced. Write the equation for the reaction.

If an aromatic compound contains saturated aliphatic side chains nitration carried out under the above conditions takes place always in the benzene nucleus and not in the side chain. Since the carbon atoms of benzene are each united directly to only one hydrogen atom, the nitro-derivatives obtained are tertiary and therefore incapable of forming salts, nitrolic acids, or pseudonitroles, as do the primary and secondary nitro-compounds.

Nitro-groups can also be introduced into side chains.[2] If toluene or ethylbenzene, for example, is heated with dilute nitric acid (d. 1·076)

[1] In large scale practice the residual nitrating acid is recovered in a similar way; 1·5 moles of nitric acid are used in the present case.

[2] Konovalov, *Ber.*, 1894, 27, Ref. 194, 468.

in a closed vessel to a temperature somewhat over 100° phenylnitromethane $C_6H_5.CH_2.NO_2$ or phenylnitroethane $C_6H_5.CH(NO_2).CH_3$ is obtained.

The same reaction can be applied, not only to the aromatic parent substances, the hydrocarbons, but also to all their derivatives, such as phenols, amines, aldehydes, acids, and so on. The nitration does not, however, always proceed with the same ease, and therefore the most favourable experimental conditions must be determined for each substance. If a substance is very easily nitrated it may be done with nitric acid sufficiently diluted with water, or else the substance to be nitrated is dissolved in a resistant solvent and is then treated with nitric acid. Glacial acetic acid is frequently used as the solvent. Substances which are less easily nitrated are dissolved in concentrated or fuming nitric acid. If the nitration proceeds with difficulty the elimination of water is facilitated by the addition of concentrated sulphuric acid to ordinary or fuming nitric acid. When nitration is carried out in sulphuric acid solution, potassium or sodium nitrate is sometimes used instead of nitric acid. The methods of nitration described may be still further modified in two ways: 1, the temperature or, 2, the amount of nitric acid used, may be varied. Thus nitration can be carried out at the temperature of a freezing mixture, at that of ice, at that of cold water, at a gentle heat, or, finally, at the boiling point. Moreover, we can either employ an excess of nitric acid or the theoretical amount. Small scale preliminary experiments will indicate which of these numerous modifications may be expected to yield the best results. Since nitro-compounds are usually insoluble or sparingly soluble in water they can be precipitated from the nitration mixture by dilution with water.

The chemical character of a compound is not fundamentally altered by the introduction of a nitro-group. Thus the ring-substituted nitro-derivatives of the hydrocarbons are neutral compounds like the hydrocarbons themselves, If, however, a nitro-group enters a substance having, for instance, an acid character, then this character is thereby intensified; the nitrophenols, for example, are more acidic than phenol. Correspondingly, the strength of bases is decreased by nitration; the nitranilines are less basic than aniline.

The great importance of the nitro-compounds depends on their behaviour on reduction. This question is discussed below.

When two nitro-groups are introduced into the benzene ring the chief product is *m*-dinitrobenzene, which conforms to the following general laws of substitution. For aromatic compounds there are three important typical reactions: 1, halogenation; 2, nitration, and 3, sulphonation. From benzene itself, of course, only one mono-halogen-, one mono-nitro-, or one mono-sulphonic acid derivative can be obtained. But in a mono-substituted benzene the halogen- or the nitro- or sulphonic acid group can take up the *o*-, *m*-, or *p*-position. Experiment has shown that there are two types of substitution reaction. In certain cases the product consists very largely of *o*- and *p*- di-substituted derivatives accompanied by only small amounts of *m*-derivative, whilst in others the *m*-derivative predominates and is accompanied by small quantities only of the *o*- and *p*-derivatives.

Substituents which direct halogens or nitro- and sulphonic acid groups—or, indeed, any other group—chiefly to the o- *and* p-*positions*, are called substituents of the *first order*. Those which direct chiefly to the m-*position* are substituents of the *second order*. Those of the first order are: the halogens, alkyl groups, hydroxyl, O-alkyl and O-acyl groups, the amino-group, etc. Substituents of the second order are: the nitro-group, the sulphonic acid group, the aldehyde group, the carboxyl, COO-alkyl and $CONH_2$ groups, the CO-alkyl group (in ketones), the $C \equiv N$ group, etc.

It follows from this enumeration that the substituents of the first order are all, according to the usual view, saturated and contain no multiple bonds, whereas the opposite holds for the substituents of the second order. It is also noteworthy that *o*- and *p*-substitutions almost always proceed more easily, *i.e.* with much greater velocity, than do *m*-substitutions.

The difficulty of nitration increases progressively with the number of nitro-groups introduced. Already the introduction of a second nitro-group into nitrobenzene requires much more powerful reagents than the nitration of benzene itself. Symmetrical *trinitrobenzene* is formed only after several days' boiling of the dinitro-compound with fuming nitric acid and even then only in poor yield.

Compare with this the greater susceptibility to substitution caused by OH and NH_2 and even by the methyl group in toluene. Trinitrotoluene is manufactured on a large scale as an explosive.

Some of the nitro-compounds are liquids, some are solids characterised by the great readiness with which they crystallise. Those nitro-compounds which distil without decomposition have higher boiling points than the parent substances.

If *ethylene* is treated with nitrating acid, *nitroethyl nitrate*, $NO_2.CH_2.CH_2ONO_2$ is produced, as has already been mentioned. The *nitroethyl alcohol* which is first formed by addition of nitric acid is fixed by esterification, whereas the addition compound with HNO_3 which is first formed at the double bond of benzene is decomposed with elimination of H_2O for reasons which have been mentioned repeatedly. This case is therefore analogous to the reactions of bromine with ethylene and with benzene (p. 93).

3. REDUCTION OF A NITRO-COMPOUND TO AN AMINE

(a) Aniline from Nitrobenzene [1]

To a mixture of 120 g. of finely granulated tin [2] and 61·5 g. (0·5 mole) of nitrobenzene in a round-bottomed flask (capacity 2 l.), 270 c.c. (320 g.) of concentrated hydrochloric acid are added in the following manner: At first only about one tenth of the acid is added, a moderately wide air condenser is then immediately attached to the flask and the mixture is shaken. After a short time the contents of the flask become warm and eventually begin to boil vigorously. Without completely suppressing the reaction, the flask is cooled in water and then the rest of the acid is added gradually, with constant shaking, in such a way that the reaction is always maintained at a good rate. Finally the mixture is heated on the water bath for one hour longer, 100 c.c. of water are added to the hot solution, and then a solution of 150 g. of commercial sodium hydroxide in 200 c.c. of water is gradually poured in until the product is strongly alkaline.[3] A long downward condenser is straightway attached to the flask and steam is passed into the hot liquid until, after the distillate has ceased to pass over as a milky fluid, a further 300 c.c. of clear liquid have distilled. Finely powdered sodium chloride (25 g. for each 100 c.c.) is dissolved in the distillate and the aniline is then extracted with ether.[4] The ethereal solution is dried over a few pieces of solid potassium hydroxide, the ether is evaporated, and the aniline is distilled. Boiling point 184°. Yield 90–100 per cent of the theoretical.

The property of being converted by energetic reduction into primary amines belongs to the nitro-compounds both of the aliphatic and of the aromatic series. Six atoms of hydrogen are required for the reduction of each nitro-group. In industry nitrobenzene is reduced, not with expensive tin, but with iron filings or iron powder according to the old method of Béchamp, which is still in use at the present time. The amount of hydrochloric acid indicated by the equation

---

[1] *Annalen*, 1842, **44**, 283.

[2] If granulated tin is not available it may be prepared by melting block-tin over a blow-pipe in a long-handled iron spoon provided with a spout, and then pouring the metal *drop by drop* from a height of 2–3 feet into a bucket filled with water.

[3] On the electrolytic removal of the tin see p. 308, footnote.

[4] On a large scale the salting out is omitted and the "aniline water" is used in raising the steam for the next batch.

$$C_6H_5.NO_2 + 3\,Fe + 6\,HCl = C_6H_5.NH_2 + 3\,FeCl_2 + 2\,H_2O \qquad (A)$$

is very much greater than that actually used on a technical scale. The process can be carried out with much less, even with about 3 per cent of the quantity calculated from the above equation. This is due to the fact that the iron is mostly oxidised to ferric hydroxide. The reaction proceeds according to the equation (B) as well as according to (A), *i.e.* $FeCl_2$ is continually re-formed.

$$C_6H_5.NO_2 + 2\,FeCl_2 + 2\,Fe + 4\,H_2O \rightarrow C_6H_5.NH_2 + 2\,FeCl_2 + 2\,Fe(OH)_3. \quad (B)$$

Ferric hydroxide is precipitated as a result of the hydrolysis of the ferric chloride, and hydrochloric acid is continuously liberated to react with further quantities of iron. The iron oxide ultimately produced is re-converted into iron powder by hydrogen at red heat, and so becomes available for the next batch.

Catalytic hydrogenation processes in which copper is the catalytic agent have also been recently introduced into industrial practice for the preparation of aniline from nitrobenzene.

For the reduction of nitro-compounds on a small scale, tin or stannous chloride and concentrated hydrochloric acid are the most suitable reagents. Solids are often difficult to reduce in the absence of a solvent and call for the addition of alcohol or glacial acetic acid. When the addition of water to a sample of the reaction mixture produces a clear solution it is known that the reduction is complete. The base will, of course, be present as hydrochloride, and hydrochlorides are, almost without exception, soluble in water. It should be noted, however, that sparingly soluble double salts with stannous chloride are often formed, but these double salts are usually soluble in boiling water.

If a double salt crystallises out in large amounts it is collected at the pump and is decomposed with alkali or, better, with hydrogen sulphide; thus the base is easily obtained in the pure state.

Of the primary monoamines, some, such as aniline, *o*-toluidine, xylidine, are colourless liquids. Others, such as *p*-toluidine, pseudocumidine and th enaphthylamines, are solids. They can be distilled without decomposition and are volatile with steam. In water they are rather sparingly soluble—a 3 per cent solution of aniline can be made. The di- and polyamines are usually solids, not volatile in steam and much more soluble in water than the monoamines. The amines are basic in character, but, as a result of the negative nature of the phenyl-group, the aromatic amines are considerably weaker bases than are the aliphatic amines. Consequently aqueous solutions of the (stoicheiometrically) neutral aniline salts are acid to litmus because of the hydrolysis which they undergo. For the same reason a small amount of the free base can be extracted with ether from an aqueous solution of an aniline salt. (Test with a solution of hydrogen chloride in ether or, after evaporation of the ether, by the reaction with bleaching powder.)

Experiments.—1. 10 c.c. of aniline water (obtained by shaking three drops of aniline with 10 c.c. of water in a test tube) are diluted with 100 c.c. of water and a small quantity of a filtered aqueous solution of *bleaching powder* is added. A violet colour is thus produced (Runge's reaction). This very sensitive reaction is given only by aqueous solutions of *free* aniline and not by those of its salts, from which it must first be isolated.

The reaction may also be used to detect small quantities of benzene or of nitrobenzene, after these have been converted in the manner just described, by reactions carried out in a test tube.

The bleaching powder reaction is peculiar to aniline; the dye which is formed is a complex quinone derivative the constitution of which is not yet known with complete certainty. The other experiments, about to be described, are group reactions for primary aromatic amines.

2. Primary and secondary amines are acylated by acid chlorides and anhydrides, in particular also by the chloride of benzene sulphonic acid (pp. 182, 190). The preparation of *acetanilide* has already been described (pp. 114, 117). The acetyl- and benzoyl-derivatives of all the simpler primary amines of the benzene and naphthalene series are known, so that these derivatives can always serve for purposes of identification.

As an exercise a primary amine should be identified in the manner described.

3. **Benzylidene-aniline.**—Aniline (1 c.c.) is heated in a test tube on the water bath with an equal quantity of benzaldehyde. The mixture becomes turbid because water separates. On cooling a solid is formed, a so-called Schiff's base (azomethine). Melting point 72°.

This feebly basic condensation product is decomposed into its components when warmed with acid. The reaction is a general one for primary amines.

4. **Isonitrile Reaction.**—Like the primary aliphatic amines of the methylamine type, aniline and its analogues give a characteristic odour with chloroform and alkali.

Aniline (2 drops) and alcohol (2 c.c.) are mixed in a test tube and 0·5 c.c. of concentrated potassium hydroxide solution along with about five drops of chloroform are added. The mixture is then gently warmed in a fume chamber.

$$C_6H_5.NH_2 + Cl_2C\langle^{Cl}_{H} + 3KOH = C_6H_5.N{:}C + 3KCl + 3H_2O.$$

In a quite analogous way ammonia yields *hydrocyanic* acid.

$$H.NH_2 + Cl_2C\begin{matrix} \diagup Cl \\ \diagdown H \end{matrix} + 3\,KOH = HN{:}C + 3\,KCl + 3\,H_2O\,.$$

The question of the constitution of hydrocyanic acid has already been considered (p. 128). Here it need only be remarked that the isonitriles are converted by hydrolysis into primary amines and formic acid; no carbon monoxide is produced, although from the formula this might be expected. The reason for this is to be sought in the fact that the first stage in the reaction consists in the addition of water to the two free valencies of the carbon atom. The reaction must therefore be formulated thus:

$$C_6H_5.N{:}C\begin{matrix} \diagup \\ \diagdown \end{matrix} + H_2O = \underset{iso\text{-form of formanilide}}{C_6H_5.N{:}C\begin{matrix} \diagup OH \\ \diagdown H \end{matrix}} = C_6H_5.NH_2 + OC\begin{matrix} \diagup OH \\ \diagdown H \end{matrix}\,.$$

It is consequently no argument against the carbimine structure of hydrocyanic acid that it likewise does not yield carbon monoxide but *formic acid* (in addition to ammonia) on decomposition by mineral acids.

In a similar way fulminic acid, another derivative of bivalent carbon, which has been accurately investigated and for which the carboxime $>C{:}NOH$ constitution has been established, is decomposed into formic acid and hydroxylamine (see experiment on pp. 150, 151).

5. Hofmann's method for the synthesis of alkylamines can be applied to the *alkylation* of aniline. The methylated anilines are especially important, and in particular the tertiary base *dimethylaniline,* which will be repeatedly used in this practical course as starting material and is also much used technically. On a large scale aniline (in the form of its hydrochloride) is methylated in an autoclave with methyl alcohol. The methyl chloride which is thus produced is the actual alkylating agent. At very high temperatures the methyl group wanders from the nitrogen to the *para*-position. This is a new example of the rearrangements which benzene derivatives undergo. Such rearrangements will be repeatedly mentioned (cf. pp. 177, 178).

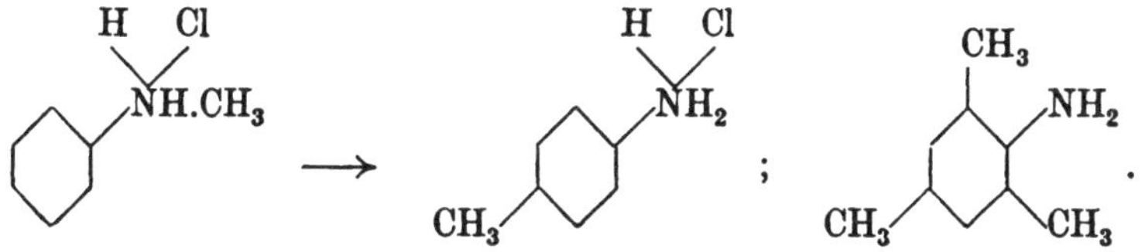

If the reaction takes place in the presence of an excess of methyl alcohol *mesidine* (formula on the right, above) is ultimately produced as a result of further methylation and wandering (A. W. Hofmann). This reaction does not proceed smoothly and has no preparative importance.

6. **Reaction with Carbon Bisulphide.**—Ammonia and the primary amines of the aliphatic series combine with carbon bisulphide to form ammonium salts of dithiocarbamic acids, *e.g.*

$$S{:}C{:}S + H_2N.CH_3 \longrightarrow S{:}C\begin{cases} SH \\ NH.CH_3 \end{cases} \xrightarrow{H_2N.CH_3} S{:}C\begin{cases} SNH_3.CH_3 \\ NH.CH_3 \end{cases}.$$

In the aromatic series, however, the reaction at once proceeds further. From the dithiocarbamate first formed, hydrogen sulphide is eliminated, and the phenyl isothiocyanate which remains combines, in its turn, with a second molecule of amine, giving a diarylthiourea.

$$\underset{\text{Dithiocarbamate}}{S{:}C\begin{cases} SH.NH_2C_6H_5 \\ NH.C_6H_5 \end{cases}} \longrightarrow \underset{\text{Phenylisothiocyanate (Mustard oil)}}{S{:}C{:}N.C_6H_5} + H_2S + H_2N.C_6H_5$$

$$\xrightarrow{C_6H_5.NH_2} \underset{\text{Diphenylthiourea}}{S{:}C\begin{cases} NH.C_6H_5 \\ NH.C_6H_5 \end{cases}}$$

In order to obtain an isothiocyanate in the aliphatic series, a dithiocarbamate must be distilled with a salt of a heavy metal ($HgCl_2$, $FeCl_3$) (A. W. Hofmann). Here, in the case of diphenylthiourea, the distillation is carried out with concentrated hydrochloric acid.

Diphenylthiourea (Thiocarbanilide).—In a round-bottomed flask provided with a long reflux condenser 20 g. of aniline, 25 g. of carbon bisulphide, 25 g. of alcohol, and 5 g. of finely powdered potassium hydroxide are kept gently boiling on the water bath for three hours. The carbon bisulphide and alcohol are then removed by distillation through a downward condenser and water is added to the residue. The crystalline material which has been produced is filtered with suction and washed successively with water, dilute hydrochloric acid, and again with water. After drying, the product weighs 15–18 g. A small amount is recrystallised from alcohol (melting point 154°) and the rest is used without further purification for the preparation of *phenylisothiocyanate.* Of the crude product 15 g. are heated on the sand bath with 60 c.c. of concentrated hydrochloric acid (d. 1·18) in a 250 c.c. flask to which a downward condenser is attached. When, as a result of distillation, the volume of the material in the flask has been reduced to 10–15 c.c., the distillate is diluted with its own volume of water and extracted with ether. The ethereal solution is shaken with a little sodium carbonate solution and dried with calcium chloride. The residue left after evaporation of the ether is distilled. Boiling point 222°. Yield almost quantitative.

Besides isothiocyanate, there is produced by the action of hydrochloric acid on thiocarbanilide *triphenylguanidine*, which separates as hydrochloride when the residue in the flask is diluted with 50 c.c. of water and allowed to stand for several hours. The free base is obtained by decomposing the salt with warm sodium hydroxide solution. Triphenylguanidine crystallises from alcohol in colourless needles which melt at 143°.

In the reaction described above the chief effect of the concentrated hydrochloric acid is to eliminate aniline :

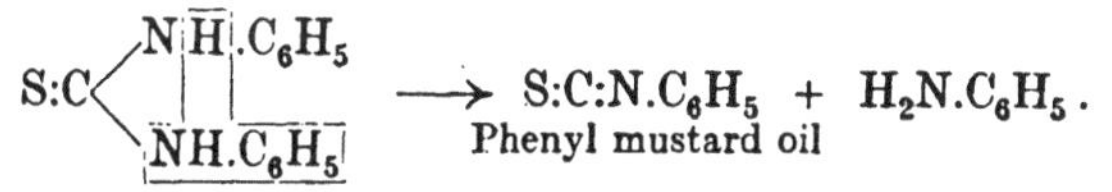

At the same time a small amount of hydrogen sulphide is also eliminated. The main product of this reaction, carbodiphenylimide (diphenylcyanamide), an extremely reactive substance, combines with the aniline present in the solution to form triphenylguanidine in the same way as, from cyanamide itself and ammonia, the unsubstituted guanidine is formed.

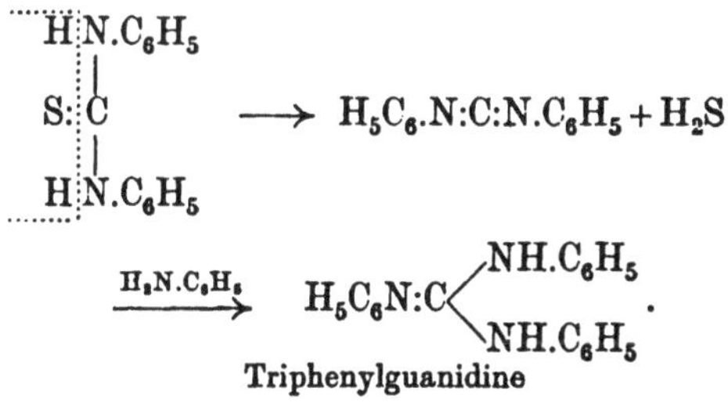

Triphenylguanidine

The isothiocyanates undergo fundamentally the same addition reactions as the isologous cyanic esters (see p. 144), *e.g.* $O:C:N.C_6H_5$.

They react, however, much more slowly. This follows already from the method of preparation of phenyl isothiocyanate (phenyl cyanate is at once decomposed by water).

The addition of aniline to phenyl isothiocyanate which leads to the re-formation of diphenylthiourea is described in the succeeding paragraph.

Phenyl isothiocyanate (5 drops) is mixed in a small test tube with an equal amount of aniline and the mixture is gently heated in a small flame. The product solidifies to crystalline *thiocarbanilide* when rubbed with a glass rod. A small sample of the material may be purified by recrystallisation from alcohol for the purpose of determining the melting point.

When phenyl isothiocyanate is heated with yellow mercuric oxide the sulphur is replaced by oxygen and the corresponding cyanate, which can be recognised by its exceptionally pungent odour, is obtained :

$$C_6H_5.NCS + HgO = \underset{\text{Phenyl cyanate}}{C_6H_5.NCO} + HgS\,.$$

Experiment.—Phenyl isothiocyanate (0·5 c.c.) is heated for some time in a test tube with an equal volume of yellow mercuric oxide. Heating is continued until the isothiocyanate boils. The yellow oxide is converted into black mercuric sulphide and at the same time the extremely pungent odour of phenyl cyanate is observed ; its vapour has a powerful lachrymatory effect.

(*b*) *m*-Nitraniline from *m*-Dinitrobenzene

The recrystallised dinitrobenzene is dissolved in hot alcohol (5 c.c. per gramme of dinitrobenzene) and the solution is rapidly cooled, which causes part of the dinitro-compound to separate again. Concentrated ammonia solution (d. 0·913 ; 0·8 g. per gramme of dinitrobenzene) is now added, the flask and contents are weighed, and the solution is saturated with hydrogen sulphide at the ordinary temperature. Then the current of hydrogen sulphide is shut off and the flask is heated for half an hour under reflux condenser on the water bath. Passing in of hydrogen sulphide in the cold and subsequent heating are repeated until, for each gramme of dinitrobenzene taken, an increase in weight of 0·6 g. has occurred. If, as a result of insufficient cooling, the necessary increase in weight is not attained, the shortage is neglected and hydrogen sulphide is passed in three times altogether. The product is diluted with water and filtered. The precipitate which has been separated is washed with water and repeatedly extracted with warm dilute hydrochloric acid. Finally, the nitraniline in the acid filtrates is liberated by neutralisation with ammonia and is recrystallised from water. Melting point 114°. Yield 70–80 per cent of the theoretical.

If it is desired to avoid working with hydrogen sulphide, commercial sodium sulphide may be used as reducing agent.[1]

[1] For details see Cobenzl, *Chem.-Ztg.*, 1913, 37, 299 ; see also Ullmann, *Enzyklopädie*, 1st edition, vol. i. p. 440.

The complete reduction of nitro-compounds which contain several nitro-groups is carried out in the same way as in the case of mononitro-derivatives. If, however, only partial reduction is desired, the use of ammonium hydrosulphide is advantageous.

$$C_6H_4\left\langle\begin{matrix}NO_2\\NO_2\end{matrix}\right. + 3NH_4SH = C_6H_4\left\langle\begin{matrix}NO_2\\NH_2\end{matrix}\right. + 2H_2O + 3S + 3NH_3.$$

For the reduction of nitro-compounds containing a group which may be attacked by nascent hydrogen, as, for example, an aldehyde group, an unsaturated side chain, and so on, special methods must be applied. In such cases ferrous hydroxide or iron powder (cf. Chap. VII. 6, arsanilic acid) are often used. The reduction is carried out thus: a weighed amount of ferrous sulphate is caused to act, in the presence of alkali (potassium or sodium hydroxide, baryta), on the substance to be reduced. In this way it is possible to reduce, for example, *o*-nitrobenzaldehyde to aminobenzaldehyde, and *o*-nitrocinnamic acid to aminocinnamic acid.

The yield of *m*-nitraniline when prepared as described above is only about four-fifths of the theoretically possible amount. This indicates, quite clearly, that the intermediate reduction products are reduced further much more rapidly than is an intact nitro-group.

When *m*-dinitrobenzene is reduced in an acid medium the product is m-*phenylenediamine*:

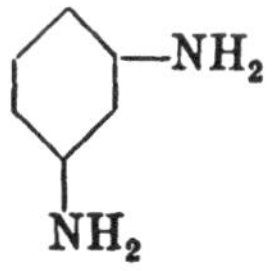

a di-acid base of technical importance. On diazotisation it yields the brown dye vesuvine or Bismarck brown. Hence *m*-phenylene diamine is used to test for traces of nitrites in water supplies.

o- and *p*-Nitranilines are prepared by nitration of aniline. Since aniline is very sensitive towards oxidising agents, the amino-group must be protected. This is most simply secured by acetylation. *Acetanilide is nitrated* and, according to the conditions chosen, a large yield of either the *o*- or the *p*-compound can be produced. The acetyl-group is subsequently removed by hydrolysis. *p*-Nitraniline is also obtained in a simpler way. *p*-Nitrochlorobenzene (which can be prepared by nitration of chlorobenzene) is caused to react with ammonia at a high temperature and under high pressure.

On the mobilising effect of nitro-groups on halogens, cf. p. 93.

The basic character of the amino-group in aniline, which is in itself feeble, is already greatly reduced by the entrance of a single nitro-group. Hence the three nitranilines are very weak bases which dissolve only in excess of acid to form salts. Whilst the nitranilines themselves are of a deep orange-yellow colour, their salts, when pure, are *colourless.* The powerful colour-deepening (bathochromic) influence which the free amino-group exerts on nitrobenzene (itself almost colourless when perfectly pure) is completely abolished by salt formation, *i.e.* by conversion of the unsaturated trivalent nitrogen into that of an ammonium compound.

Of the three nitranilines the *o*-compound is the most feebly basic, then comes the *p*- and finally the *m*-compound.

These facts illustrate relationships which are of great importance throughout the whole chemistry of aromatic compounds, and have already been mentioned above in connection with the mobilisation of halogens by nitro-groups in the *o*- and *p*-positions. Altogether two substituents in ortho- or in para-positions have a much stronger influence on each other than is exerted mutually by two substituents which are meta to each other. A satisfactory explanation of these facts has not yet been found although the close relationships between the 1 and 2 and the 1 and 4 positions (Thiele) throw some light on the question.

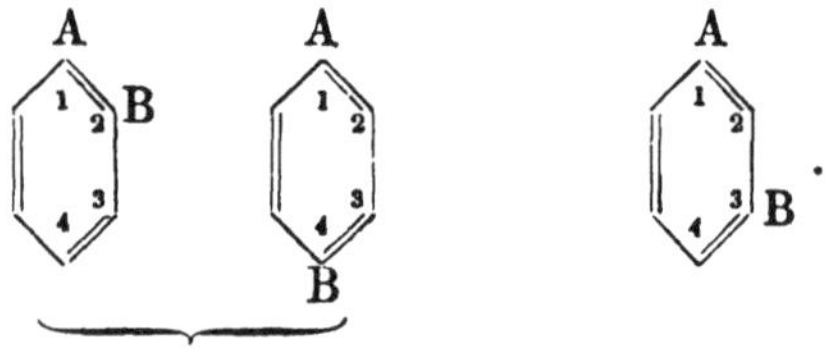

The unequal basicities of the three nitranilines can be illustrated by the following experiment. It is a general property of the salts of weak bases—as well as of weak acids—that in aqueous solution they are stable only if an excess of acid (or alkali) is present. When such solutions are diluted with water *hydrolysis* occurs as a result of the operation of the law of mass action. In the present case this phenomenon shows itself in the appearance of the yellow colour characteristic of the bases and finally, since the nitranilines are sparingly soluble in water, in their precipitation in crystalline form. The weaker the base the smaller is the amount of water which must be added in order to make the hydrolysis perceptible.

Experiment.—Of each of the three nitranilines—they can be procured in any laboratory—0·5 g. is separately dissolved by stirring with glass rods in three test tubes, each containing 3 c.c. of concentrated sulphuric acid. The colourless solutions thus prepared are then poured, separately, into beakers each containing 200 c.c. of water. *o*-Nitraniline is partially precipitated from the yellow solution produced, the somewhat more basic *p*-compound remains in solution but is yellow in colour, whilst the solution of *m*-nitraniline sulphate remains colourless.

4. PHENYLHYDROXYLAMINE[1]

A solution of 20 g. of ammonium chloride in 400 c.c. of water is mixed in a thick-walled filter jar (capacity about 2 l.) with 40 g. of freshly distilled nitrobenzene. To this mixture, which is continuously and vigorously stirred (preferably with a wooden rod with wide end), 60 g. of zinc dust (purity at least 75 per cent) are added during the course of forty minutes. The temperature is maintained at or below 10° by throwing in small pieces of ice.

After all the zinc has been added, stirring is continued for ten minutes—the odour of the nitrobenzene should then have disappeared—and the zinc hydroxide is at once removed by filtration with suction on a Büchner funnel. The filtrate (solution I) is poured into a beaker. The sludge of zinc hydroxide is washed on the funnel with 400 c.c. of water at 45° in the following manner: the funnel is filled with water which is then carefully mixed with the hydroxide by stirring, while as yet no suction is applied; then, by gentle suction, the solution is slowly filtered, and finally, with the pump full on, the residue in the funnel is pressed with a glass stopper (solution II). In each of the two aqueous solutions 120 g. of finely powdered sodium chloride are *completely* dissolved, which precipitates the phenylhydroxylamine in fine crystalline flocks. The suspension is allowed to stand in ice for half an hour and is then filtered at the pump as dry as possible. The phenylhydroxylamine is pressed on porous plate and can be recrystallised from a little benzene to which petrol ether is added. In this way the substance is obtained in the form of soft felted glistening needles which are quite pure and can be preserved for some time. Melting point 81°. For further work no special purification is necessary, however. Yield of *dry* material 75–80 per cent of the theoretical.

The material from solution I is usually purer than that from the other filtrate, but otherwise there is no reason why the two solutions should not be combined. Phenylhydroxylamine, which has not been recrystallised, can only be preserved undecomposed for a few days at the most.

Care should be taken not to allow phenylhydroxylamine, especially when in solution, to come into contact with the skin. In many individuals it produces severe eczema, whilst others are quite unaffected.

Ammonium hydrogen sulphide is also a very suitable reagent for

[1] E. Bamberger, *Ber.*, 1890, 27, 1347; A. Wohl, *Ber.*, 1890, 27, 1432.

the reduction of nitrocompounds to arylhydroxylamines ; it is used in alcoholic solution in the cold.[1]

An aluminium-mercury couple acts in the same way as zinc dust. This aluminium amalgam, prepared by the action of mercuric chloride on aluminium (preferably granulated), is also suitable for reducing substances dissolved in ether or alcohol ; the water which is required is slowly added drop by drop. (The method is that of H. Wislicenus.) The extent of the reduction varies according to the nitro-compound concerned, but corresponds approximately to the effect of zinc dust in a neutral medium. Consequently reduction usually does not proceed beyond the hydroxylamine stage.

Phenylhydroxylamine, especially when impure, is very unstable. It darkens and decomposes even in a closed container when kept for a short time. The pure material can be kept for a considerable time in a desiccator.

In contrast to the parent substance of the series *p*-tolylhydroxylamine, which is prepared by the reduction with zinc dust of *p*-nitrotoluene in boiling alcohol, is a quite stable compound.

The arylhydroxylamines are weak bases which dissolve in dilute acids to form salts. The instability of phenylhydroxylamine has three causes : the effect of atmospheric oxygen, of alkalis, and of acids. On exposure to air the compound, especially if impure, is oxidised to nitrosobenzene, which can be detected in decomposing phenylhydroxylamine by its pungent odour. As is often the case, the velocity of this process, which is known as autoxidation, is increased by alkalis ; it is accompanied by the production of hydrogen peroxide according to the equation :

$$C_6H_5.NHOH + O_2 \longrightarrow C_6H_5.NO + H_2O_2 .$$

Since nitrosobenzene condenses with phenylhydroxylamine forming azoxybenzene, the latter substance occurs amongst the decomposition products of phenylhydroxylamine. Also, by the action of alkalis, water is eliminated and *azobenzene* is produced. The action of acids will be discussed below.

Like hydroxylamine itself, all its derivatives of the form $R.N{<}^{OH}_{H}$ are *reducing agents.*

Experiment.—Test the reducing effect of phenylhydroxylamine by dissolving material on the point of a knife in 2 c.c. of warm water and adding to the solution a few drops of ammoniacal silver nitrate solution.

Experiment.—*Isomeric change of phenylhydroxylamine to* p-*aminophenol.*[2] The base (2·2 g. ; 0·02 mole) is added in small

[1] Willstätter, *Ber.*, 1908, 41, 1936. [2] Bamberger, *Ber.*, 1890, 27, 1552.

portions to a mixture of 10 c.c. of concentrated sulphuric acid and 30 g. of ice, externally cooled by more ice.

The solution is diluted with 200 c.c. of water and boiled until a sample, when mixed with dichromate solution, no longer smells of nitrosobenzene but of *quinone* (ten to fifteen minutes). To the cooled solution dichromate dissolved in 2 c.c. of water is added, a downward condenser is attached to the flask containing the mixture, and steam is passed through. The quinone is carried over with the steam. On the mechanism of its formation in this reaction compare p. 301. Test the residue in the flask for ammonia.

The change which arylhydroxylamines undergo by the action of mineral acids, especially when warm, is worthy of special note. If the position *para* to the NHOH-group is free, a rearrangement takes place to the isomeric p-*aminophenol*, *e.g.* in the case of phenylhydroxylamine according to the equation :

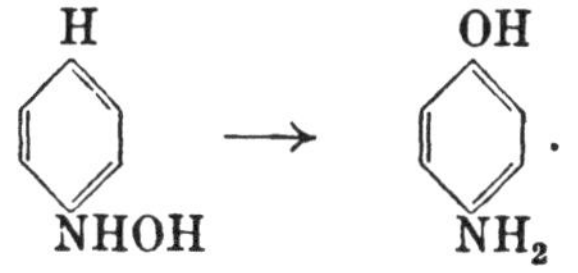

A process for the direct reduction of nitrobenzene to p-*aminophenol*, an important intermediate for the production of dyes, depends on the above interesting transformation. Nitrobenzene in alcoholic solution is mixed with concentrated sulphuric acid and electrolysed with a lead cathode. This process proves that phenylhydroxylamine is also an intermediate in the reduction of nitrobenzene in *acid* solution, as was mentioned above. Here, as a result of the rapidity of the rearrangement which takes place, it is not converted into aniline.

In the experiment described above, the *p*-aminophenol is not isolated but converted by oxidation into quinone.

If the position *para* to the NHOH-group is occupied, as, for example, in *p*-tolylhydroxylamine, the catalytic action of strong acids produces a different result. The hydroxyl group is indeed shifted to the carbon atom in the *p*-position, but this leads to the production of a quinone-like substance :

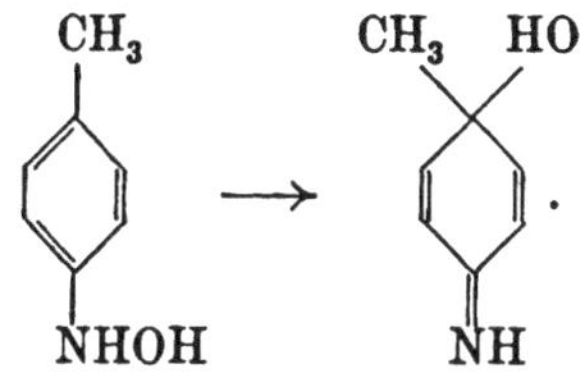

This " *quinolimine* " contains the quinonoid imino-group, which is very unstable towards acids and is eliminated hydrolytically in the form of ammonia, being replaced by oxygen :

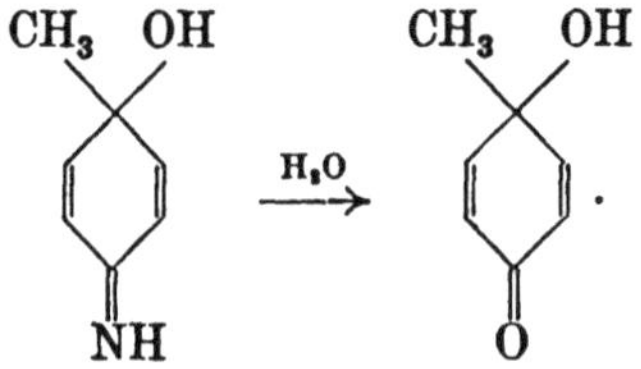

There is thus formed as final product of the rearrangement the simplest of the *quinols*, a substance which is very soluble in water, is colourless, in contrast to quinone, and is difficult to prepare (Bamberger).

Experiment. — *Nitrosophenylhydroxylamine.*[1] Phenylhydroxylamine (2·2 g.) is dissolved in 20 c.c. of *N*-hydrochloric acid and an aqueous solution of 1·4 g. of sodium nitrite is rather quickly added with strong cooling in ice. White needles are at once precipitated. They are filtered with suction, washed with ice-cold water, and dried on porous plate. Melting point 59°.

The substance is dissolved in ether and by passing in dry ammonia gas the ammonium salt is precipitated. From this salt the iron and copper salts, insoluble in water, are prepared by precipitation with Fe^{+++} and Cu^{++} (" Cupferron ").

In this reaction phenylhydroxylamine behaves like a secondary amine. To the class of nitrosohydroxylamines there belong also the so-called *isonitramines* and the compound of nitric oxide and sodium sulphite.

$$KO_3S{-}\underset{\displaystyle NO}{\underset{|}{N}}{-}OK \; .$$

The formation of the ammonium salt and the analytical application of nitrosophenylhydroxylamine as a reagent for the determination of iron and copper show that compounds of this kind are acidic.

Phenylhydroxylamine condenses with aldehydes, *e.g.* benzaldehyde, in the same way as it does with nitrosobenzene (cf. the following section, 5) :

$$C_6H_5.N\begin{matrix}\diagup H\\ \diagdown OH\end{matrix} + O{:}\underset{H}{C}.C_6H_5 \longrightarrow \left[C_6H_5.\underset{\displaystyle OH}{\underset{|}{N}}{-}\underset{\displaystyle OH}{\underset{|}{\overset{H}{C}}}.C_6H_5\right]$$

$$\xrightarrow{-H_2O} \underset{\displaystyle O}{\underset{\|}{C_6H_5.N}}{=}CH.C_6H_5 \; .$$

[1] Bamberger, *Ber.*, 1890, 27, 1552.

Substances of this kind are called "*nitrones*". Their method of formation is quite analogous to that of the oximes from aldehydes and hydroxylamine.

These nitrones are identical with the N-ethers of the aldoximes and are also formed—when an alkyl group occupies the place of C_6H_5—by the action of alkyl halides on the stereoisomeric β-aldoximes:

$$\begin{matrix} R.CH \\ \| \\ HON \end{matrix} + CH_3Br \longrightarrow \begin{matrix} R.CH \\ \| \\ ON—CH_3 \end{matrix} + HBr.$$

The nitrone from phenylhydroxylamine and benzaldehyde can easily be prepared in beautiful crystals from alcoholic solutions of these components.

Finally, it is interesting to note that the reduction of *nitroethylene*, an olefinic nitro-compound having a structure analogous to that of nitrobenzene, leads to acetaldoxime.

$$H_2C{:}CH.NO_2 + 4H = H_3C.\underset{H}{C}{:}NOH + H_2O.$$

Phenylnitroethylene, which is much more easily obtained (see p. 151), reacts in an analogous manner, yielding phenylacetaldoxime $C_6H_5.CH_2.\underset{H}{C}{=}NOH$ (Bouveault). In any case both these nitroethylenes first yield derivatives corresponding to phenylhydroxylamine. These derivatives, however, undergo immediate rearrangement to the stable oxime form:

$$R.CH{=}\underset{H}{C}.NHOH \longrightarrow R.CH_2.\underset{H}{C}{=}NOH.$$

Such a rearrangement does not take place in the benzene ring, where the three neighbouring double linkages provide the most complete expression of saturation. Here the NHOH-group remains excluded from the ring, and the "aromatic" constitution of the nucleus is preserved.

Similar considerations explain why, so far, it has been impossible to obtain an "aliphatic aniline" of the type $R.CH{:}C{<}_{NH_2}$.

5. NITROSOBENZENE

Freshly prepared phenylhydroxylamine (12 g.) is dissolved as rapidly as possible by gradual addition to an ice-cold mixture of 50 c.c. of concentrated sulphuric acid and 250 c.c. of water. Ice-water (500 c.c.) is now added to the solution, which is cooled to 0°, and then a likewise cold solution of 12 g. of sodium dichromate in 200 c.c. of water is run in rather quickly from a dropping funnel, while the reaction flask is further cooled and shaken. The nitrosobenzene soon separates in yellow crystalline flocks. These are filtered with suction on a small Büchner funnel, washed twice with water, and transferred with the filter paper to a round-bottomed flask from which the easily volatile nitrosobenzene is distilled with steam. The green vapour which passes over deposits an almost colourless crystalline crust already in the condenser. Towards the end of the distillation the current of water through the latter is shut off and the nitrosobenzene which has been deposited is cautiously melted with steam and run into the receiver. By filtration of the distillate the nitrosobenzene is separated, pressed on porous plate, and dried in a vacuum desiccator over calcium chloride (*not* over sulphuric acid). A sample of the dried material is washed in a test tube with a little ether (the solution is green) and again dried. This sample is used for determination of the melting point. Nitrosobenzene melts to a *green* liquid at 68°.

It can be obtained in an absolutely pure, stable condition by recrystallisation from twice its weight of alcohol.

Aromatic nitroso-compounds can also be obtained by oxidation of primary amines, but only one oxidising agent is known with which the process can satisfactorily be carried out. This reagent is *monopersulphuric* acid (Caro's acid):

$$C_6H_5.NH_2 + 2O \longrightarrow C_6H_5.NO + H_2O.$$

Experiment.[1] — Powdered potassium persulphate (18 g.) is thoroughly ground in a mortar, well cooled in ice, with 15 c.c. of concentrated sulphuric acid. The mixture, after standing for one hour, is poured on to 100 g. of ice and while being cooled is neutralised with crystalline sodium carbonate. Into this neutral solution 100 c.c. of aniline water (2·8 g. of aniline in 100 c.c. of water) are poured. After a short time nitrosobenzene separates in yellow flocks. The mixture is stirred and then, after the solid material has settled and the supernatant liquid is clear, filtered with suction. The nitrosobenzene is distilled with steam. The yield is equivalent to rather more than half the aniline taken.

Apart from rare exceptions, the only nitroso-compounds known are those in which the NO-group is united to a *tertiary* carbon atom, as in nitrosobenzene. Nitrosoisobutane $(H_3C)_3:C.NO$, for example, is a representative from the aliphatic series.

[1] Caro, *Z. angew. Chem.*, 1898, 11, 845; Baeyer, *Ber.*, 1900, 33, 124; *Ber.*, 1901, 34, 855.

In the solid state almost all nitroso-compounds are *colourless*,[1] but when *fused* or in *solution* they are *blue* or *green*. Determinations of the molecular weight of nitrosobenzene in liquid hydrogen cyanide have shown (Piloty) that the colourless form is bimolecular. The NO-groups of two molecules are loosely united in one of the ways indicated by the following formulae:

$$\begin{array}{c} C_6H_5.N{=}O \\ \| \\ C_6H_5.N{=}O \end{array} \quad \text{or} \quad \begin{array}{l} C_6H_5.N{-}O \\ \quad\;\; | \quad\; | \\ \quad\;\; O{-}N.C_6H_5 \end{array} .$$

When the crystal structure is destroyed by fusion or dissolution, a dissociation into single coloured molecules occurs to an extent which increases with the temperature. This behaviour is exactly similar to that so well known in the case of nitrogen peroxide:

$$(C_6H_5.NO)_2 \rightleftarrows 2\,C_6H_5.NO\,; \quad (NO_2)_2 \rightleftarrows 2\,NO_2\,.$$

The NO-group is the most active colour-producing (chromophoric) group known. With a radical such as isobutyl which is of no account for the absorption of light, it produces a blue nitrosohydrocarbon. In spite of their intense coloration the nitroso-compounds are not dyes, since they lack the "auxochromic" groups (*e.g.* NH_2 or OH) necessary for combination with textile fibres.

In many respects the nitroso-group resembles the aldehyde group, *i.e.* the reactions which, in the aldehydes, are due to the reactivity of the $>C{=}O$ double bond can, for the most part, be reproduced with the nitroso-compounds in virtue of the reactivity of the $-N{=}O$ double bond.

The condensation of nitrosobenzene with phenylhydroxylamine which is described below is an example of this similarity. Hydroxylamine and phenylhydrazine also react with nitrosobenzene, but the details of these reactions cannot be given here.

Aldehydes react with primary amines to give the so-called *azomethines* (Schiff's bases), by elimination of water (p. 158), *e.g.*

$$\underset{H}{C_6H_5.C{:}O} + H_2N.C_6H_5 \longrightarrow \underset{H}{C_6H_5.C{:}N.C_6H_5} + H_2O\,.$$

Similarly, nitrosobenzene and aniline give *azobenzene*:

$$C_6H_5.NO + H_2N.C_6H_5 \longrightarrow C_6H_5.N{:}N.C_6H_5 + H_2O\,.$$

Experiment.[2]—Nitrosobenzene (1 g. in 10 c.c. of alcohol) is added to a solution of 1 c.c. of aniline in 3 c.c. of glacial acetic acid. On

[1] *p*-Nitrosodimethylaniline, 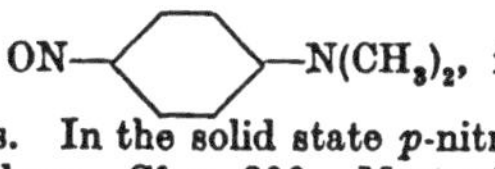

for example, and the bases related to it are exceptions. In the solid state *p*-nitrosodimethylaniline has a magnificent emerald-green colour. Cf. p. 306. Most of the pseudonitrols are also exceptions.

[2] A. Baeyer, *Ber.*, 1874, 7, 1638.

gentle warming the colour changes to dark orange. After heating for ten minutes longer on the boiling water bath, a few cubic centimetres of water are added. On cooling the solution azobenzene crystallises in orange-red platelets. Washed on the filter with 50 per cent alcohol and dried on porous plate, it melts at 68°. Azobenzene can very readily be recrystallised from little alcohol.

In this way mixed (asymmetrical) azo-compounds can conveniently be prepared in good yield. Prepare p-*methylazobenzene*, for example, from nitrosobenzene and *p*-toluidine according to the procedure given above.

Aldehydes condense with compounds containing a reactive methyl- or methylene-group to form unsaturated ketones, *e.g.*

$$C_6H_5.\underset{H}{C}{:}O + H_3C.CO.CH_3 \longrightarrow C_6H_5.\underset{H}{C}{:}CH.CO.CH_3 .$$

Benzylideneacetone

An analogous reaction is known with aromatic nitroso-compounds, but for it an exceptionally mobile hydrogen atom must be present in the ketone and hence no condensation occurs with simple ketones such as acetone. The products of the reaction are, of course, *azomethines*. This condensation has made possible the synthesis of 1:2:3-triketones (F. Sachs), *e.g.*

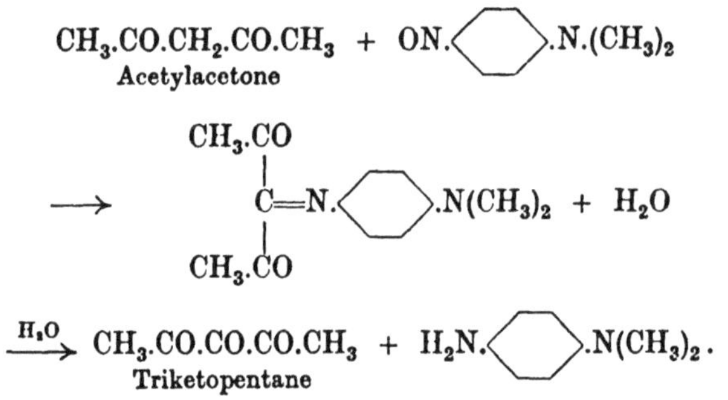

The last phase of the reaction depends on the fact that azomethines are easily decomposed by acids into carbonyl compounds and primary base.

The practical effect of the condensation consists, therefore, in the conversion of methylene into $>C{=}O$. The same result is attained in a quite similar reaction by the action of nitrous acid on ketones (cf. the synthesis of diacetyl from methylethyl ketone).

Finally, nitrosobenzene reacts with Grignard reagents. With phenylmagnesium bromide, in the usual way there is produced *diphenylhydroxylamine*, an exceptionally reactive substance :

$$C_6H_5.N{:}O + Br.Mg.C_6H_5 \longrightarrow C_6H_5.N\begin{matrix}\diagup C_6H_5 \\ \diagdown OMgBr\end{matrix} \xrightarrow{+H_2O}$$

$$\underset{\displaystyle\overset{\bullet}{O}H}{C_6H_5.N}.C_6H_5 + MgBr(OH).$$

Diphenylhydroxylamine, like phenylhydroxylamine, can best be dehydrogenated with silver oxide. Here only one H-atom, that from the OH-group, can be removed and the red crystalline substance which is thus produced contains *quadrivalent* nitrogen. Like nitrogen peroxide, therefore, this dehydrogenation product reacts like a free radicle. As its formula indicates, it is derived from nitrogen peroxide by the substitution of two C_6H_5-groups for one O.

$$\begin{matrix}H_5C_6\diagdown \\ H_5C_6\diagup\end{matrix}N{=}O \quad \textit{Diphenylnitrogen oxide.}$$

Experiment.—*Azoxybenzene from phenylhydroxylamine and nitrosobenzene.*—Phenylhydroxylamine (1 g.) is added to a solution of 1 g. of nitrosobenzene in 10 c.c. of alcohol. The mixture is shaken while a few drops of concentrated potassium hydroxide solution (1 : 1) are added, and is then warmed on the water bath for a few minutes. The yellowish-red solution thus formed deposits yellow crystals of the reaction-product when cooled and rubbed with a glass rod. Since azoxybenzene melts at 36°, it has a great tendency to separate from a supersaturated solution in the form of an oil. By recrystallisation from a little alcohol or from petrol ether (retain a few crystals for inoculation) the compound is obtained as a pale yellow or almost colourless solid.

The feeble colour of azoxybenzene, in contrast to the red of azobenzene, would appear to be more readily intelligible on the basis of the old formula, I, than on that by which Angeli has replaced it, II :

$$\text{I}\quad H_5C_6.\underset{\diagdown\ \diagup}{N{-}N}.C_6H_5,\ \ \ (\text{O bridging both N}) \qquad \text{II}\quad {}^{\bullet}C_6H_5.\underset{\displaystyle\overset{\|}{O}}{N}{=}N.C_6H_5.$$

Nevertheless, the existence of asymmetrical azoxybenzenes in two isomeric forms (Angeli),

$$R.\underset{\displaystyle\overset{\|}{O}}{N}{=}N.R' \quad \text{and} \quad R.N{=}\underset{\displaystyle\overset{\|}{O}}{N}.R'$$

convincingly supports formula II.

The mechanism of the condensation described is plain, and corresponds entirely to the production of nitrones from phenylhydroxylamine and aldehydes (pp. 168, 169):

$$C_6H_5.\underset{|\atop OH}{NH} + ON.C_6H_5 \longrightarrow C_6H_5.\underset{|\atop OH}{N}-\underset{|\atop OH}{N}.C_6H_5 \xrightarrow{-H_2O} C_6H_5.N:\underset{\|\atop O}{N}.C_6H_5 \, .$$

The relationships of azoxybenzene to azo- and hydrazobenzene are discussed in the explanations given for the next preparation.

The interesting rearrangement which azoxybenzene undergoes as a result of the action of concentrated sulphuric acid may also be mentioned here. p-*Hydroxyazobenzene*, the parent substance of the acid azo-dyes, is formed by this rearrangement (Wallach).

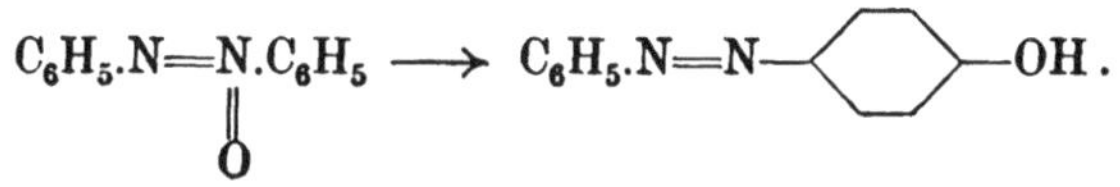

6. HYDRAZOBENZENE AND AZOBENZENE

(a) Hydrazobenzene

A round-bottomed flask (capacity 1 l.) is provided with a well-fitting thin-walled double neck attachment (Fig. 32). An ascending Liebig condenser is clamped obliquely and its inner tube is joined by a short wide rubber tube to the oblique neck in such a way that the flask can be vigorously shaken without trouble. The upright neck, which serves for the introduction of the zinc dust required for the reduction, is closed with a cork.

Sodium hydroxide (50 g.) is dissolved in 150 c.c. of water, and the solution, while still warm, is poured into the flask, along with 50 c.c. of *alcohol* and 41 g. (0·33 mole) of *nitrobenzene*. Zinc dust (6–8 g.) is now added with vigorous shaking which is continued until the reaction, at first rather violent, has ceased. The reaction mixture is then kept boiling by continuous addition of zinc dust. The reaction should not be allowed to become too violent, but, on the other hand, it should not be interrupted by cooling.

The contents of the flask become at first red (azobenzene), but finally turn pale yellow when the necessary amount of reducing agent has had its effect. About 100 to 120 g. of 75 per cent zinc dust are required. If the reaction should cease prematurely, the flask is heated on a vigorously boiling water bath.

It is essential that the contents of the flask be continually agitated by vigorous shaking in order that the heavy dust may remain in contact with the organic substance.

When the reduction is complete, the mixture is heated on the water bath and 500 c.c. of alcohol are added. The precipitated hydrazobenzene dissolves in this alcohol at the boiling point. The whole contents of the flask are filtered while boiling hot through a Büchner funnel (*flames in the vicinity must first be extinguished*), the flask is at once washed out with 50 c.c. of hot alcohol, and this alcohol is used to wash the excess of zinc dust on the funnel. The filtrate is allowed to cool in the closed filter flask, which is placed in a freezing mixture in order to accelerate crystallisation. After an hour the almost colourless reaction product is thoroughly separated by filtration at the pump, and is washed several times with 50 per cent alcohol, to which a small quantity of aqueous sulphurous acid has been added, until the filtrate is no longer alkaline. By rapid recrystallisation from a not too large volume of hot alcohol the hydrazobenzene is obtained quite colourless and pure. Melting point (with production of a yellow colour) 124°. Because of its great tendency to undergo autoxidation hydrazobenzene should be thoroughly dried *in vacuo*, and then can only be preserved for any length of time in a colourless condition if kept in well-closed containers in an atmosphere of CO_2 or N_2 or, better, in sealed tubes. This tendency to oxidation also requires the preparation being carried through without interruption.

The yield of crude product, which can be used directly for further preparations, amounts to 20–25 g.

(*b*) Azobenzene from Hydrazobenzene

1. **By Dehydrogenation.**—Bromine (10 g., 3·2 c.c.) is dropped into a solution of 6·0 g. of sodium hydroxide in 75 c.c. of water (75 c.c. of 2*N*-sodium hydroxide solution) which is kept cool in ice. Hydrazobenzene (9·2 g., 0·05 mole) in 60 c.c. of ether is shaken for ten minutes in a small separating funnel with this hypobromite solution, the ether layer is separated from the aqueous and the ether is distilled. Orange-red platelets of azobenzene, which melt at 68° when recrystallised from a little alcohol, are thus obtained. Yield quantitative.

Azobenzene is also obtained in good yield by drawing air through an alkaline alcoholic solution of hydrazobenzene for several hours.

2. **By Dismutation.**—Hydrazobenzene (1–2 g.) is melted in a test tube over a small flame. The orange-red liquid thus produced is

carefully heated until the aniline which has been formed begins to boil. On cooling, a mixture of red azobenzene and aniline is obtained. The aniline can be shaken out with water and identified by means of the bleaching powder reaction. The azobenzene may be recrystallised from alcohol as described above. If it is desired to isolate the aniline also, when larger amounts of hydrazobenzene are used, the base is separated from the azobenzene by means of dilute acetic acid. From the solution of its acetate the aniline is then liberated with concentrated alkali hydroxide solution, extracted with ether, and purified in the manner already described.

Azobenzene, which contains the chromophore group —N=N—, and is the parent substance of the azo-dyes, is a very stable compound, capable of being distilled without decomposition. The group N=N between the two aromatic nuclei is very firmly bound, although in most other azo-compounds this is not the case. The fastness of the azo-dyes is thus explained.

With concentrated mineral acids azobenzene gives red salts, as may be shown by pouring hydrochloric acid on it. Oxygen is added on and the azoxy-compound formed by the action of hydrogen peroxide or nitric acid. The synthesis of asymmetrical aromatic azo-compounds from nitroso-compounds and primary amines was discussed above.

At its melting point hydrazobenzene decomposes into *azobenzene* and *aniline* in the manner shown in the equation :

$$\begin{array}{ll} H_5C_6\text{—}N\vdots H & HN\text{—}C_6H_5 \\ \vert \;\;\; \rightrightarrows & \vert \\ H_5C_6\text{—}N\vdots H & HN\text{—}C_6H_5 \end{array} \longrightarrow \begin{array}{c} H_5C_6.N \\ \Vert \\ H_5C_6.N \end{array} + \begin{array}{c} H_2N.C_6H_5 \\ \\ H_2N.C_6H_5 \end{array}.$$

An exactly similar reaction of *phenylhydrazine* will be discussed later (p. 289). A simple example of this type is provided by the *spontaneous decomposition* of *hydrogen peroxide* into oxygen and water :

$$\begin{array}{ll} O\vdots H & OH \\ \vert \;\; \rightrightarrows & \vert \\ O\vdots H & OH \end{array} \longrightarrow \begin{array}{c} O \\ \Vert \\ O \end{array} + \begin{array}{c} HOH \\ \\ HOH \end{array}.$$

The spontaneous decomposition of hydrazobenzene, like that of hydrogen peroxide, is catalytically accelerated by metallic platinum.

(c) Benzidine from Hydrazobenzene

Hydrazobenzene (9·2 g.) dissolved in the minimum quantity of ether is added drop by drop with shaking to 100 c.c. of ice-cold, approximately 7 *N*-hydrochloric acid (concentrated acid diluted with an equal volume of water). Crystalline benzidine hydrochloride separates and, after 50 c.c. of concentrated hydrochloric acid have been added and the mixture has stood for half an hour, the hydrochloride is filtered at the pump, and washed, first with 7 *N*-hydrochloric acid and then with a little ether. Yield 9–10 g. The hydrochloride can be recrystallised by dissolution in hot water and addition of concentrated hydrochloric acid to the slightly cooled solution.

To obtain the *free base* a slight excess of concentrated sodium hydroxide solution is added to a not too concentrated solution of the salt prepared by dissolving the latter in warm water containing a little hydrochloric acid, and cooling rapidly to 15°–20°. The base, which separates in crystalline form, is filtered with suction and thoroughly washed with water. Before the alkali is added the solution of the salt must be clear; any hydrochloride which has been precipitated must be removed by filtration.

The free benzidine can be recrystallised from hot water or else from a little alcohol. Melting point 122°.

The conversion of hydrazobenzene into the isomeric benzidine—discovered by the Russian chemist Zinin in the year 1846—is started catalytically by mineral acids and results from the tendency of the molecule to pass into a form possessing less energy, *i.e.* into a more saturated condition. The reaction is suitably classified with those of which the chief characteristic is that a substituent united to nitrogen exchanges its point of attachment with an H-atom of the nucleus—usually an H-atom in the *p*-position. To this class belong the conversion of phenylsulphaminic acid into sulphanilic acid (p. 189), of phenylhydroxylamine into *p*-aminophenol (p. 167), and also of acetanilide into *p*-aminoacetophenone and of N-chloroacetanilide into *p*-chloroacetanilide:

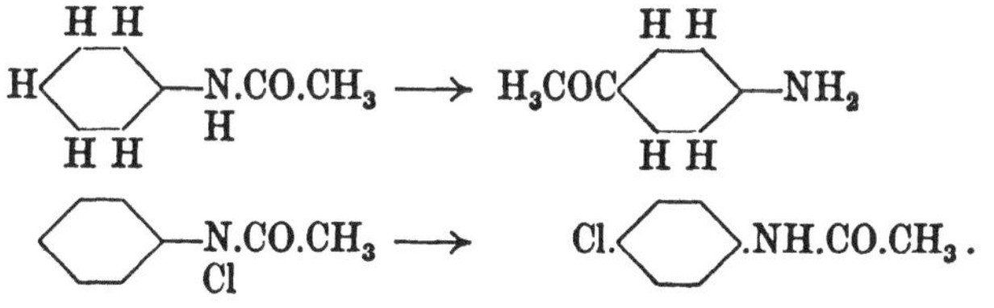

The rearrangement of aromatic nitrosamines to be discussed later is also of this type, *e.g.*

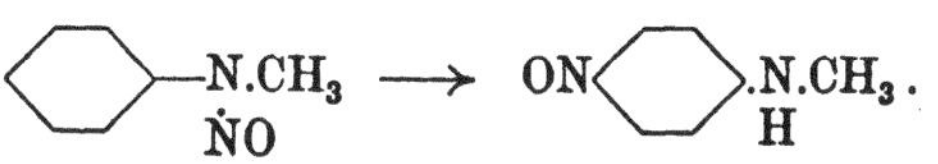

In the same way, in the benzidine reaction, the group $HN.C_6H_5$ becomes separated from nitrogen and attaches itself as $H_2N.C_6H_5$ to the *p*-position left free by the hydrogen atom.

N

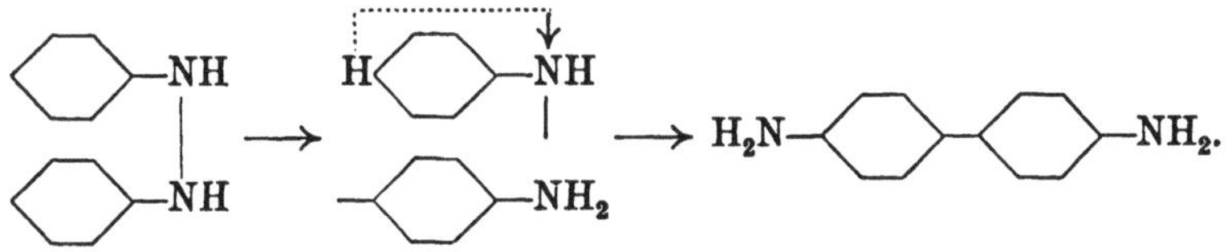

Special attention must be drawn to the fact that the radicles which wander do not do so as " free radicles ", but that the movements of these groups take place within the range of the molecular forces.

The similarity of the rearrangement of aromatic hydrazo-compounds to the exchange reactions with which they were compared above becomes more marked in those cases where the *p*-positions of the two benzene buclei are occupied. Then, as a rule, a diphenyl base is not produced, but the radicle which separates moves so that its nitrogen atom takes up the *o*-position with respect to the other nitrogen atom ; derivatives of *o*-aminodiphenylamine are thus produced, *e.g.*

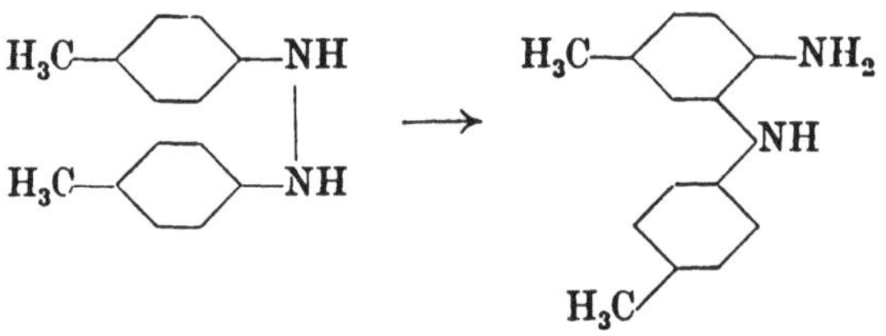

This form of the isomerisation is known as the *semidine transformation* (P. Jacobson).

Benzidine, and the diphenyl bases *tolidine* and *dianisidine* which are produced in the same way from *o*-nitrotoluene and *o*-nitroanisole, are manufactured on a large scale in the dye industry as important intermediates for the preparation of substantive azo-dyes (which dye cotton directly, cf. in this connection pp. 292, 294).

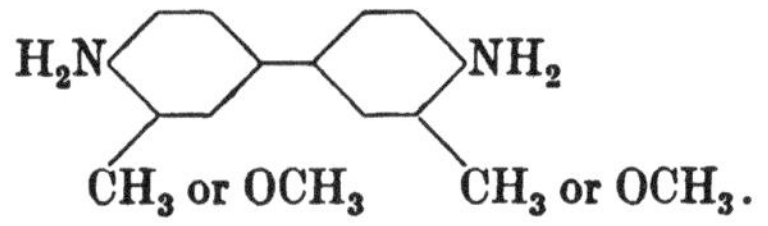

On the Mechanism of the Reduction of Nitrobenzene

The reduction of aromatic nitro-compounds is of exceptionally great interest, not only scientifically, but also technically. The conversion of the hydrocarbons of coal tar into useful products began with the discovery of the nitration process ; the conversion, on the technical scale, of the nitro-group of nitrobenzene into the amino-group gave aniline, the starting material for the preparation of innumerable dyes and pharmaceutical products; to aniline were added the homologous toluidines, xylidines, naphthylamines, and so on.

The production of aniline from nitrobenzene proceeds in such a way that the reactive hydrogen is added to the nitro-group, the oxygen is eliminated as water, and finally hydrogen is again added on. The process is not a simple one and involves a series of intermediate stages:

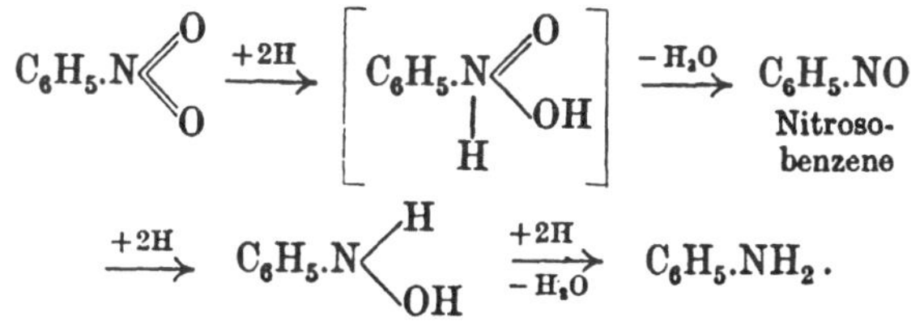

β-Phenylhydroxylamine

Under the conditions prevailing during the production of aniline neither nitrosobenzene nor phenylhydroxylamine is encountered. The reason for this is that the rate of reduction of these intermediate products is much greater than that of the nitrobenzene itself (F. Haber).

In neutral or alkaline solution the conditions are altered so as to favour the immediate precursor of the final product of hydrogenation, namely, *phenylhydroxylamine.* This compound is obtained from nitrobenzene, suspended in ammonium chloride solution, by reduction with zinc dust. Zinc dust can decompose water with the formation of $Zn(OH)_2$ if a substance is present which takes up the liberated hydrogen. Molecular, *i.e.* ordinary, oxygen is capable of doing this and is thereby converted into *hydrogen peroxide* (M. Traube):

$$Zn + 2\,H_2O + O_2 \longrightarrow Zn(OH)_2 + H_2O_2 .$$

In the case under discussion the nitrobenzene takes the place of the oxygen. (Write the equation.) If the experiment is properly carried out, the reduction is limited in this way to the phenylhydroxylamine stage.

If the reduction takes place in an alkaline medium, products derived from two molecules of nitrobenzene are formed. In these products the unaltered portions of the nitrobenzene molecules are united by means of nitrogen atoms. These substances are:

$$C_6H_5.\underset{\ddot{O}}{N}{:}N.C_6H_5 \qquad \text{Azoxybenzene,}$$

$$C_6H_5.N{:}N.C_6H_5 \qquad \text{Azobenzene,}$$

$$C_6H_5.NH.NH.C_6H_5 \qquad \text{Hydrazobenzene.}$$

The least powerful method of reduction, boiling nitrobenzene with sodium methoxide in solution in methyl alcohol, provides azoxybenzene in excellent yield (Zinin); the methoxide is converted into formate. (Write the equation.)

Since azoxybenzene is attacked by more powerful reducing agents, *e.g.* zinc dust and sodium hydroxide solution or ammonia, the use of such agents converts nitrobenzene to azobenzene and hydrazobenzene, by passing at once beyond the azoxybenzene stage. The three reduction products with " paired " nitrogen atoms, therefore, stand in very close genetic relation to each other.

Experiment. Reduction of Azoxybenzene to Hydrazobenzene.— Azoxybenzene (1 g.) is dissolved in 5 c.c. of alcohol, the solution is heated to boiling, and 3 c.c. of 50 per cent sodium hydroxide solution and 2–3 g. of zinc dust are added with shaking. At first the mixture becomes red, because of the formation of azobenzene, but on more prolonged boiling a colourless solution is obtained just as in the reduction of nitrobenzene. When this stage has been reached, the mixture is filtered with suction through a small Büchner funnel and the hydrazobenzene is finally isolated in the manner described on p. 175.

Thus the joining of the two molecules by nitrogen takes place when azoxybenzene is formed, and the experiment described on p. 173 shows quite definitely that this substance is produced with extraordinary ease from phenylhydroxylamine and nitrosobenzene in the presence of alkali, that is to say, under the conditions which prevail during the production of the whole series. Nitrosobenzene is the first stage, but cannot be isolated, for in the course of the reaction it is trapped by the phenylhydroxylamine as soon as it is formed.

An explanation is thus provided of the otherwise puzzling formation of important products with paired nitrogen atoms during the reduction of aromatic nitro-compounds. The technical importance of the process concerns the synthesis of benzidine and of analogous bases.

The electrolytic reduction of nitrobenzene can be carried out conveniently according to the methods described by K. Elbs in *Übungsbeispiele für die elektrolytische Darstellung chemischer Präparate* (Halle a. S., 1911).

CHAPTER IV

SULPHONIC ACIDS

1. BENZENE MONOSULPHONIC ACID FROM BENZENE AND SULPHURIC ACID

Benzene (45 c.c. = 0·5 mole) is gradually added in small portions to 150 g. of liquid fuming sulphuric acid containing 5–8 per cent of anhydride ; no fresh portion is run in until the previous one, which first floats on the surface of the acid, has been dissolved by shaking. The acid is contained in a 200 c.c. flask which is kept cool with water and is shaken well throughout. The time required for the sulphonation is about ten to fifteen minutes. The reaction mixture is run slowly from a dropping funnel into a beaker containing three to four times its volume of cold saturated brine kept cool with ice and stirred while the mixture is added. After some time the sodium benzene sulphonate separates in the form of lustrous nacrous plates and on prolonged standing forms a thick crystalline sludge. (Crystallisation may be started by scratching the beaker with a glass rod.) The sludge is filtered at the pump and the crystals are pressed with a cork or glass stopper and then washed with a little saturated sodium chloride solution. Finally, the salt is dried in air on filter paper or on porous plate, powdered, and heated to 110° in a drying oven until it forms a dry dust. The yield is about 100 g., but the material contains sodium chloride.

Of the crude product 5 g. may be purified by recrystallisation from absolute alcohol. (The admixed sodium chloride is insoluble in alcohol.)

To isolate the *diphenylsulphone*, which is produced as a by-product, 30 g. of the powdered salt are warmed with 50 c.c. of ether, the mixture is filtered with suction while hot and the undissolved material is washed with ether. The ether leaves on evaporation a small quantity of a crystalline residue which is recrystallised from ligroin in a test tube. Melting point 129°.

To prepare *benzenesulphonyl chloride* 60 g. (0·33 mole) of the sodium salt are heated for a quarter to half an hour on a vigorously boiling water bath in a fume chamber with 80 g. of finely powdered phosphorus pentachloride. The reaction product is cooled and then poured slowly into a separating funnel containing 600 c.c. of ice-water. In order to decompose the phosphorus oxychloride, the funnel is repeatedly shaken, and after standing for one hour the sulphonyl chloride is extracted with ether. The ethereal solution is dried over a little calcium chloride, the ether is evaporated, and the residue is distilled *in vacuo*. The bulk passes over at 120°–124°/12 mm. Pure benzenesulphonyl chloride solidifies when cooled in ice-water.

Benzenesulphonamide.—Finely powdered ammonium carbonate (10 g.) and benzenesulphonyl chloride (about 1 c.c.) are ground in a porcelain basin which is then warmed over a small flame until the odour of the sulphochloride has disappeared; the mixture is well stirred meanwhile. After cooling, water is added and the product is collected at the pump, washed several times with water, and then crystallised from alcohol by adding hot water until a turbidity appears. Melting point 156°.

Benzenesulphohydroxamic Acid.[1]—Hydroxylamine hydrochloride (10 g.) is boiled under reflux condenser with just enough methyl alcohol to dissolve it, and when still hot is decomposed by a solution of 3 g. of sodium in 60 c.c. of ethyl alcohol, which should not be added too quickly. After the mixture has been cooled, precipitated sodium chloride is removed at the pump and 8·5 g. of benzene-sulphonyl chloride are then added in small portions to the solution of free hydroxylamine. Most of the alcohol is now removed by distillation from the water bath, the hydroxylamine hydrochloride which has separated is removed by filtration, and the filtrate is evaporated to dryness *in vacuo* at a moderate temperature. The residue is extracted three times with 15 c.c. portions of boiling absolute ether. Evaporation of the combined ethereal extracts in an open dish yields the benzene sulphohydroxamic acid in the form of a mass of crystalline plates which are digested with cold chloroform and filtered with suction. Yield 5–6 g. Melting point 126°.

The homologous tolyl compound is prepared in a similar way from

[1] Piloty, *Ber.*, 1896, 29, 1559.

toluene *p*-sulphochloride, which is a cheap commercial by-product in the manufacture of saccharin.

The most important reaction of benzenesulphohydroxamic acid is its decomposition by alkalis. This decomposition does not consist in a reversal of the process of formation (*i.e.* conversion into benzenesulphonic acid and hydroxylamine). An exchange of the state of oxidation takes place : *benzenesulphinic* acid and *nitroxyl* are produced :

$$\left.\begin{array}{ll} & C_6H_5.SO_2.NHOH \\ \text{or} & C_6H_5.\underset{\displaystyle |\atop \displaystyle OH}{S}(O):NOH \end{array}\right\} \longrightarrow C_6H_5.SO_2H + O{:}NH\,.$$

Advantage is taken of this decomposition in carrying out the Angeli-Rimini test for aldehydes (p. 205).

2. *p*-TOLUENESULPHONIC ACID [1]

In the processes described under 1 and 3 the sulphonating agent used is concentrated sulphuric acid in excess and therefore the product of the reaction is isolated in the form of its sodium salt. The method now to be described, on the other hand, permits of the direct isolation of the *free sulphonic* acid. Such isolation is possible because of the fact that the water formed in the reaction is removed by distillation in an ingenious apparatus (Fig. 48). When stoicheiometrical amounts of sulphuric acid are used, its sulphonating action is soon stopped by this water. (Hence the excess of acid in the other methods.) By the use of an excess of toluene the whole of the sulphuric acid is consumed. In the flask (capacity 0·5 l.) shown in the figure 40 c.c. of concentrated sulphuric acid (d. 1·8) and 200 c.c. of toluene are heated to boiling on the sand bath. The hydrocarbon which distils is condensed in the condenser K, drops through a small funnel into the water-trap H which separates the flask from the condenser and is provided with a drain cock ; after the lower portion of the trap has filled, the toluene flows back into the flask. The capacity of the portion of the trap below the delivery tube is 10–15 c.c. The water produced in the reaction distils with the toluene vapour and after condensation collects under the toluene in H. To keep the lower portion of the trap cool it is enclosed in a lead coil through which flows a current of water. From time to time the water which collects is run into a small measuring cylinder. After the toluene-

[1] H. Meyer, *Annalen*, 1923, 433, 331.

sulphuric acid mixture has boiled for five hours, about 18 c.c. of water are obtained, partly from the sulphuric acid and partly produced in the reaction (12·5 c.c.).

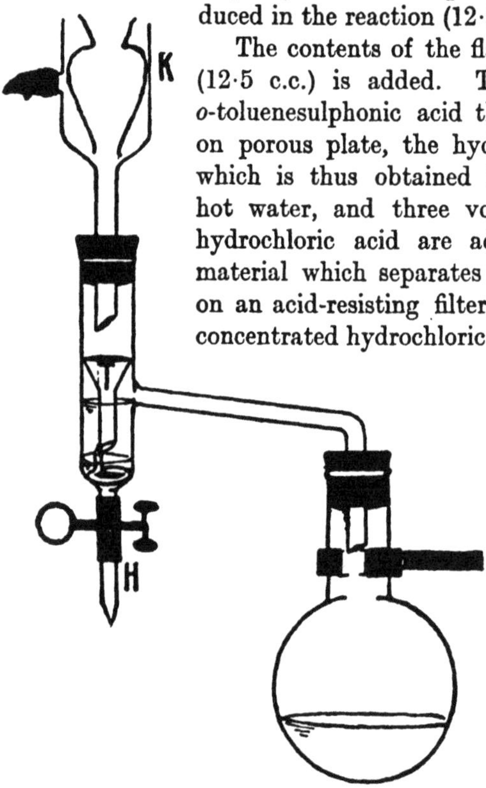

FIG. 48.

The contents of the flask solidify when water (12·5 c.c.) is added. To remove toluene and *o*-toluenesulphonic acid the solid is pressed well on porous plate, the hydrated *p*-sulphonic acid which is thus obtained is dissolved in a little hot water, and three volumes of concentrated hydrochloric acid are added. The crystalline material which separates is filtered with suction on an acid-resisting filter, washed with ice-cold concentrated hydrochloric acid, and then twice recrystallised in the same way. Finally, the acid is dried in a desiccator over potassium hydroxide until completely free from hydrochloric acid. (Test a sample.) Owing to the presence of some carbon particles the crystals are slightly grey in colour. Melting point 104°–105°. Yield after three recrystallisations about 50 g.

The specimen is used in the determination of acetyl (p. 70).

3. β-NAPHTHALENESULPHONIC ACID

A mixture of 64 g. of naphthalene and 45 c.c. (80 g.) of pure concentrated sulphuric acid is heated in an open flask in an oil bath at 170°–180° for four hours. The solution is cooled somewhat and is then cautiously poured with stirring into a litre of water and neutralised while boiling in a large basin with a sludge of lime (from about 70 g. of slaked lime). The sludge should not be too thin. The neutralised material is filtered as hot as possible through a large Büchner funnel into a previously warmed filter flask. The solid on the funnel is washed three times with hot water and the filtrate (clarified, if necessary, through a folded filter paper) is evaporated in a basin over a free flame until a sample solidifies to a crystalline paste when rubbed with a glass rod. The concentrated solution is allowed to stand over night and the calcium naphthalene-β-sulphonate which has separated is filtered with suction, pressed down well on the funnel, and washed with a little water. To obtain the sodium salt concentrated sodium carbonate solution is added to the hot solution of the calcium salt until the mixture is just permanently alkaline. The precipitated calcium carbonate is removed at the filter pump while the mixture is still warm, and is washed with water. The filtrate is evaporated in a basin over a naked flame until crystals begin to separate from the hot liquid. After standing for several hours in the cold, the crystals are collected, the mother liquor is further concentrated, and after standing for a long time is again filtered. The two crops of crystals are mixed and dried on the water bath. Yield 75–85 g.

A very elegant process for the direct preparation of the free β-naphthalenesulphonic acid from its components has been described by O. N. Witt (*Ber.*, 1915, **48**, 751). It is specially recommended as an alternative to the process described above.

4. SULPHANILIC ACID FROM ANILINE AND SULPHURIC ACID

Pure concentrated sulphuric acid (100 g.) and freshly distilled aniline (31 g. = 0·33 mole) are gradually mixed with shaking in a dry flask and the mixture is heated at 180°–190° in an oil bath until sodium hydroxide no longer liberates aniline from a sample diluted with water (four to five hours). The reaction mixture is cooled somewhat and then poured with stirring into cold water. Sulphanilic acid crystallises, is collected at the filter pump, washed with water, and recrystallised from water with animal charcoal. Yield 30–35 g.

On a technical scale aniline is heated with *one* mole only of H_2SO_4, *i.e.* as acid sulphate; the temperature is about the same as that given above. (The "baking" process.) Compare this process, which can also be used in the laboratory, with that here described.

5. 2:4-DINITRO-α-NAPHTHOL-7-SULPHONIC ACID [1]

(Naphthol Yellow *S*)

Finely powdered α-naphthol (50 g.) is added gradually to 200 g. of 25 per cent oleum and brought into solution by continuous shaking. The solution is then heated at 125° in an oil bath for one hour. In order to ascertain whether the α-naphthol has been completely converted into the 2:4:7-trisulphonic acid, a sample of the solution is mixed with about 10 c.c. of water in a test tube, 10 c.c. of concentrated nitric acid are added, and the mixture is heated nearly to boiling. If the yellow solution which is thus formed neither becomes turbid nor deposits flocks on cooling, the melt can be worked up for naphthol yellow *S*. Otherwise the conversion of the α-naphthol into trisulphonic acid must be completed by the addition of a more concentrated oleum and renewed heating.

The cooled melt is gradually mixed with 500 g. of crushed ice, the liquid thus obtained is filtered and 120 g. of nitric acid (d. 1·4) are added to the brown filtrate, which is then heated at 50° for half an hour. After the solution has stood for twelve hours at ordinary temperature, the greater part of the dinitronaphtholsulphonic acid produced will have separated. It is filtered off and crystallises from hot dilute hydrochloric acid in small yellow needles which are dried first on porous plate and then in a desiccator over sulphuric acid and potassium hydroxide. Melting point 151°. Yield about 85 per cent of theoretical.

Naphthol yellow *S*, called by A. Kossel " flavianic acid ", is used in the isolation of arginine (p. 392).

Explanations

The technical method for the sulphonation of aromatic compounds is the exact counterpart of the nitration process. In both cases the OH group of the acid, along with a hydrogen atom of the benzene ring, is eliminated and in the position vacated by this hydrogen atom the groups $—NO_2$ and $—SO_3H$ enter. For various reasons (refer to p. 93) it is probable that an addition reaction takes place somewhat as follows, one double bond of the benzene ring being involved :

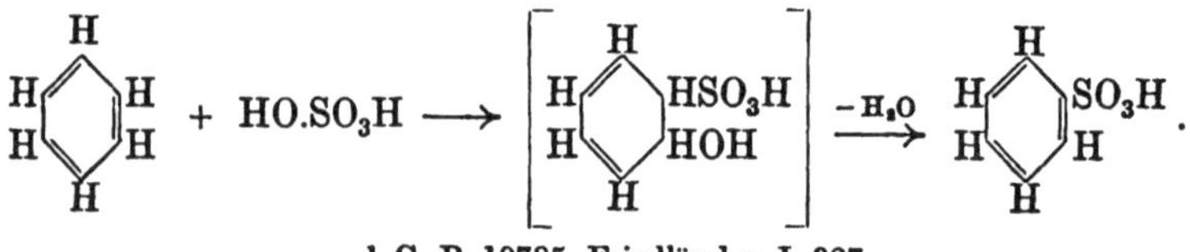

[1] G. P. 10785, Friedländer, I, 327.

Because of the tendency to revert to the stable aromatic ring system, the intermediate product (shown in brackets) will lose water and change into benzenesulphonic acid.

A comparison should always be instituted between the benzene series and the aliphatic olefines; in this case they behave essentially in the same way. Ethylene indeed forms an addition compound, ethyl sulphuric acid, with concentrated sulphuric acid at a low temperature (about 50°).

$$CH_2{=}CH_2 \longrightarrow CH_3.CH_2.O.SO_3H,$$

hence the H_2SO_4 molecule is here broken up in a different way. For obvious reasons this reaction cannot take place with benzene, since it must be very readily reversible. If, however, ethylene is subjected to the action of *fuming* sulphuric acid, a sulphonation product, so-called *carbyl sulphate*, is formed just as in the case of benzene. Carbyl sulphate is derived from the alcohol sulphonic acid which is first formed by esterification with sulphuric acid and subsequent elimination of water.

$$CH_2{=}CH_2 \longrightarrow CH_2OH{-}CH_2SO_3H \longrightarrow \begin{array}{l}CH_2{-}CH_2.SO_3H\\ |\\ O{-}SO_3H\end{array}$$

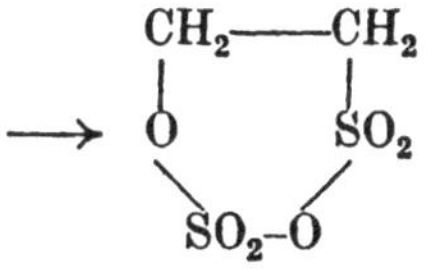

(Carbyl sulphate).

In a quite analogous way ethylene reacts with nitric-sulphuric acid, nitroethyl nitrate (p. 155) being produced. How are alkyl sulphonic acids prepared ?

The ease with which the sulphonic group enters into aromatic compounds depends on the nature of the substituents present, just as it does in nitration. Benzene is rather difficult to sulphonate, toluene and naphthalene are somewhat more easy, phenols and amines very easy. The sulphonation of nitrobenzene or the further sulphonation of the benzene sulphonic acids proceeds with more difficulty, and the action of the sulphuric acid must here be intensified by increasing its SO_3-content.

Since NO_2 and SO_3H are substituents of the second order, a second substituent enters in the *m*-position. Oleum containing a high percentage of sulphur trioxide finally converts benzene into benzene trisulphonic acid. Chlorosulphonic acid condenses with aromatic hydrocarbons, giving aryl sulphochlorides.

From naphthalene two monosulphonic acids are obtained, namely, α- and β-naphthalenesulphonic acids :

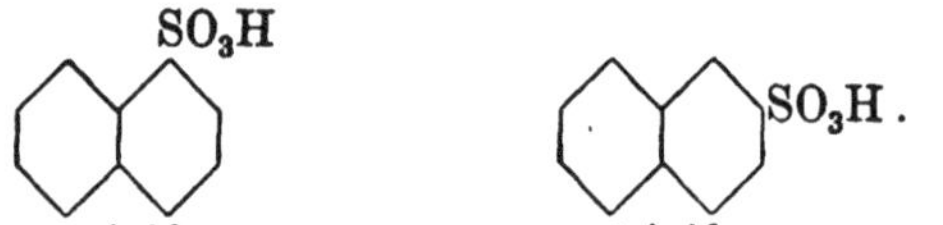

α-Acid β-Acid

Substitution reactions in the naphthalene ring take place, without exception, in the α-position, which is characterised by its high reactivity. The introduction of halogen and of the nitro-group leads exclusively to the α-derivative. This is also the case with the sulphonic group. When naphthalene is sulphonated at a low temperature, such as that mentioned above, the α-sulphonic acid is produced; it can thus be prepared also on a technical scale. The β-sulphonic acid, on the other hand, is only formed at higher temperatures when the α-acid is, to a large extent, decomposed hydrolytically into naphthalene and sulphuric acid. The equilibrium between sulphonation and hydrolysis at the temperature (170°–180°) here used lies rather to the left in the case of the α-acid, and far to the right in that of the β-acid.

$$\text{Naphthalene} + H_2SO_4 \rightleftarrows \text{Sulphonic acid} + H_2O\,.$$

But since some α-acid is always found in the reaction mixture, the β-acid must also be subject to a hydrolytic equilibrium.

Actually when naphthalene β-sulphonic acid is melted with sulphuric acid (containing water) small amounts of the isomeric α-acid are obtained.

A similar behaviour is observed in the case of the *phenolsulphonic* acids and in particular in that of *anthraquinone*, which, in its substitution reactions, is extraordinarily like naphthalene. Anthraquinone is sulphonated with more difficulty than is naphthalene, and in consequence the conditions of increased temperature which must be applied bring about the formation of the β-acid, the important starting point for the synthesis of alizarin. In industrial practice, however, ways and means have been found for producing also anthraquinone α-sulphonic acid, which was formerly not readily obtainable. α-Substitution takes place when the sulphonation is catalysed by mercury [1] (R. E. Schmidt).

Aniline is very easily sulphonated, for example, by heating its sulphate ("baking" process). This change recalls that of aniline acetate into acetanilide. Actually, it is very probable that an analogous product acylated at the amino-group, a *sulphaminic acid*, is first formed, since examples of this type of change are known, *e.g.* the conversion of phenylhydroxylamine to *p*-aminophenol and of phenylnitramine to

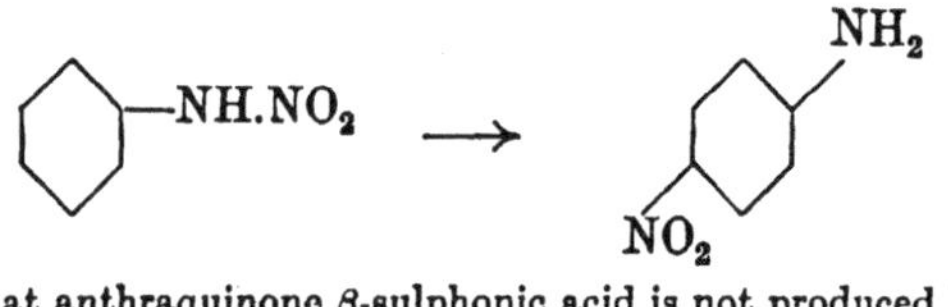

[1] It seems that anthraquinone β-sulphonic acid is not produced by rearrangement of the α-acid.

p-nitraniline. The sulphonic group may be presumed to wander to the p-position.

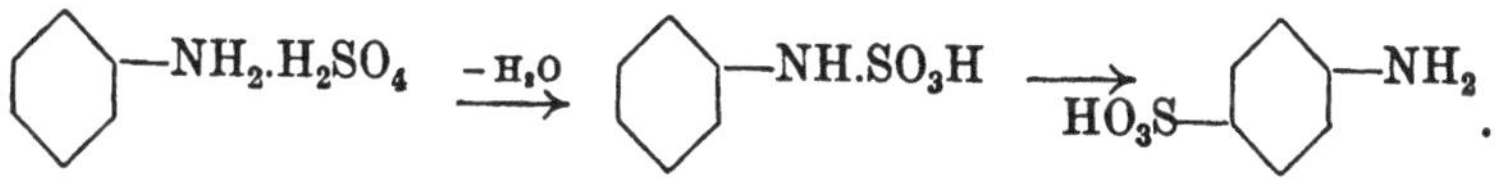

Primary aniline sulphate   Phenylsulphaminic acid   Sulphanilic acid

That this is the course of the reaction is proved by the possibility of isolating the sulphaminic acid corresponding to *a-naphthylamine* if mild experimental conditions are chosen. At a higher temperature the naphthylsulphaminic acid is converted into 1-naphthylamine-4-sulphonic acid (*naphthionic acid*).

In the sulphonation of aniline small amounts of the o-compound are produced along with sulphanilic acid. Aniline o-sulphonic acid, however, is of no further interest. Metanilic acid, on the other hand, is also manufactured as an intermediate in the azo-dye industry. It is obtained from nitrobenzene m-sulphonic acid by reduction. The *amino- (and hydroxy-) sulphonic acids of the naphthalene series* are of the greatest technical importance. They are either diazotised themselves or serve for the coupling with other diazo-compounds. In this way the most important azo-dyes are produced.

The phenols are as easily sulphonated as are the aromatic amines. When polynitrophenols have to be prepared, sulphonic groups are often first introduced and are then easily eliminated and replaced by NO_2 by the action of nitric acid. This method is used, for example, in preparing picric acid.

In the sulphonation of α-naphthol, SO_3H-groups are easily introduced into the 2- and 4-positions of the ring containing the OH-group. The SO_3H-groups in these positions, moreover, can be replaced by NO_2. The procedure described in section 5 of the present chapter shows how a third sulphonic group is introduced. The solubility in water of the dinitronaphthol is greatly increased by the presence of this group. 2:4-Dinitronaphthol (Martius' yellow) and its 7-sulphonic acid were formerly amongst the most important yellow dyes for wool.

The sulphonic acids of aromatic hydrocarbons are difficult to isolate, at least those of the simpler members, because they are generally easily soluble in water and do not crystallise readily. They are among the strongest organic acids, and in their affinity constants differ only very little from the strong mineral acids. Their alkaline earth salts are, in general, soluble in water, and on this property depends the method used for removing the excess of sulphuric acid which is always present in the usual process employed for preparing the acids. In the preparation of naphthalene β-sulphonic acid described above, the excess of sulphuric acid is removed in this manner. From the alkaline earth salts the alkali salts are then easily obtained in all cases by boiling with solutions of alkali carbonate. As will be shown in a preparation to be described later, these alkali salts are converted into phenols by fusion with alkali. The aromatic sulphonic acids are decomposed into hydrocarbons and sulphuric acid by hot dilute mineral acids or even by superheated steam.

In example 2 (above) a direct process for obtaining free sulphonic acids is described.

In the aminosulphonic acids the electrochemical character of both substituents is appreciably weakened, as it is in aminocarboxylic acids like glycine or anthranilic acid. The influence of the sulphonic group, however, preponderates over that of the amino-group, which is already quite feeble in aniline itself. Sulphanilic acid no longer forms salts with aqueous mineral acids, but does so easily with alkalis, against which it can be sharply titrated in aqueous solution. In the case of aliphatic amino-acids such titration is only possible in alcoholic solution (Willstätter).

In the preparation of benzenesulphonic acid, *diphenylsulphone* is produced as a by-product. The benzene sulphonic acid which is first formed itself acts on benzene, just as sulphuric acid does, and water is eliminated:

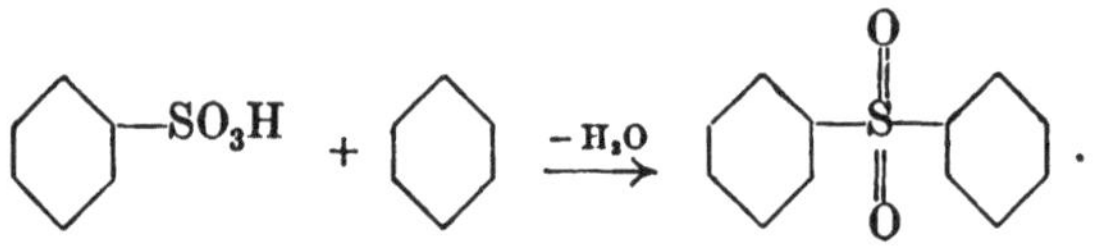

The sulphone may be regarded as a product of further sulphonation.

The sulphones are neutral, crystalline, indifferent substances. They are also produced by energetic oxidation of sulphides. The hypnotic sulphonal (write its formula) is a double sulphone.

On Sulphonyl Chlorides and Sulphonamides.—The conversion of benzene sulphonic acid into its chloride and amide shows that derivatives of sulphonic acids, analogous to those of carboxylic acids, can be obtained. The sulpho-chlorides are much less reactive than are the chlorides of the carboxylic acids; benzene sulphochloride, for example, can be distilled in steam almost without decomposition.

Amines are frequently characterised by conversion to sulphonamides, since the latter are notable for the great ease with which they crystallise. For preparative purposes the various amines may be separated by taking advantage of their behaviour towards benzene sulphochloride. Only the primary and secondary amines react to form amides, and of these only the sulphonamides from the primary amines are soluble in alkali, so that they can be separated; by energetic hydrolysis they are subsequently converted into sulphonic acid and amine (Hinsberg). The sulphonamides from primary bases are, therefore, acids. The salts of these amides are probably derived from an "enol" form, and salt formation is favoured by the strongly negative character of the $—SO_2$-group.

$C_6H_5.SO_2.NH.C_6H_5$,
Benzenesulphanilide

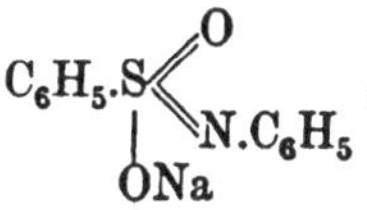

Sodium salt

$C_6H_5.SO_2.N(CH_3).C_6H_5$.
Benzenesulpho-*N*-methylanilide (insoluble in alkali)

The sweetening agent saccharin is also derived from a sulphonamide; it is prepared from toluene-*o*-sulphonamide by oxidising the CH_3-group to carboxyl with permanganate; ring closure is subsequently brought about by the action of concentrated hydrochloric acid:

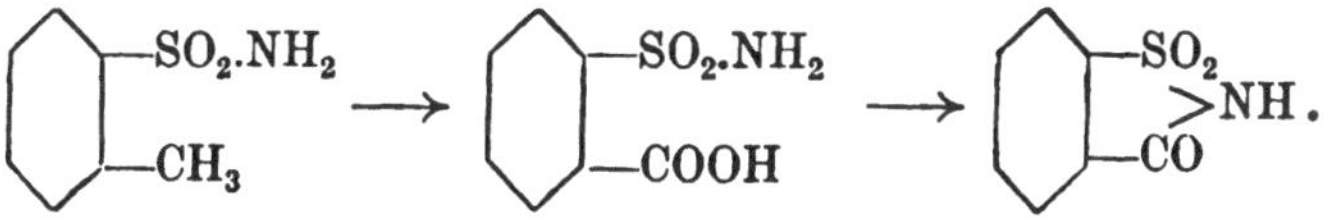

Saccharin, as its formula shows, is acidic in virtue of its imino hydrogen atom. The soluble sweetening agent is the sodium salt.

If an alkali salt of a sulphonic acid is subjected to dry distillation with potassium cyanide or potassium ferrocyanide a nitrile is obtained, *e.g.*

$$C_6H_5.SO_3K + KCN = C_6H_5.CN + SO_3K_2.$$

Whilst the sulphonic acids themselves cannot, in practice, be reduced, their sulphochlorides may be brought to a lower state of oxidation, that of the *sulphinic acids*, by treatment with metals, preferably with zinc. In this way the zinc salts of these acids are produced directly:

$$2\,C_6H_5.SO_2Cl + 2\,Zn \longrightarrow (C_6H_5.SO_2)_2Zn + ZnCl_2.$$

The decomposition by alkalis of benzene sulphohydroxamic acid—from benzene sulphochloride and hydroxylamine—likewise leads, as was already mentioned, to the sulphinic acid (Piloty).

From the sulphochlorides, by energetic reduction with nascent hydrogen, the corresponding mercaptans are produced:

$$C_6H_5.SO_2Cl + 6\,H \longrightarrow C_6H_5.SH + 2\,H_2O + HCl.$$

The sulphinic acids can also be reduced in this way.

Thiophenol.—Benzene sulphochloride (8 g.) is allowed to run in small portions from a dropping funnel into a round-bottomed flask containing 20 g. of finely granulated tin and 50 c.c. of concentrated hydrochloric acid. During the addition of the sulphochloride and until most of the tin has dissolved, the flask, which is fitted with a double neck attachment carrying a reflux condenser, is heated on the boiling water bath. The mercaptan produced is distilled with steam and extracted from the distillate with ether. The ethereal solution is dried over sodium sulphate, the ether is removed by distillation, and the thiophenol which forms the residue is likewise distilled. It passes over as an almost colourless liquid at 173°.

This extremely malodorous substance should only be prepared and used under a well-ventilated hood or in a fume chamber. In particular, the substance must not be allowed to come into contact with the hands or the clothing, since the odour clings for days.

The thioalcohols have a pronounced acidic character. The alkali salts of the aliphatic mercaptans are, indeed, very largely hydrolysed by water, but the aromatic compounds, on the other hand, can be exactly titrated with alkali and phenolphthalein. The yellow lead salts and the colourless mercury salts are characteristic.

Experiment.—A few drops of thiophenol are added to alcoholic solutions of lead acetate and mercuric chloride.

The ease with which the hydrogen atom is detached from the sulphur is noteworthy (analogy with H_2S) ; even atmospheric oxygen gradually converts the mercaptans into aryl or alkyl disulphides, and mild oxidising agents do so at once :

$$2\,R.SH \longrightarrow R.S{-}S.R\,.$$

Experiment.—A few drops of thiophenol on a watch-glass are slowly evaporated to dryness on the water bath with a few cubic centimetres of very dilute ammonia solution (*fume chamber*). An oil remains which sets to a crystalline solid on cooling. It is phenyldisulphide, melting point 61°.

On reduction the disulphides take up hydrogen and are reconverted into mercaptans.

A biological example of this relationship is provided by *cysteine* and *cystine* :

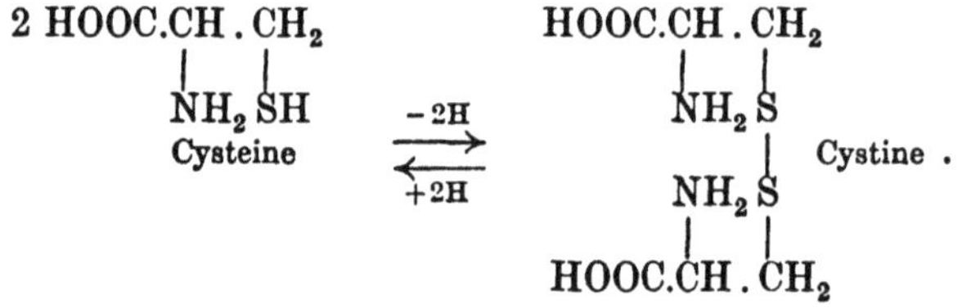

Another is the behaviour of "*glutathione*" (Hopkins), a tripeptide in which the NH_2-group of cystine (or of cysteine) is combined with a carboxyl group [1] of glutamic acid to form an amide grouping, and the carboxyl group of cystine is likewise combined with the NH_2-group of glycine.

[1] The one which is further from the NH_2-group.

Cystine is one of the units from which the proteins are built up. It is prepared by acid hydrolysis of keratin (from hair or horns).

The aromatic mercaptans react with chlorine to form aryl sulphur chlorides (Zincke); phenyl sulphur chloride is a deep red, very reactive liquid (Lecher):

$$C_6H_5.SH + Cl_2 \longrightarrow C_6H_5SCl + HCl\,.$$

From the mercaptans the sulphonic acids are re-formed by energetic oxidation.

CHAPTER V

ALDEHYDES

1. FORMALDEHYDE[1]

THE side tube (length about 10 cm.) of a distilling flask (capacity 250 c.c.) is bent upwards at the junction with the neck of the flask. The end of the side tube, drawn out into a capillary (internal diameter 1·0–1·5 mm.), is then inserted through a cork into a piece of combustion tubing about 30 cm. long (Fig. 49). Within the tubing

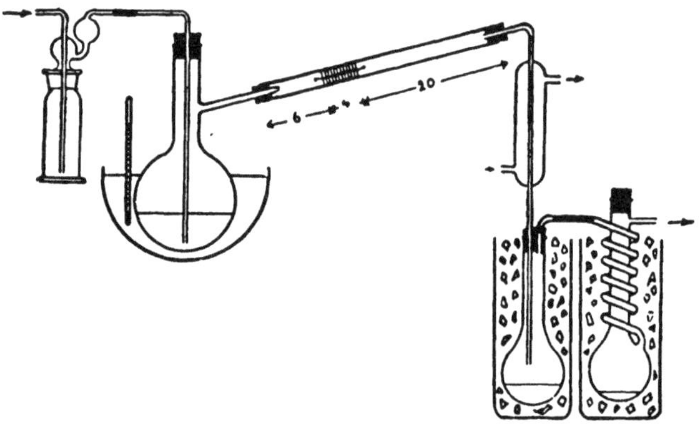

FIG. 49.

and about 6 cm. from the point of the capillary is a copper spiral 4 cm. long. The tubing slopes upwards at a small angle and its upper end is connected with a vertical condenser, preferably of the coil type. To the lower end of the condenser there are attached two communicating receivers which, during the experiment, are almost completely immersed in a freezing mixture. The short side tube of the second receiver is connected to the pump. Into the distilling

[1] Tollens, *Ber.*, 1886, 19, 2133. The explanatory remarks on the preparations 1–3 are collected along with the experiments on p. 201 *et seq.*

flask, which is lowered as deeply as possible into a water bath kept exactly at 46°–47°, 100 c.c. of methyl alcohol are poured. The flask is then closed with a rubber stopper, through which is inserted a glass tube reaching nearly to the bottom. Through this tube air is drawn in, and when the air is passing, the copper spiral is warmed in the flame, cautiously at first, until, when red heat is reached, the reaction sets in. The air current must now be so regulated that the spiral continues to glow quite feebly without further application of heat. If the experiment is carried out in this way there will be complete freedom from explosions. The region within which methyl alcohol-air mixtures explode is indeed reached when the temperature of the bath is too low (42°–44°), but the flame strikes back no further than the capillary tube, since the rapid current in the latter prevents further striking back. (Compare the Bunsen burner which also only strikes back if the current of gas is too slow.)

The two receivers contain 110–115 c.c. of a 30 to 32 per cent formaldehyde solution after all the methyl alcohol has been evaporated. A further small quantity of formaldehyde may still be collected in a third receiver containing a little water.

The following paragraph contains some points which should be considered in carrying out gaseous reactions.

In order to dehydrogenate one mole of methyl alcohol 0·5 mole of oxygen is required, and hence for one volume of the alcohol half as much oxygen or two and a half times as much air. The stoicheiometrical mixture must therefore contain methyl alcohol and air in the proportions (by volume) 1 : 2·5, *i.e.* 28·5 per cent of methyl alcohol. Since the volumes vary as the partial pressures the temperature of evaporation (of the alcohol) must be so chosen that its vapour pressure shall be 28·5 per cent of the atmospheric pressure, *i.e.* about 210 mm. of mercury. The vapour pressures of better known substances at various temperatures are to be found in Landolt-Börnstein, *Physikal-chem. Tabellen*, 5th Edition, 1923.

With the simple type of apparatus here described complete saturation of the air with methyl alcohol vapour is not reached, and hence a temperature somewhat higher than the theoretical is used.

Determination of the Formaldehyde Content of the Solution obtained.—With a pipette 5 c.c. of the solution are run into a measuring flask and diluted to 50 c.c. with water. Of the diluted solution 20 c.c. are placed in a conical flask (capacity 250 c.c.), 30 c.c. of about 3 per cent hydrogen peroxide solution, previously made exactly neutral to phenolphthalein, and then 30 c.c. of *N*-sodium hydroxide solution are added and the mixture is shaken.

After a short time evolution of hydrogen sets in and becomes very vigorous; it is accompanied by the liberation of heat, and is finally brought to completion by heating for a short time. The cooled solution is then titrated with *N*-hydrochloric acid solution after adding more phenolphthalein. The amount of alkali consumed gives the content of formaldehyde according to the equation:

$$2\,CH_2O + H_2O_2 + 2\,NaOH \longrightarrow 2\,H.COONa + H_2 + 2\,H_2O.$$

Thus if, for example, 22·5 c.c. of sodium hydroxide solution were used in the reaction, then the 20 c.c. (=2 c.c. of the original solution) contain 22·5 × 30 mg. = 0·675 g. of formaldehyde, *i.e.* the content of formaldehyde was 33·8 per cent.

In this very remarkable reaction, by the addition of two moles of formaldehyde to hydrogen peroxide, an intermediate product having the constitution $HOH_2C.O.O.CH_2OH$, *dihydroxymethyl peroxide*, is formed and can indeed be isolated in the crystalline state. It breaks up, however, extremely easily by the action of alkali into formate and hydrogen:

$$HOH_2C.O.O.CH_2OH + 2\,NaOH \longrightarrow 2\,H.COONa + H_2 + 2\,H_2O.$$

[For details of this reaction see *Annalen*, 1923, **431**, 301.]

On the determination of formaldehyde with ammonia, cf. p. 206.

The simple aldehydes react with neutral sulphites to form salts of aldehydesulphurous acids. In this reaction free alkali is formed, and the aldehyde content of a solution can also be determined in this way if the alkali is titrated.

2. ACETALDEHYDE

(a) From Ethyl Alcohol

The most suitable flask to use for this preparation is one with a capacity of 1·5 l. having a wide double neck of the type shown in the figure. A round flask with a not too short neck and a *tightly* fitting branched tube attachment will also serve. To the oblique branch a large condenser is fitted (likewise in an oblique position) and into the vertical neck a two-holed rubber stopper carrying a long dropping funnel and a delivery tube (0·4 cm. internal diameter) is inserted. The delivery tube, which reaches to the bottom of the flask, supplies carbon dioxide from a steel cylinder provided with a reducing valve. To the upper end of the condenser a medium sized calcium chloride U-tube is attached, and the free end of this U-tube is then connected with a condenser of the coil type (Fig. 50). At the lower end of this second condenser two connected receivers each containing 100 c.c. of absolute ether are fixed. These receivers stand in a freshly prepared ice and salt mixture, and during the experiment their temperature should not rise above $-10°$. Through the first condenser a very slow current of water at about 26° is passed, and by means of a cotton thread tightly held by the upper rubber stopper a thermometer is maintained in such a position inside the central tube

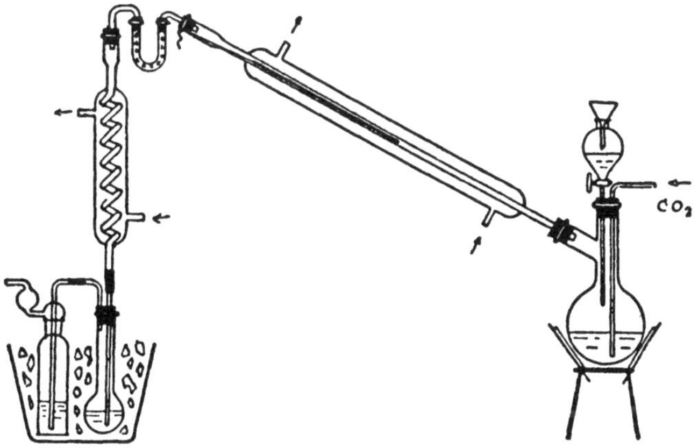

FIG. 50.

of the condenser that the bulb is midway between the ends. *All connections in the apparatus must be completely gas-tight.* During the experiment the coil condenser is supplied with water at 5°–10°. In winter such water may be taken direct from the tap, but in warm weather it must first be passed through an ice-cooled coil.

Alcohol (125 c.c.; 2 moles C_2H_5OH) is now poured into the flask along with about one-third of a mixture of 150 c.c. (270 g.) of sulphuric acid and 250 c.c. of water and this alcohol-sulphuric acid-water mixture is heated to boiling. To the main portion of the diluted sulphuric acid 100 c.c. of water are added, 200 g. of sodium dichromate are dissolved in the liquid so prepared, and the solution is poured while still warm into the dropping funnel, the delivery tube of which is completely filled. After the funnel has been fixed in position its contents are allowed to run at a moderate speed into the just-boiling mixture in the flask; the heat produced by the reaction makes further application of external heat unnecessary. By means of a moderately rapid current of carbon dioxide the aldehyde produced is carried out of the boiling solution and so protected from further oxidation. (The vapour pressure of the acetaldehyde at the temperature of the first condenser is still sufficiently high to prevent condensation when the current of gas is rapid. Condensation takes place only in the strongly cooled receivers.)

The rate of addition of the dichromate solution is so regulated that the mixture in the flask is kept constantly boiling and that the thermometer in the condenser registers 25°–30°.

If the head of dichromate solution is not sufficient to overcome the pressure in the apparatus a T-tube is inserted in the carbon dioxide delivery circuit, and by means of a piece of rubber tubing a by-pass, leading into the opening of the dropping funnel, is constructed. The rubber tubing is provided with a screw-clip to be used when refilling the dropping funnel.

In place of a steel cylinder a freshly filled Kipp apparatus may be used, but must be fitted with a tall tube filled with dilute hydrochloric acid in order to overcome the pressure in the rest of the apparatus.

The addition of the dichromate solution is complete in half an hour, and a further period of ten minutes suffices for the complete removal of aldehyde from the flask. The receivers are then disconnected, and not till then is the valve of the CO_2-cylinder closed.

Since the acetaldehyde cannot be separated by fractional distillation from the ether in which it has been collected, it is converted into the crystalline *aldehyde-ammonia*. The contents of the two receivers are transferred to a small filter jar which is cooled in a freezing mixture, and ammonia gas is led in from a cylinder. The delivery tube for the current of ammonia is a straight calcium chloride tube with its wide end deeply submerged in the liquid. To break up the crystalline mass which is formed, this tube is frequently moved to and fro. The jar is kept covered with a clock-glass (bored), piece of cardboard, or copper gauze. *On account of the ether which evaporates, all flames in the vicinity must be extinguished.*

If an ammonia cylinder is not available the gas is obtained by heating concentrated ammonia solution in a round-bottomed flask over a screened flame. The gas must be dried by passage through a tower containing potassium hydroxide and good quicklime. The amount of ammonia required is calculated as follows (small excess).

In order to avoid using too much ammonia a calculation must be made of how many litres of the gas will be required to combine with the highest yield of aldehyde which can be expected (60 g.). This weight of aldehyde is equal to 60/44 = 1·36 mole, and corresponds to 30 litres. (Why ?) If the effects of pressure and temperature, which on the average cause an alteration of 10 per cent, are taken into account it is evident that about 33 litres of ammonia will be required. The rate at which the ammonia escapes from the cylinder is approximately standardised once for all by determining exactly with a watch the time required for a current of reproducible velocity (fixed by counting the number of bubbles passing in unit time through a bubbler containing concentrated potassium hydroxide solution or mercury) to neutralise 42 c.c. of *N*-hydrochloric acid (measuring cylinder; methyl orange). This time is the period in which, at the rate of bubbling measured, 1 litre of ammonia is delivered from the cylinder.

Ammonia is passed into the ethereal solution of aldehyde for about thirty times this litre period (determined above), and the mixture is left for one hour for complete crystallisation; a sample of the liquid portion is then treated in a test tube with more ammonia gas in order to ascertain whether further precipitation occurs. If this is not so the aldehyde-ammonia is separated at the pump, washed a few times with absolute ether, and dried first on filter paper and then in a non-evacuated desiccator over sulphuric acid. If kept in a well-closed container the dry substance can be preserved for a long time; impure material turns brown and decomposes in a few days. Yield 50–60 g.

To obtain the *free aldehyde* 25 g. of the aldehyde ammonia are dissolved in water, a cooled mixture of water (40 c.c.) and concentrated sulphuric acid is added, and the acetaldehyde liberated is distilled from the water bath through a calcium chloride U-tube (gently warmed if the external temperature is low) and through an efficient coil condenser. In order to prevent autoxidation of the acetaldehyde the apparatus is filled with carbon dioxide before distillation, and, since the vapour pressure of the aldehyde is high, a current of carbon dioxide is passed again, for a short time only, at the end of the distillation. On this occasion also a slow current is employed. Since acetaldehyde boils at 21° the receiver, which is attached to the condenser by means of a cork stopper, must be well cooled in an ice-salt freezing mixture.

For the experiments to be carried out with acetaldehyde see pp. 202-210.

(*b*) From Acetylene

Although the process (*a*) is the only one which need be considered for laboratory purposes, the technically much more important hydration of acetylene is described here. From time to time this second process should be carried out in the laboratory.[1]

Mercuric oxide (5 g.) is dissolved for the most part in a still warm mixture of 110 c.c. of water and 50 c.c. of concentrated sulphuric acid. The mixture is brought into a large hydrogenation flask (Fig. 54, p. 368) and shaken for some time with acetylene prepared from calcium carbide, purified with acid solutions of dichromate and copper nitrate, and collected over saturated sodium chloride solution in a glass gas-holder (capacity 10–15 litres). Before shaking is begun the air present must be displaced by the hydrocarbon.

If necessary the absorption can also be brought about in a thick-walled flask fixed in an upright position on a shaking machine; after the air has been driven out by the acetylene the rubber stopper through which the delivery tube passes is fastened down with wire.

In from two to four hours up to 10 l. of acetylene are taken up. The colourless intermediate mercury compound very soon begins to separate. After the passing in of acetylene has ceased the whole of the reaction mixture is transferred to a round-bottomed flask and heated on a conical (Babo) air bath, while steam is passed through to decompose the mercury compound. The acetaldehyde liberated distils with the steam. An apparatus similar to that described under (*a*) is used; one receiver containing ether and cooled in a freezing mixture is sufficient. The aldehyde is precipitated from the ethereal solution as aldehyde-ammonia in the manner described above. Yield of aldehyde-ammonia 5–6 g.

3. BENZALDEHYDE FROM BENZYLIDENE CHLORIDE

Dry chlorine is led into 50 g. of boiling toluene until its weight has increased by 40 g., and the boiling point has risen to 187°; the operation is carried out in the same way as described for the preparation of benzyl chloride (p. 87). We are merely concerned with a continuation of this same reaction.

The crude benzylidene chloride so obtained is boiled for four hours in a round-bottomed flask under an efficient reflux condenser with 500 c.c. of water and 150 g. of precipitated calcium carbonate (or prepared chalk or *finely* powdered marble). An oil bath at 130°

[1] The procedure here described was worked out in the chemical laboratory of the University of Göttingen.

(thermometer in the oil) is used to heat the flask, and a gentle current of carbon dioxide is passed through the apparatus. The flask is then removed from the oil bath and steam is at once passed through the hot liquid to remove the benzaldehyde formed.

The *residue* in the flask is then filtered at the pump while hot, and the filtrate is strongly acidified with concentrated hydrochloric acid. When the acid filtrate is cooled the by-product of the reaction, *benzoic acid*, separates in glistening plates. These are filtered with suction, and recrystallised from boiling water. Melting point 121°. Benzoic acid is somewhat volatile with steam.

The distillate from the steam distillation is twice shaken with not too much ether, and the ethereal extract, if necessary after concentration, is transferred to a wide-mouthed bottle, into which technical sodium bisulphite solution is poured in small portions with stirring (a glass rod is used) so that the aldehyde addition compound formed sets to a thick paste. The bottle is then stoppered and vigorously shaken; the stopper is removed from time to time until all the benzaldehyde has entered into combination. (Odour!) The paste is now filtered with suction, and the solid on the funnel, after washing with ether, is at once decomposed by mixing it with an excess of sodium carbonate solution; the liberated aldehyde is removed without delay by steam distillation. The distillate is extracted with ether, the extract is dried over a little calcium chloride, the ether is removed by distillation, and the benzaldehyde which remains is likewise distilled. Boiling point 179°. Yield 35–40 g. (70 per cent of the theoretical).

Since benzaldehyde is very sensitive to the action of oxygen, all stages of the preparation must be performed in rapid succession.

Explanations and Experiments Relating to Chapter V., Sections 1–3

The lowest members of the series of the aldehydes are colourless liquids of pungent odour and miscible with water; the middle members are also liquids, but do not dissolve easily in water; the aldehydes of high molecular weight are crystalline solids. The boiling points of the aldehydes are considerably lower than those of the corresponding alcohols:

{ $CH_3.CHO$	Boiling point 21°
{ $CH_3.CH_2OH$	„ „ 78°
{ $CH_3.CH_2.CHO$	„ „ 50°
{ $CH_3.CH_2.CH_2OH$	„ „ 97°.

The aromatic aldehydes have a pleasant odour. (Oil of bitter almonds, vanillin, piperonal.)

The most general method for the preparation of aldehydes consists in removing two hydrogen atoms from a primary alcohol (*al*cohol *dehyd*-rogenatus); from secondary alcohols, in the same way, ketones are formed. Since the hydrogen is generally removed by an oxidising agent the process is regarded as an oxidation. Alcohols can, however, also be decomposed catalytically into aldehyde and hydrogen. Such decomposition is brought about by palladium black in the cold or by copper at higher temperatures. The part played by the copper in the preparation of formaldehyde (according to O. Loew) consists in the separation of hydrogen (dehydrogenation), and the purpose of the air which is mixed with the methyl alcohol is to burn the hydrogen and upset the equilibrium:

$$CH_3OH \rightleftarrows H_2.C:O + 2H.$$

Substances having triple linkages when dissolved in sulphuric acid take up the elements of water, particularly if mercuric salts are present. The simplest example of this reaction is the preparation of *acetaldehyde* from *acetylene* itself in the manner described above.

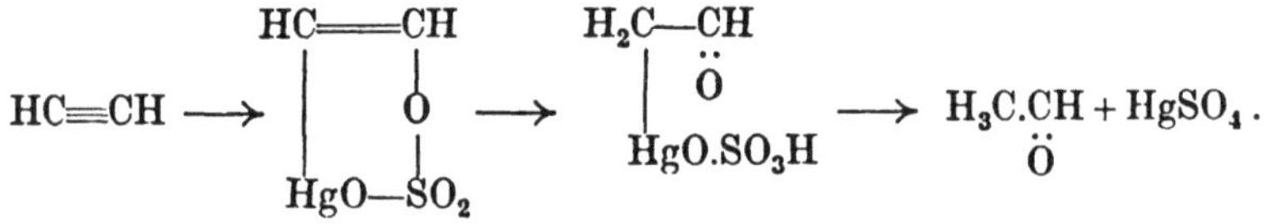

This process has great technical importance for the synthesis of acetic acid.

The aldehydes are readily oxidised and therefore behave towards ammoniacal silver solution and towards Fehling's solution as reducing agents.

Experiment 1.—Dilute a few drops of formaldehyde or acetaldehyde with a few c.c. of water, add a small amount of ammoniacal silver solution, and divide the mixture between two test tubes. Into one test tube run a few drops of sodium hydroxide solution; an immediate separation of metallic silver takes place. From the other solution after standing for some time in the cold, or more quickly on warming, the silver separates. Thus the oxidising action of ammoniacal silver solution is very considerably increased by sodium hydroxide (Tollens). Also test the reducing action of the aldehydes on Fehling's solution.

By oxidation aldehydes are converted into *carboxylic acids*. This process is indeed a direct continuation of the dehydrogenation of the alcohols. It is not the aldehyde itself but the *aldehyde hydrate* formed by the addition of water which actually reacts with the oxidising agent, *e.g.*

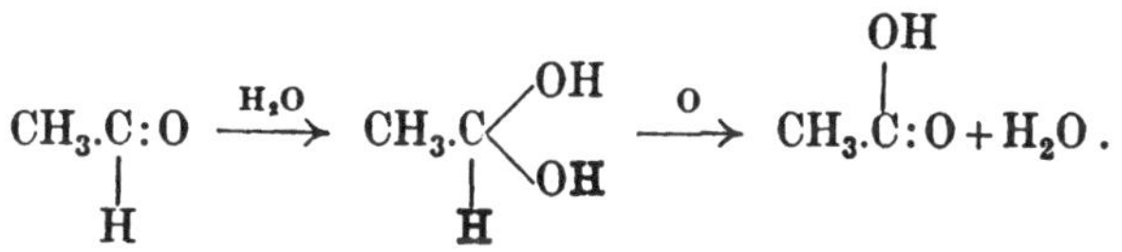

The conversion of ethyl alcohol by way of acetaldehyde into acetic acid is the chemical expression equivalent to *acetic fermentation.* In this process the acetic bacteria utilise atmospheric oxygen in order to bind the hydrogen. That the hydrogen which has to be removed is activated, and not the oxygen (as was formerly thought), is shown by experiments in which *oxygen is excluded* and replaced by *quinone*; the bacteria produce acetic acid from alcohol as before and the quinone is reduced to hydroquinone.

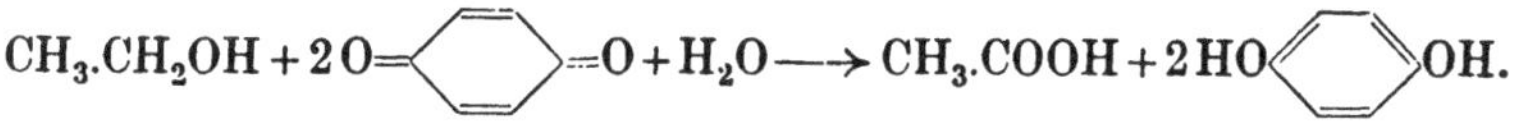

The important part which acetaldehyde plays in alcoholic fermentation (C. Neuburg) is shown by the fact that it is formed by decarboxylation of the intermediate product, *pyruvic acid*:

$$CH_3.CO.COOH \longrightarrow CH_3.CHO + \mathbf{CO_2}.$$

In the manner expressed by the fundamental equation:

$$\underset{\text{Methylglyoxal (hydrate)}}{CH_3.CO.CH(OH)_2} + CH_3.CHO \longrightarrow CH_3.CO.COOH + \mathbf{CH_3.CH_2OH}$$

ethyl alcohol and fresh pyruvic acid are formed by mutual hydrogenation and dehydrogenation. Compare, in this connection, p. 391.

The aldehydes are liable to a pronounced extent to undergo *autoxidation, i.e.* to combine with molecular oxygen. Here it is the genuine aldehyde structure $-\overset{H}{C}{=}O$ which first adds on the unsaturated oxygen molecule at the reactive double linkage, so that a per-acid is produced:

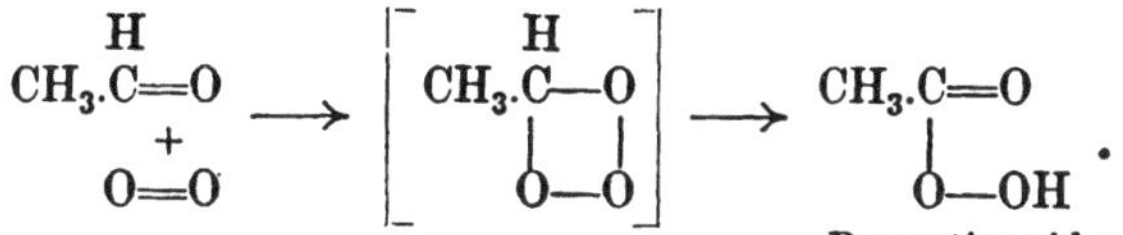

Peracetic acid

The per-acids are strong oxidising agents and react with a second molecule of aldehyde to give two molecules of acid:

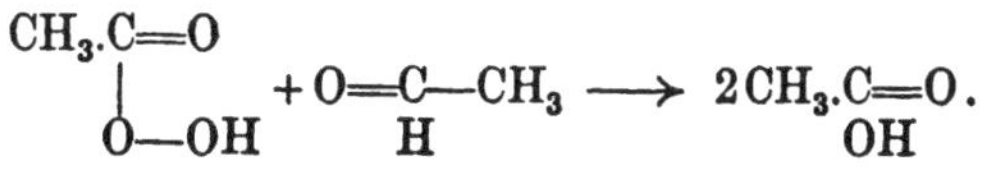

Thus the autoxidation of the aldehydes leads finally to acids. That a per-acid is first formed can be very easily shown in the case of acetaldehyde by the immediate liberation of iodine from potassium iodide solution which is caused by this strong oxidising agent. In the case of benzaldehyde, which combines exceptionally rapidly with oxygen, it has been possible to trap the per-acid with acetic anhydride as benzoyl-acetyl peroxide (Nef):

$$\begin{matrix} C_6H_5.C:O \\ | \\ O{-}OH \end{matrix} + (CH_3CO)_2O \longrightarrow \begin{matrix} C_6H_5.C{=}O \\ | \\ O{-}O{-}CO.CH_3 \end{matrix} + CH_3.COOH\,.$$

Experiment 2.—One c.c. of freshly prepared acetaldehyde is shaken for a few minutes in a long glass cylinder provided with a tightly fitting rubber stopper. Half of the liquid is then poured into a little dilute potassium iodide solution and half into two or three volumes of water which is then tested with litmus paper to show the presence of the acetic acid formed. It will be found that the aldehyde which has been dissolved in water hardly liberates any iodine from potassium iodide solution after standing for some time.

Experiment 3.—Two drops of benzaldehyde are exposed to the air on a watch-glass for one hour.

For preparative purposes the method of obtaining aldehydes from the primary alcohols is preferable by far, at least in the aliphatic series. The simple aromatic aldehydes can be obtained by alkaline hydrolysis of the arylidene chlorides, $R.CHCl_2$, which are produced from the hydrocarbons by substitution with chlorine (technical method for the preparation of benzaldehyde). In addition to these methods the elegant synthesis of Gattermann and Koch should be mentioned here. This synthesis, which proceeds like that of Friedel-Crafts, consists in acting on the aromatic hydrocarbon with carbon monoxide and hydrogen chloride in the presence of aluminium chloride and cuprous chloride (cf. Chap. IX. No. 5, on p. 338).

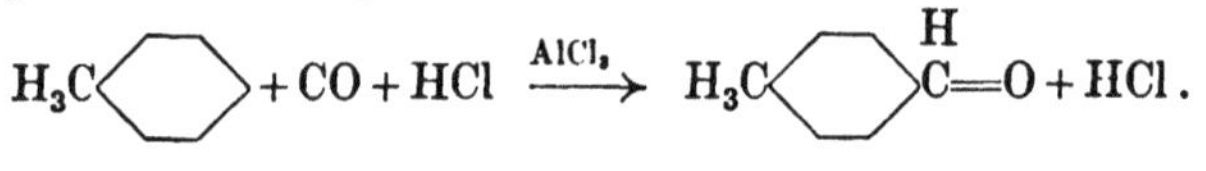

The carbon monoxide may be replaced by hydrogen cyanide (Gattermann) or by fulminic acid (in the form of mercury fulminate, Scholl), the first product then being respectively an imine or an oxime.

There is no absolutely general method for converting carboxylic acids into aldehydes; in many cases, however, the chlorine of acid chlorides can be replaced by hydrogen activated by means of palladium (Rosenmund).

$$C_6H_5.\underset{Cl}{C}{=}O \xrightarrow{2H} C_6H_5.\underset{H}{C}{=}O + HCl\,.$$

Another possibility, which has already been of great use in many syntheses, consists in the energetic reduction of hot esters with a little alcohol and much metallic sodium to the corresponding alcohols (Bouveault); these can then be oxidised to aldehydes in the usual way.

Experiment 4.—Colour reaction with fuchsine and sulphurous acid. A fragment of fuchsine is dissolved in a large volume of hot water to form an approximately 0·2 per cent solution, and after this has been cooled, concentrated aqueous sulphurous acid is added gradually until, after standing for some time, decolorisation has taken place. The sensitivity of the reaction should be tested with progressively diluted solutions of formaldehyde and acetaldehyde. When testing aldehydes which, like benzaldehyde, are sparingly soluble in water, some alcohol is added. The alcohol must first be tested since, on long standing, particularly when exposed to light, it may contain detectable amounts of acetaldehyde. In the case of formaldehyde the colour developed with the fuchsine solution becomes pure blue by the action of concentrated hydrochloric acid whilst with other aldehydes, under the same conditions, the colour almost completely disappears. (Distinction between formaldehyde and acetaldehyde.)

The colour reaction with fuchsine-sulphurous acid provides a means of distinguishing sharply between aldehydes and ketones. In dilute aqueous solution dextrose reacts negatively. Concerning the mechanism of the colour reaction see *Ber.*, 1921, **54**, 2527.

Experiment 5. Angeli-Rimini Reaction.—A few drops of an aldehyde (any of those prepared) are dissolved in aldehyde-free[1] alcohol and about the same amount of benzene sulphohydroxamic acid (for the preparation of which see p. 182) is added; in the case of aliphatic substances, twice as much of the acid is used. To this mixture, kept cool and shaken, 2 *N*-sodium hydroxide is added, in an amount judged to be about two molecular proportions. After standing for fifteen minutes the alkaline mixture is made just acid to Congo red and finally a drop of ferric chloride solution is added. An intense red colour is produced.

It was mentioned on p. 183 that benzene sulphohydroxamic acid is decomposed by alkali into benzene sulphinic acid and the very unstable compound nitroxyl, O=NH. If nitroxyl is produced in the presence of an aldehyde, it adds itself to the carbonyl group and a hydroxamic

[1] This is naturally only important when, in actual practice, an unknown substance is to be tested as to its aldehyde character.

acid results, the presence of which—and consequently also that of the aldehyde—is indicated by the intense colour reaction with ferric chloride.

$$C_6H_5.\underset{H}{C}:O + HN:O \longrightarrow \left[C_6H_5.\underset{H}{C}\begin{matrix} \diagup OH \\ \diagdown NO \end{matrix} \right] \longrightarrow C_6H_5.\overset{OH}{\overset{|}{C}}:NOH.$$

If the nitroxyl is formulated as hydrate, *i.e.* as dihydroxyammonia $N\begin{matrix} \diagup OH \\ - OH \\ \diagdown H \end{matrix}$, the similarity of this reaction to that which occurs when aldoximes are formed from aldehydes and hydroxylamine becomes still clearer. Nitroxyl is also formed in the alkaline hydrolysis of the sodium salt of *aci*-nitrohydroxylamine (Angeli):

$$O:N:\underset{\underset{NaO}{|}}{N}OH \longrightarrow O:N.ONa + O:NH.$$

The other reactions of the aldehydes, which are extraordinarily reactive substances, need only be mentioned here. Such reactions are: reduction to alcohols, formation of hydrazones, oximes, semicarbazones, bisulphite compounds, acetals and cyanohydrins (by addition of hydrogen cyanide).

Experiment 6. Reaction with Ammonia.—Of the formaldehyde prepared 10 c.c. are mixed with a small excess of ammonia and the mixture is evaporated in a small glass dish on the water bath. The colourless crystals which remain consist of *hexamethylenetetramine* $(CH_4)_6N_2$ (urotropine). This reaction proceeds quantitatively. It should be so carried out and the result compared with that obtained by titration.

Acetaldehyde combines with ammonia to form *aldehyde-ammonia*, as was shown by the method of preparation described.

$$\left(CH_3.\underset{H}{C}\begin{matrix} \diagup OH \\ \diagdown NH_2 \end{matrix} \right)_3$$

Benzaldehyde gives so-called *hydrobenzamide*

$$\begin{matrix} C_6H_5.CH:N \diagdown \\ C_6H_5.CH:N \diagup \end{matrix} CH.C_6H_5.$$

The products of the reaction with ammonia are therefore fundamentally different for each of the three aldehydes, but in each case the reaction begins with an addition:

$$\overset{\text{H}}{\text{—C:O}} + \text{NH}_3 \longrightarrow \text{—}\overset{\text{H}}{\text{C}}\begin{matrix}\diagup \text{OH} \\ \diagdown \text{NH}_2\end{matrix} .$$

The reaction with acetaldehyde does not proceed further, but with the other two water is eliminated and further changes take place.

Experiment 7.—A few drops of benzaldehyde are vigorously shaken in a test tube with three parts of commercial bisulphite solution. The crystals which separate are the sodium bisulphite compound of the benzaldehyde.

The. bisulphite compounds are formed according to the following equation :

$$\text{R—}\underset{\text{H}}{\text{C}}\text{:O} + \text{HSO}_3\text{Na} \longrightarrow \text{R—}\underset{\text{H}}{\text{C}}\begin{matrix}\diagup \text{OH} \\ \diagdown \text{SO}_3\text{Na}\end{matrix} .$$

The problem of the constitution of the bisulphite compounds of the aldehydes and ketones seems to have been solved in accordance with the above formula by the investigations of Raschig and Prahl.[1] It follows from their work, that these compounds are salts of *a-hydroxy-sulphonic* acids, the sulphonic groups of which are mobilised by the adjacent hydroxyl group. G. Schroeter,[2] however, has obtained from esters of dimethylmethanedisulphonic acid by hydrolytic elimination of one sulphonic group in the manner indicated below the salt of the a-hydroxy*iso*propylsulphonic acid and this salt is not identical with acetone sodium bisulphite.

$$\begin{matrix}\text{H}_3\text{C} \diagdown \\ \text{H}_3\text{C} \diagup\end{matrix}\text{C}\begin{matrix}\diagup \text{SO}_3\text{R} \\ \diagdown \text{SO}_3\text{R}\end{matrix} + 3\,\text{NaOH} \longrightarrow \begin{matrix}\text{H}_3\text{C} \diagdown \\ \text{H}_3\text{C} \diagup\end{matrix}\text{C}\begin{matrix}\diagup \text{OH} \\ \diagdown \text{SO}_3\text{Na}\end{matrix} + \text{Na}_2\text{SO}_3 + 2\,\text{ROH} .$$

This contradiction has yet to be explained.

Since the bisulphite compounds are decomposed into their constituents by warming with sodium carbonate solution or with dilute acids, they are particularly useful for enabling aldehydes (and ketones) to be separated from mixtures with other substances.

Polymerisation.—The simple aldehydes polymerise very easily. Anhydrous formaldehyde, indeed, cannot be kept for long, but changes very rapidly into an amorphous solid of high molecular weight $(CH_2O)_n$, *paraformaldehyde*. At ordinary temperature this substance breaks down again slowly to the simple molecule, a change which takes place more rapidly on heating. From the aqueous solution (formalin), in which form it is usually prepared, the anhydrous aldehyde cannot be obtained, since it only begins to distil with water vapour when the solution

[1] *Annalen*, 1926, 448, 265 ; *Ber.*, 1928, 61, 179.
[2] *Ber.*, 1926, 59, 2341 ; 1928, 61, 1616.

is boiled, and even then distils very slowly. This is due to the fact that, for the most part, it is present in solution as a hydrate, $H_2C{<}{OH \atop OH}$.

Acetaldehyde also polymerises gradually on keeping, yielding the liquid *paraldehyde* $(CH_3.CHO)_3$, which boils without decomposition at 124°. If a little hydrogen chloride is passed, with cooling, into acetaldehyde the likewise polymeric *metaldehyde* crystallises out.

Experiment 8.—To 5 c.c. of freshly distilled acetaldehyde in a moderate-sized conical flask one drop of concentrated sulphuric acid is added with cooling. When the vigorous reaction is over, the paraldehyde produced is shaken in a small separating funnel with water in order to remove the sulphuric acid, and the polymeride, which is insoluble in water, is separated. After being dried with a little calcium chloride the substance is distilled from a small distilling flask. Boiling point 124°.

Conversely paraldehyde can be reconverted into acetaldehyde by adding a few drops of concentrated sulphuric acid and distilling from the water bath through a column.

By this means fresh acetaldehyde can be prepared whenever required.

Experiment 9.—Pure paraldehyde is tested by the previously described aldehyde reactions: with ammoniacal silver nitrate, fuchsine-sulphurous acid, and bisulphite solutions. All are negative.

Vapour density determinations show that paraldehyde is formed by the union of three molecules of acetaldehyde. Since the substance has no aldehydic properties the following structure, that of a cyclic triacetal, is properly given to it:

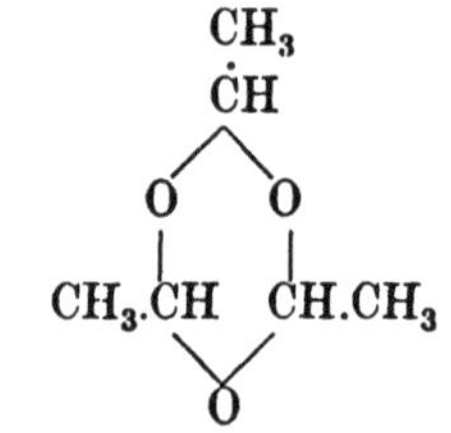

(Compare with this the polymerisation of acetylene to benzene.)

As was seen above, a small amount of concentrated sulphuric acid effects both the polymerisation of acetaldehyde and the depolymerisation of paraldehyde. An equilibrium occurs and its attainment is catalytically accelerated by the concentrated sulphuric acid:

$$3\,CH_3.CHO \rightleftarrows (CH_3.CHO)_3.$$

Moderate temperatures greatly favour the displacement of the equilibrium to the right. That, nevertheless, as was mentioned above, paraldehyde can be depolymerised to acetaldehyde by sulphuric acid is due to the fact that the equilibrium is shifted in accordance with the law of mass action. Thus the denominator of the fraction in the equation

$$\frac{\text{Concentration of paraldehyde}}{(\text{Concentration of acetaldehyde})^3} = K$$

always tends to decrease, owing to the volatilisation of the acetaldehyde present in small quantity; the re-establishment of the equilibrium brings about a decrease in the concentration of the paraldehyde and so favours its depolymerisation. Although equilibrium is reached at a point at which the mixture consists almost entirely of paraldehyde, yet because of the high vapour pressure of the monomer almost complete depolymerisation can be brought about in practice.

At low temperatures a second polymeric form of acetaldehyde, the beautifully crystalline *metaldehyde*, is formed.

Experiment 10.—A few bubbles of hydrogen chloride are passed into a solution of a few cubic centimetres of acetaldehyde in twice as much absolute ether kept cold in a freezing mixture. After a short time the *metaldehyde* separates in magnificent needle-shaped crystals which are filtered with suction and washed with ether. The filtrate yields a second crop on repeating the treatment.

Like paraldehyde, metaldehyde can be preserved, and, when freshly prepared, is odourless. On keeping, however, a distinct odour of acetaldehyde becomes evident—a sign that here also an equilibrium is slowly being established. Metaldehyde can be completely depolymerised by heating. Molecular weight determinations (in phenol) show that metaldehyde is tetramolecular (Hantzsch); the examination of the space lattice of crystals by the method of Laue and Bragg points to the same conclusion (Mark).

Metaldehyde is prepared on a technical scale for use as fuel (" solid methylated spirits ").

These *reversible polymerisations* of aldehydes are to be distinguished from the *condensations* which they can also undergo. Thus formaldehyde is converted by quite weak alkalis ($Ca(OH)_2$, $CaCO_3$) into *glycollic aldehyde* and *glyceraldehyde*, and further into a mixture of *hexoses* (Butlerow, O. Loew) from which E. Fischer isolated the so-called *α-acrose* (*dl*-fructose). In such condensations several molecules combine with the formation of new carbon-carbon linkages. Recall the part played by formaldehyde in the assimilation of carbon dioxide.

The so-called *aldol condensation*, which all aldehydes of the formula $^{R}_{R}\!>$CH.CHO undergo under the influence of dilute alkalis or acids, also involves the formation of new carbon-carbon linkages. The hydrogen atom in the α-position possesses mobility induced by the adjacent CO-group, and this mobilised hydrogen atom combines with the likewise very reactive C═O-group of a second molecule :

$$\overset{\text{H}}{\text{O═CH.CH}_2} + \text{O═CH.CH}_3 \longrightarrow \underset{\text{Aldol}}{\text{O═CH.CH}_2.\overset{\text{OH}}{\text{CH}}.\text{CH}_3}.$$

The aldols are β-hydroxyaldehydes and, like all β-hydroxycarbonyl-compounds, they easily lose water and become converted into α-β-unsaturated aldehydes. From aldol as starting material a technical route to butadiene is provided and possibly, in the future, to a synthetic rubber.

Experiment 11.—A few drops of acetaldehyde, dissolved in about 2 c.c. of water, are heated in a test tube with 0·5 c.c. of dilute sodium hydroxide solution. A yellow colour develops and the acetaldehyde is converted by way of aldol into *crotonaldehyde*, which can be recognised in the boiling solution by its pungent odour. If acetaldehyde is heated with concentrated alkali solution yellow *aldehyde resin* is precipitated as a result of further condensation.

The brown colour which develops in solutions of ethoxides and of potassium hydroxide in ethyl alcohol is to be attributed to the formation of similar substances, following oxidation of the alcohol.

The benzoin reaction and the reaction of Cannizzaro, which are discussed later, likewise take place because of the tendency of the aldehydes to undergo condensation. The specific catalyst determines in each case the particular way along which the condensation will proceed.

Experiment 12. Schardinger's Reaction.—Of two 25 c.c. portions of fresh milk one is boiled for a short time and cooled ; then 1 c.c. of the formaldehyde (prepared in an earlier experiment) and a few drops of an aqueous solution of methylene blue are added to each portion. When the two samples are now warmed to about 50° the dye in the unboiled milk is very rapidly decolorised and the same happens to further amounts added. In the boiled milk the colour remains unaltered.

Fresh cows' milk contains an enzyme which very greatly accelerates the reduction by aldehyde of methylene blue to its leuco-compound. This reduction does not show itself if the enzyme is absent. Two H-atoms of the hydrated aldehyde $\text{R—C}\begin{cases}\text{OH}\\ \textbf{OH}\\ \textbf{H}\end{cases}$ are "activated" by the enzyme in such a way that the aldehyde acts as a reducing agent and is thereby itself converted into acid. Finely divided metallic platinum can produce the same result as does the enzyme (Bredig). Heat destroys the effect of Schardinger's aldehyde-dehydrase. For details see *Ber.*, 1914, **47**, 2085; *Annalen*, 1929, **477**, 32.

Technical Importance of the Aldehydes.—Formalin solution is used as a disinfectant and as a preservative. Caseinogen hardened with formaldehyde (*galalith*) yields a widely used substitute for vulcanite, as does also the synthetic resin *bakelite*, which is produced by the condensation of formaldehyde with phenol (p. 234).

Sodium hyposulphite (hydrosulphite) is converted by aldehydes into *aldehyde bisulphite* compound and *aldehyde sulphoxylate*:

$$(SO_2Na)_2 + 2R.CHO + H_2O \longrightarrow R.C\begin{cases}H\\OH\\SO_3Na\end{cases} + R.C\begin{cases}H\\OH\\SO_2Na\end{cases}.$$

The sulphoxylate obtained from formaldehyde is much used under the name of "rongalite" as a reducing agent in discharge printing in dyeworks.

Small amounts of *acetaldehyde* (from acetylene) are converted industrially into alcohol by catalytic hydrogenation, and large amounts are transformed into *acetic acid* by catalysed autoxidation (with oxides of manganese).

Benzaldehyde is an important intermediate for dyes (see malachite green); many other aldehydes (phenylacetaldehyde, vanillin, piperonal, citral, etc.) are used in *perfumery* or as flavouring agents.

4. CANNIZZARO'S REACTION. BENZOIC ACID AND BENZYL ALCOHOL FROM BENZALDEHYDE [1]

Freshly distilled benzaldehyde (20 g.) is shaken in a stoppered cylinder or in a thick-walled flask with a cold aqueous solution of potassium hydroxide (18 g. in 12 g. of water) until a permanent emulsion is formed. The container is corked and left over night. To the crystalline paste (potassium benzoate) which separates, just sufficient water [2] is added to enable the benzyl alcohol to be extracted by repeated shaking with ether. The combined ether extracts are concentrated to 30–40 c.c. and are then *thoroughly* shaken twice in a separating funnel with 5 c.c. portions of commercial (40 per cent) bisulphite solution. After the ethereal solution has been separated, freed from dissolved sulphurous acid by shaking

[1] *Ber.*, 1881, **14**, 2394.

[2] If too much is added it is difficult to extract the (water-soluble) benzyl alcohol completely.

with a few cubic centimetres of sodium carbonate solution, and dried with anhydrous sodium sulphate, the ether is removed by distillation and the benzyl alcohol then passes over at 206°. Yield of benzyl alcohol about 8 g.

The aqueous alkaline liquid (from which the benzyl alcohol was removed with ether) is acidified with hydrochloric acid. Benzoic acid is precipitated and, when cold, is filtered off with suction. It is recrystallised from boiling water without previous washing. Melting point 121°. Yield 9–10 g.

Cannizzaro's reaction probably proceeds in such a way that two molecules of aldehyde condense to form an ester which is then hydrolysed to alcohol and acid:

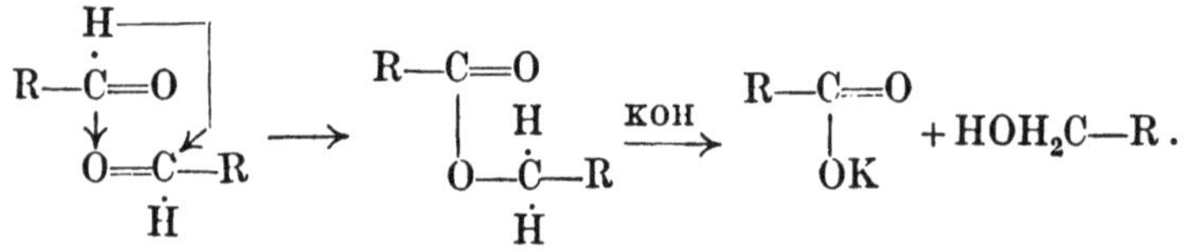

This view is supported by the fact that aldehydes are actually condensed to esters by aluminium ethoxide (Tistschenko).

The dismutation of aldehyde to acid and alcohol also plays an important part in cell metabolism, particularly in alcoholic fermentation (p. 391) (Mechanism?) although the chemical process is certainly different in this case.

The Cannizzaro reaction is by no means confined to aromatic aldehydes. *Formaldehyde* undergoes the same change, yielding *formic acid* and *methyl alcohol.* That the aliphatic aldehydes from acetaldehyde upwards do not undergo the reaction is due to the fact that the aldol condensation (mentioned above), in virtue of its much greater velocity, takes precedence over the Cannizzaro reaction.

In the case of tertiary aldehydes, which cannot undergo the aldol condensation, the Cannizzaro reaction replaces it among the aliphatic series. Thus *glyoxylic* acid is dismuted into *glycollic* and *oxalic* acids.

Related to the Cannizzaro reaction there is a reaction discovered by Meerwein,[1] in which an aldehyde is converted into an alcohol by the action of aluminium ethoxide.

Aldehydes R.CHO react with $Al(OC_2H_5)_3$ in such a way that addition takes place and $R.C\begin{cases}OC_2H_5\\ Oal\\ H\end{cases}$ is formed.[2] In the second stage of the reaction this substance is decomposed into $R.C\begin{cases}H\\ Oal\\ H\end{cases} + OHC.CH_3$.

[1] *Annalen*, 1925, 444, 221. [2] al = Al/3.

By means of this reduction process it is possible to obtain, from the corresponding aldehydes, alcohols such as trichloroethyl alcohol or cinnamyl alcohol, which are not otherwise readily accessible.

5. ACYLOIN CONDENSATION. BENZOIN FROM BENZALDEHYDE

Freshly distilled benzaldehyde (10 g.) is mixed with 25 c.c. of alcohol and a solution of 2 g. of potassium cyanide in 5 c.c. of water. The mixture is boiled on the water bath under reflux for one hour. Then the product is allowed to cool slowly, and the crystals which form, after being separated by filtration and washed with a little alcohol, are dried on the water bath. In order to obtain some quite pure benzoin a small sample of the crude product is recrystallised from a little alcohol. Melting point 134°. Yield about 90 per cent of the theoretical.

Benzil from Benzoin

The crude benzoin, prepared as above, after being dried and finely powdered, is heated (with frequent shaking) in an open flask on a vigorously boiling water bath for 1·5 to 2·0 hours with twice its weight of pure concentrated nitric acid. The reaction mixture is then diluted with cold water and when the material which separates has solidified the liquid is poured off, and the solid, after being dried on porous plate, is crystallised from alcohol. The crystals, after filtration, are dried in air on filter paper. Melting point 95°. Yield about 90 per cent of the theoretical.

The so-called acyloin or benzoin condensation is a further interesting aldehyde reaction. In the aromatic series it takes place as a result of the action of potassium cyanide, and it is very probable that the potassium compound of the cyanohydrin is formed as an intermediate product. As in the case of benzyl cyanide (p. 251), there is here a labile H-atom which, in an alkaline medium, is capable of undergoing a species of aldol condensation with a second molecule of aldehyde:

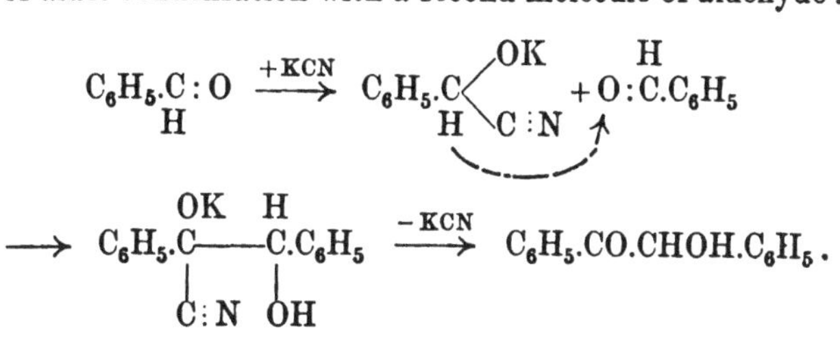

The condensation product is then converted, with elimination of potassium cyanide, into benzoin, The catalytic participation of the potassium cyanide is obvious. The distinction between this reaction and the cyanohydrin synthesis should be thoroughly grasped.

Substitution products react like benzaldehyde itself (anisaldehyde gives anisoin). Similarly furfural gives furoin. (Write out the equation.)

The reason why the acyloin synthesis is especially characteristic of aromatic aldehydes, depends on the circumstance that in the aromatic series the tertiary carbon atom in the ring does not allow of the aldol condensation, a reaction for which conditions are otherwise much more favourable. The simplest example of the acyloin condensation, moreover, was already encountered in the case of formaldehyde (p. 209); glycollic aldehyde is the simplest acyloin. Acyloin compounds are also produced, in the aliphatic series, by the action of sodium or potassium on esters, and hence are also formed as by-products in the acetoacetic ester synthesis (Bouveault, Scheibler).

Finally, agents which assist the acyloin synthesis have been discovered in the living cell; these enzymes (so-called *carboligases*) bring about the union of two molecules of aldehyde in the manner of the acyloin synthesis. Thus in fermenting yeast added benzaldehyde is caused to combine with acetaldehyde, the intermediate product of the fermentation, to form $C_6H_5.CHOH.CO.CH_3$, the (optically active) *benzacetoin*. If acetaldehyde itself is added (instead of benzaldehyde) acetoin is formed (Neuberg). The acyloins, being α-hydroxyketones, are to some extent related to the ketoses. Thus they also reduce Fehling's solution and are converted by phenylhydrazine into osazones.

Experiment.—Benzoin (1 g.) in concentrated alcoholic solution is boiled for some time on the water bath with 1·5 c.c. of phenylhydrazine. On cooling the osazone of benzil crystallises. Melting point 225°.

The production of the ammonia during the reaction should be demonstrated and the equation for the process written out.

An identical compound is formed from benzil by the action of phenylhydrazine, and from benzaldehyde phenylhydrazone by autoxidation (Busch). The formation of osazones from α-hydroxyketones (and α-hydroxyaldehydes) will be discussed later (p. 288).

The preparative importance of the acyloins depends on the fact that they are intermediate products, from which many 1:2-diketones can be obtained. The simplest aromatic member of this group is *benzil* (anisil and furil are analogous); like its aliphatic prototype *diacetyl* $CH_3.CO.CO.CH_3$ (and like anhydrous glyoxal) it is yellow in colour. Diacetyl is obtained from methyl-ethyl ketone via the monoxime of the former compound (von Pechmann). It is remarkable that diacetyl condenses to *p*-xyloquinone. (Formulate).

That the two C:O-groups of these diketones are adjacent is proved by the fact that they are capable of condensing with *o*-phenylenediamine (Quinoxalines, Hinsberg).

Experiment.—About 0·1 g. each of benzil and benzoin are dissolved in a test tube in 10 c.c. of alcohol and a few drops of alkaline hydroxide solution are added in the cold. A fine red colour is at once produced and disappears when the liquid is shaken with air. The colour reappears after a short time and can be caused to disappear by renewed shaking; these changes may be brought about repeatedly. When, after a few more drops of alkali have been added, the colour does not reappear there is no more benzoin left in the solution. Quite pure benzil does not give the colour.

This remarkable reaction occurs because alkalis (potassium hydroxide) convert benzoin partly into the di-enol form, *i.e.* into the potassium derivative of *stilbenediol* $C_6H_5.COK:COK.C_6H_5$.[1]

If water is excluded this potassium salt can be isolated in the form of orange-yellow crystals which, with benzil, yield the red solution sensitive to the action of air. The solution probably contains the *potassium-benzil radicle* which is also obtained by the addition of metallic potassium to benzil (Beckmann and Paul,[2] Schlenk[3]):

$$\underset{\substack{|\\OK}}{C_6H_5.C}{=\!=}\underset{\substack{|\\OK}}{C}.C_6H_5 + C_6H_5.CO.CO.C_6H_5 \longrightarrow 2\,C_6H_5.CO{-}\underset{\substack{|\\OK}}{C}.C_6H_5\ .$$

The radicle is converted by autoxidation partly into benzil, partly into benzoic acid.[4]

The most important reaction of benzil and related compounds is the *benzilic acid rearrangement* already discovered by J. von Liebig.

Experiment.[5]—Benzil (5 g.) is heated for ten minutes to boiling on the water bath with 15 c.c. of alcohol and a solution of 5 g. of potassium hydroxide in 10 c.c. of water. After cooling, the suspension of potassium benzilate crystals is filtered as dry as possible by suction, and the salt, after being washed with a little alcohol, is dissolved in 20–30 c.c. of cold water. The solution is filtered and

[1] Scheuing, *Annalen*, 1924, 440, 72.
[2] *Annalen*, 1891, 266, 23. [3] *Ber.*, 1913, 46, 2840.
[4] Compare, in this connection, A. Weissberger, H. Mainz, and E. Strasser, *Ber.*, 1929, 62, 1942.
[5] After H. von Liebig, *Ber.*, 1908, 41, 1644.

dilute sulphuric acid is added, at the boiling point, to the clear filtrate. The free acid is precipitated partly in the form of crystals. It is separated by filtration with suction while hot and washed with hot water. It can then be recrystallised at once from a large volume of hot water or, after drying, from benzene. Yield about 4 g.

The first stage of the rearrangement, which proceeds according to the equation

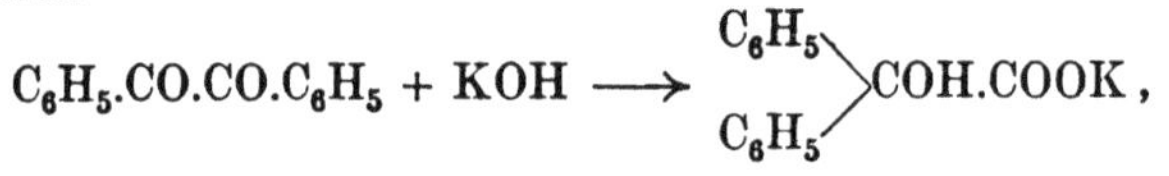

consists in the addition of a molecule of alkali hydroxide to the benzil (Scheuing). In this addition product, evidently because of the tendency of the potassium to become neutral, the C_6H_5- and OK-groups exchange places:

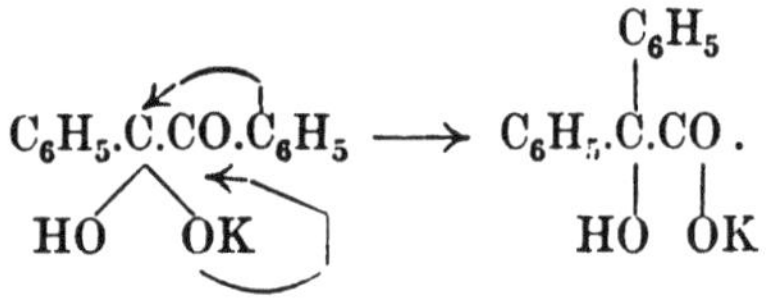

By a similar reaction phenanthraquinone yields biphenyleneglycollic acid. (The equation should be written.) The benzilic acid rearrangement also plays a part in many other reactions (croconic acid, purpurogallin).

The so-called *pinacoline rearrangement* is closely related:

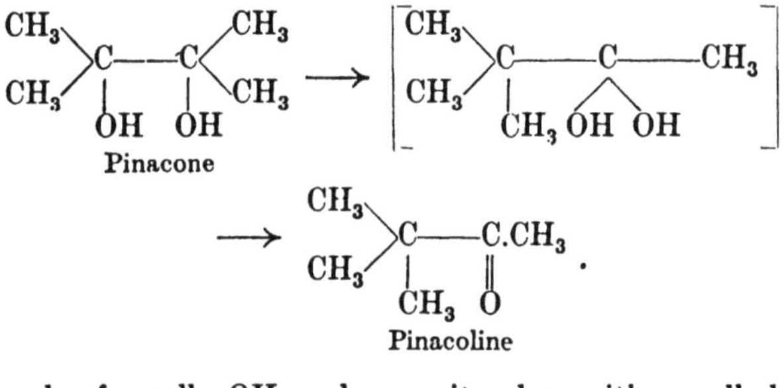

Here also formally OH exchanges its place with an alkyl radicle, CH_3, although actually—since concentrated sulphuric acid is used—the elimination of water between the two OH-groups provokes the wandering of a methyl group.

A rearrangement which takes place in similar compounds, and has been much studied recently, may be briefly mentioned here. This rearrangement has been named—not quite correctly—the *retro-pinacoline rearrangement.*

It consists in the conversion of *pinacolyl alcohol* into *tetramethylethylene* with elimination of water :

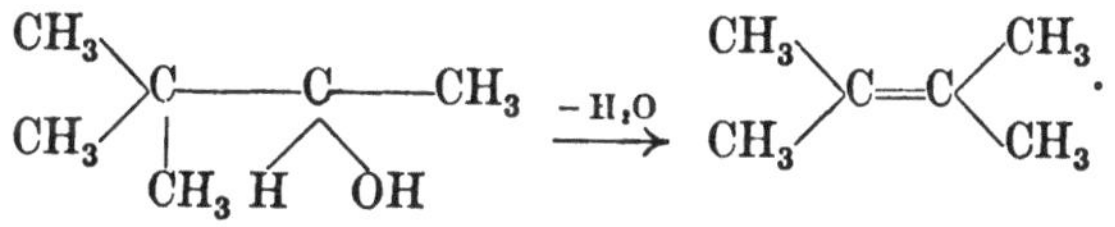

Closely related to this change is the conversion of borneol and its derivatives into compounds of the camphene type :

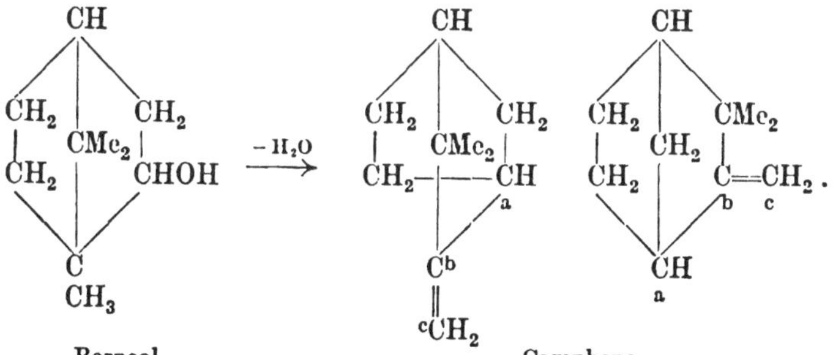

Borneol Camphene

The only difference between the two reactions evidently consists in the shift of the double bond in respect of the methyl group, from *a*, *b*, to *b*, *c*. For between *a* and *b*, through stereochemical reasons, no double bond can exist, since, in accordance with Bredt's rule, none of the C-atoms which are common to both rings of a bicyclic system of the camphane type can take part in an unsaturated linkage.

Further inspection shows the second of the above camphene formulae to be merely another and more convenient way of writing this hydrocarbon.

Information about this important work, which can be considered only briefly here, is to be derived from the publications of H. Meerwein. A clear and comprehensive review of our knowledge of molecular rearrangements is to be found in: Henrich, *Theorien der organischen Chemie*, 5th Ed., 1924, Chap. XVII.

Only one more rearrangement of benzil, leading to an elegant preparative method, need be mentioned here, namely its conversion into *diphenylketene* by the method of G. Schröter (*Ber.*, 1909, **42**, 2346). The *hydrazone* of benzil is dehydrogenated with mercuric oxide (preferably prepared by the experimenter himself) to yield the diazo-compound, the so-called " azibenzil " (Curtius, Staudinger) :

When this substance is heated in benzene with the exclusion of air and moisture, nitrogen is eliminated and the rest of the molecule undergoes rearrangement to *diphenylketene* :

$$C_6H_5.CO.\underset{\wedge}{C}.C_6H_5 \longrightarrow O{=}C{=}C\Big\langle\begin{matrix}C_6H_5\\ C_6H_5\end{matrix}.$$

This interesting derivative of ketene is also obtained from benzilic acid by the old process of Staudinger : the acid is converted, by the action of phosphorus pentachloride, into diphenylchloracetyl chloride from which the two chlorine atoms are removed by means of zinc. (Formulate this equation.) For information about the ketenes see H. Staudinger, *Die Ketene*, Stuttgart, 1912.

6. ADDITION OF HYDROGEN CYANIDE TO AN ALDEHYDE. MANDELIC ACID FROM BENZALDEHYDE

Mandelic Nitrile.—Freshly distilled benzaldehyde (15 g.) is stirred with a glass rod in a cylinder with about 50 c.c. of a concentrated solution of sodium bisulphite until a paste of the addition compound $C_6H_5.C\Big\langle\begin{matrix}H\\ OH\\ SO_3Na\end{matrix}$ has formed. The cylinder is then closed with a rubber stopper and vigorously shaken. The bisulphite compound is now separated by filtration at the pump, pressed well together on the filter funnel, and washed several times with *small* quantities of ice-water. It is stirred to a thick paste with water, and a cooled solution of 12 g. of pure potassium cyanide in 25 c.c. of water is added. After a short time, and especially if the mixture is stirred, the crystalline material dissolves and the mandelic nitrile appears as an oil which is separated in a tap funnel from the aqueous solution and immediately used for the next stage of the preparation.

Hydrolysis of the Nitrile.—The nitrile, mixed in a porcelain basin with four times its volume of concentrated hydrochloric acid, is heated on the water bath until an abundant separation of crystals begins to take place on the surface of the liquid. The reaction mixture is then allowed to stand over night in a cool place, and the crystals which have been deposited, after being rubbed with a little water, are separated at the filter pump and washed with water (not too much). A further quantity of the acid is obtained from the filtrate by extraction with ether. The crude mandelic acid is pressed on a porous plate, dried, and purified by crystallisation from benzene. Melting point 118°. Yield about 10–15 g.

Resolution of Inactive Mandelic Acid into its Active Components.[1]— A mixture of 10 g. of crystallised mandelic acid and 20 g. of crystallised cinchonine are heated with 500 c.c. of water in an open flask, which is very frequently shaken, on a vigorously boiling water bath for one hour. After cooling, undissolved material is separated by filtration (but not subsequently washed with water). A few crystals of cinchonine *d*-mandelate (see below) are then dropped into the clear solution (*a*), which is now allowed to stand in a cool place (6°–8°; in summer in the ice-chest, in winter conveniently in a cellar) for one or several days as required. The crude cinchonine *d*-mandelate which thus separates is collected by filtration with suction (filtrate A, to be kept) and recrystallised from 20 parts of hot water. If the solution is seeded with a few crystals of cinchonine *d*-mandelate the salt crystallises in a purer form under the same conditions as those described above. In order to obtain the free *d*-mandelic acid the purified salt is dissolved in water (not too much) and a slight excess of ammonia is added. Cinchonine is precipitated and, after being filtered off, can be recrystallised from dilute alcohol and used again for another resolution. The filtrate, which contains ammonium *d*-mandelate, is acidified with hydrochloric acid and shaken with ether. When the material left on evaporation of the ether is heated for some time on a watch-glass on the water bath, it solidifies on cooling to crystals of *d*-mandelic acid, which are pressed on a porous plate and recrystallised from benzene. Melting point 133°–134°.

Pure *l*-mandelic acid is not easily obtained from small amounts of *dl*-mandelic acid. However, a feebly laevo-rotatory preparation is obtained as follows: The filtrate A is treated as has just been described for pure cinchonine *d*-mandelate, and the free acid thus obtained must be laevo-rotatory since part of *d*-modification has been removed.

Of the three products so obtained, namely: (1) inactive racemic acid, (2) pure *d*-acid, and (3) impure *l*-acid, aqueous solutions of suitable concentration should be prepared. The rotations of the solutions should then be determined with a polarimeter.

If cinchonine *d*-mandelate is not available a specimen suitable for the inoculation in the first experiment is prepared as follows: To a few cubic centimetres of the solution (*a*) saturated aqueous sodium chloride

[1] Cf. *Ber.*, 1883, 16, 1773 and 1899, 32, 2385.

solution is added drop by drop until slight precipitation occurs. The mixture is then heated until the precipitated material has dissolved, after which the solution is allowed to stand until crystals have been deposited, which may require a day. The crystals consist of cinchonine hydrochloride on which a small amount of cinchonine *d*-mandelate is deposited, enough, however, to induce the separation of further quantities of the *d*-salt.

7. ALANINE [1]

Freshly distilled acetaldehyde (13·2 g.; 0·3 mole) dissolved in 100 c.c. of ether is poured into a pressure bottle (Fig. 30) containing a cold saturated aqueous solution of 18 g. of ammonium chloride; then a solution of 20 g. of sodium cyanide in 30 c.c. of water is added slowly drop by drop from a tap funnel with shaking and ice cooling. The closed bottle is then agitated at ordinary temperature for three hours on a shaking machine. The contents of the bottle are next transferred to a round-bottomed flask (capacity 0·5 l.) which is kept cool in ice, while 100 c.c. of concentrated hydrochloric acid are added in small portions. (Fume chamber! Free hydrocyanic acid!) The ether is removed by distillation through a downward condenser and the aqueous solution, after being heated for one hour longer on the boiling water bath, is transferred to a basin and evaporated to dryness. (A brown colour develops in the aqueous solution after the ether has been removed.) When the material in the basin is completely dry and free from hydrochloric acid (test the odour) it is extracted twice with 100 c.c. of boiling alcohol in a small round-bottomed flask. The filtered alcoholic extracts are evaporated to dryness and the extracted material is finally dried *in vacuo* on the water bath. The alanine hydrochloride is now freed from admixed sodium chloride by dissolution in 100 c.c. of hot absolute alcohol to which 5 c.c. of ether have been added and subsequent evaporation to dryness, once again, of the alcoholic solution of the alanine salt. This salt, which is difficult to prepare in the crystalline condition, is converted into the free acid in the following way:

The hydrochloride is washed into a beaker with 100 c.c. of water and boiled with 40–50 g. of litharge, added in small portions, until no more ammonia is evolved (about ten to fifteen minutes). The

[1] A. Strecker, *Annalen*, 1850, 75, 30; Zelinsky and Stadnikov, *Ber.*, 1908, 41, 2061.

ammonia is derived from ammonium chloride, of which a little is present in the solution. The mixture is now filtered with suction while hot, the material on the filter being washed with 20–30 c.c. of hot water, and the clear brown filtrate is freed from lead by passing hydrogen sulphide into the hot liquid. The lead sulphide is separated at the pump, and the filtrate, while still lukewarm, is shaken in a glass-stoppered bottle with about 3 g. of freshly precipitated and carefully washed silver carbonate in order to remove all chlorine ions (derived from the slightly soluble lead chloride. (A sample of the treated solution should be tested for chlorine.) During the shaking the stopper of the bottle should be removed from time to time. The filtered solution, into which hydrogen sulphide is again passed, leaves alanine in the form of a dark-coloured syrup when evaporated. When this syrup is rubbed with absolute alcohol it crystallises ; the crystals, after standing for some time, are separated at the pump, washed successively with a little absolute alcohol and absolute ether, and dried in a vacuum desiccator. Yield 15–20 g. Alanine can be recrystallised, but only with great loss, from its own weight of water. It is better to dissolve it in the minimum amount of boiling water and to add alcohol at the boiling point until crystallisation begins. Melting point 264° decomp.

Remarks on Sections 6 and 7. The method here described for the synthesis of cyanohydrins—treatment of the bisulphite compound of the aldehyde with potassium cyanide—cannot be used in all cases. Concentrated solutions of hydrocyanic acid or anhydrous hydrogen cyanide are often used. The general method for the synthesis of α-amino-acids, the nitriles of which are formed by the union of ammonium cyanide with aldehydes or ketones (Strecker), is to be contrasted with that for the synthesis of α-hydroxy acids. For additional amino-acid syntheses see Chap. VII. 2, p. 265.

The amygdalin of bitter almonds and of other stone-fruits is the glucoside formed by the combination of *l*-mandelic nitrile and gentiobiose (see p. 387). This glucoside belongs to the class of β-glucosides, since it is hydrolysed by the enzyme emulsin yielding two molecules of dextrose, benzaldehyde, and hydrocyanic acid. Natural *l*-mandelic acid was first obtained by Liebig, who subjected amygdalin to acid hydrolysis.

The cyanohydrin synthesis has been applied in the study of the sugars by H. Kiliani, who used it in the synthesis of higher members of the class. The carboxylic acids which result from the hydrolysis of the nitriles can be reduced, in the form of their lactones, to the corresponding aldehydes.

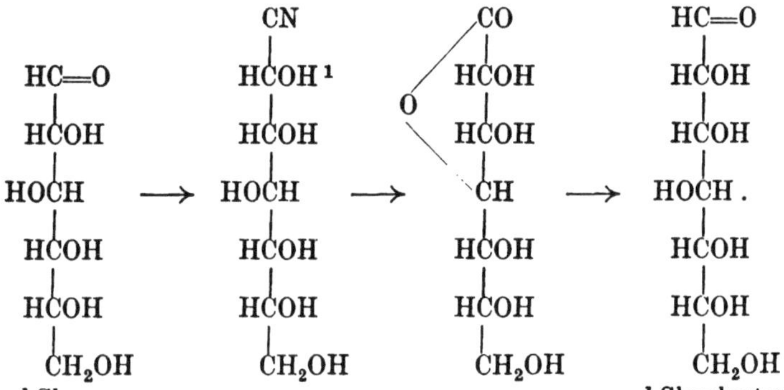

d-Glucose *d*-Glucoheptose

In the synthesis of mandelic nitrile described above asymmetry is produced in one carbon atom, but exactly equal numbers of molecules of the two antipodes are formed, since there is just as much chance that one should be formed as the other.

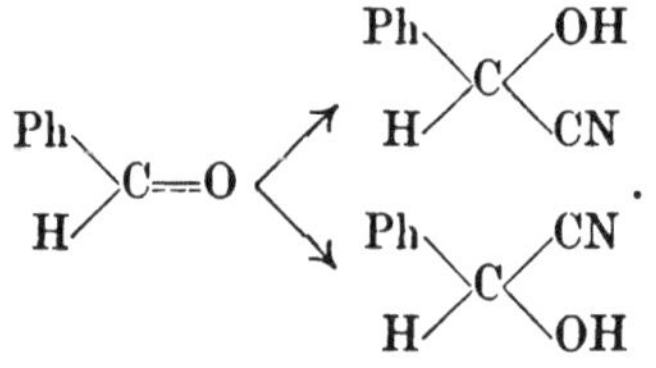

The *d*- and *l*-varieties differ only in crystal form and in direction of the rotation; in all other physical and in all chemical properties they behave exactly alike. The salts which the two forms yield with an optically active base, however, are no longer related to each other as object and mirror-image, and hence differences in physical constants, *e.g.* in solubility, appear. How are racemic bases resolved?

Enzymes exhibit specificity as regards their behaviour towards stereo-isomerides. Pasteur's biological resolution of racemic acid with moulds depends on this fact, and such moulds may also be used for the resolution of racemic mandelic acid, for the partial digestion of racemic polypeptides according to E. Fischer's method, and in numerous other processes.

[1] The compound having the opposite configuration HOCH (with CN above) is produced at the same time, but not, as in the production of a simple racemic substance, in equal amount. The asymmetry already present exerts a directing influence which favours one of the two configurations.

8. PERKIN'S SYNTHESIS

Cinnamic Acid from Benzaldehyde and Acetic Anhydride

In a flask to which a long (about 80 cm.) wide air condenser is fitted, freshly distilled benzaldehyde (21 g., 0·2 mole), acetic anhydride (30 g.), and powdered anhydrous sodium acetate (cf. p. 116) are heated at 180° for eight hours (beginning in the morning) in an oil bath. When the reaction has ceased the hot mixture is poured into a large flask and washed out with water; steam is then blown through until benzaldehyde ceases to be carried over. Sufficient water is now added to dissolve the cinnamic acid. A small amount of oily impurity remains undissolved. The solution is boiled for a short time with animal charcoal and filtered. On cooling, cinnamic acid separates from the filtrate in shining plates. If the melting point of the material thus obtained is not correct, recrystallisation from hot water is necessary. Melting point 133°. Yield about 15 g.

Perkin's synthesis proceeds according to the general principle of aldehyde condensations, namely, with the elimination of water formed from the carbonyl oxygen and two hydrogen atoms from a methylene or methyl group. The conditions for the synthesis depend on the reactivity of this group: with acid or alkali as catalyst, aldehydes react even in the cold with aldehydes or ketones, but if the aldehyde is to react with an acid anhydride a high temperature is required, with an alkali salt as condensing agent.[1]

The difference in the conditions under which the reaction takes place is to be traced to the slight reactivity of the methyl (or α-methylene) group in the acid anhydride.

Both CH_2-groups of succinic acid are capable of taking part in condensations. If unsaturated aldehydes such as cinnamaldehyde are used, with lead oxide as catalyst, highly unsaturated dicarboxylic acids are obtained which are transformed by loss of $2CO_2$ into coloured "*polyenes*" (R. Kuhn[2]).

$$2C_6H_5.CH:CH.CHO + \begin{array}{l} H_2C.CO_2H \\ \quad | \\ H_2C.CO_2H \end{array} \longrightarrow \begin{array}{l} C_6H_5.CH:CH.CH:C.\boxed{CO_2}H \\ \qquad\qquad\qquad\quad\ | \\ C_6H_5.CH:CH.CH:C.\boxed{CO_2}H \end{array}.$$

When the trebly unsaturated aldehyde

$$C_6H_5.CH{=}CH.CH{=}CH.CH{=}CH.CHO$$

[1] See Kalmin, *Helv. Chim. Acta*, 1928, 11, 977.
[2] *Helv. Chim. Acta*, 1928, 11, 87.

is used, this synthesis leads to the *copper-coloured* hydrocarbon $C_{28}H_{26}$, which contains eight conjugated double bonds.

The interesting group of compounds called *carotinoids* consists of purely aliphatic polyenes. Carotin itself, $C_{40}H_{56}$, the colouring matter of carrots, contains eleven double bonds, probably conjugated, and its isomeride *lycopin* from tomatoes and hawthorn has thirteen. *Xanthophyll* $C_{40}H_{56}O_2$ (Willstätter), which occurs in nature along with chlorophyll, and the isomeride of xanthophyll, *zeaxanthin* (Karrer), the yellow colouring matter of maize, are closely related to these hydrocarbons. Other members of this class are carboxylic acids of the polyenes, such as the *crocetins* of saffron (Karrer) and *bixin*, the red compound from *Bixa orellana* (Kuhn). *Physaliene*, the red colouring matter of the Cape gooseberry and winter cherry, is dipalmityl zeaxanthin (Kuhn, Zechmeister). It is a true "*lipochrome*".

As a result of the important investigations of Euler and Karrer carotin seems to be very closely related to the fat soluble growth-promoting vitamin-A.[1]

In the preparation described above we have an important means of making *β-arylacrylic* acids, and from them, by hydrogenation, *β-arylpropionic* acids. The method is used in the investigation of alkaloids and in the synthesis of coumarin from salicylaldehyde.

The methylene groups of *hippuric* acid and *malonic* acid are much more reactive than that of acetic acid. They may be caused, therefore, to condense with aldehydes under much milder conditions, *e.g.* by the action of pyridine. The use of malonic acid forms an extension of Perkin's reaction to the aliphatic series (Doebner), *e.g.*

$$CH_3.CHO + CH_2.(COOH)_2 \longrightarrow CH_3.CH{=}C\begin{cases}COOH\\COOH\end{cases}$$

$$\longrightarrow CH_3.CH{=}CH.COOH + CO_2.$$

Crotonic Acid

Cinnamic acid exists in two different configurations, as *cis* and *trans* isomerides. Ordinary cinnamic acid has the *trans*-form (fumaric acid); *cis*-cinnamic acid is called *allo*-cinnamic acid, and was first found, along with its isomer, in the vegetable kingdom. Synthetically, it is obtained from phenylpropiolic acid $C_6H_5.C \equiv C.COOH$ by partial catalytic hydrogenation (Paal). Phenylpropiolic acid can be obtained from the dibromide of cinnamic acid by the elimination of two molecules of HBr. Two other cinnamic acids, melting at 42° and 58° respectively, are regarded as crystallographically different polymorphs of *allo*-cinnamic acid (Biilmann).

Experiment. Hydrogenation of Cinnamic Acid.—In a glass-stoppered bottle (capacity 250 c.c.) 10 g. of cinnamic acid are

[1] See Moore, *Biochem. J.* 1929, 23, 803; Javillier, *Bull. Soc. Chim.*, 1930, 47, 489.

dissolved by shaking with about 100 c.c. of dilute sodium hydroxide solution, which is added little by little until the reaction, when tested with phenolphthalein paper—spot test with drops of the two solutions on filter paper—is just alkaline. Two per cent sodium amalgam (freshly prepared as described below) is then added in small portions with continual shaking and frequent removal of the stopper. Altogether about 250 g. of amalgam are used. To complete the reaction the mixture is warmed on the water bath (placed in warm water, which is then heated) until all the mercury has separated in liquid form. After the liquid has cooled, the metal is separated in a funnel and the aqueous layer is acidified with hydrochloric acid. Hydrocinnamic acid is precipitated, at first in the form of an oil, which solidifies on cooling and rubbing. It is collected at the pump and crystallised from hot water containing some hydrochloric acid. Because of the low melting point of the acid (47°) the solution must be allowed to cool slowly. Compare pp. 6, 7. Yield 8 g.

A sample of the material, dissolved in a little dilute sodium carbonate solution, should not decolorise a drop of permanganate solution.

Cinnamic acid may also be hydrogenated catalytically (p. 368). If the sodium amalgam method is chosen, the catalytic method should be practised with linolenic acid or phenol.

Sodium Amalgam.—Mercury and sodium react with violence, sparks and flames being produced; *the amalgam must therefore be made inside a fume chamber and goggles must be worn.* In a mortar of moderate size 300 g. of mercury are warmed to 30°–40° and the sodium (6·5 g. in all), cut into small cubes, is introduced on the point of a glass rod about 30 cm. long; the pieces are pushed under the surface of the liquid in rapid succession. A porous plate is used as a cover for the mortar in order to provide protection against spurting. While still warm the solidified amalgam is broken into small pieces and is kept in a well-stoppered bottle.

Large amounts of sodium amalgam are best prepared in a fire-clay crucible; larger pieces of sodium are fixed on the tip of a long, heavy glass rod and allowed to slide by gravity into the mercury. In order to obtain an amalgam of more than 3 per cent the crucible must be heated with a flame.

$\alpha\beta$-Unsaturated carbonyl compounds (aldehydes, ketones, esters, and acids) can be hydrogenated by means of the usual reducing agents such as zinc and acid, sodium amalgam, sodium and alcohol, whilst the carbon-carbon double bond of olefines can only be saturated by hydrogen, when activated by a catalyst. According to Thiele, the explanation of this fact is provided by assuming a 1:4-addition for which the doubly bound oxygen offers a suitable point of attack to the hydrogen :

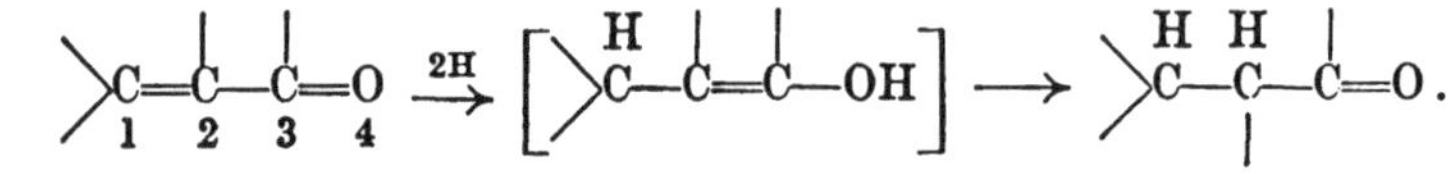

The double bonds of the benzene ring also enter, in this way, into a relationship of conjugation with carboxyl groups which are directly united to the ring. *Benzoic* acid and the *phthalic* acids are hydrogenated by sodium in alcohol (Baeyer). But *hydrocarbons of the styrene type* also can be hydrogenated by the same reducing agent (Klages) *e.g.*

$$C_6H_5.CH{=}CH{-}CH_3 \xrightarrow{2H} C_6H_5.CH_2.CH_2.CH_3 .$$

9. THE REIMER-TIEMANN SYNTHESIS

Salicylaldehyde from Phenol and Chloroform [1]

In a round-bottomed flask (capacity 1 l.) sodium hydroxide (80 g.) is dissolved by heating in water (80 c.c.), pure *phenol* (25 g.) is added to the hot solution, and the mixture is cooled to 60°–65° by dipping the flask in cold water, without shaking, however, so that separation of crystalline sodium phenoxide is avoided. By means of a two-holed cork the flask is then fitted with an efficient reflux condenser and a thermometer, the bulb of which dips into the liquid. *Chloroform* (20 g.) is next poured in through the condenser, and the contents of the flask are gently shaken ; a transient fuchsine-red colour is developed in the liquid. After a period of about ten minutes, during which the temperature of the mixture is maintained between 65° and 70° by dipping the flask in cold or hot water, a second 20 g. portion of chloroform is added and once again, after an interval of fifteen minutes, a third portion of 20 g. At this stage the flask should frequently be shaken. To complete the reaction the flask is heated on the water bath for one hour and steam is then blown through the reaction product until chloroform ceases to pass over. The reaction mixture is now allowed to cool somewhat, and dilute sulphuric acid is carefully added until the orange-coloured alkaline liquid becomes acid and almost colourless. Steam is then passed through again until no more drops of oil come over with the water.

[1] *Ber.*, 1876, 9, 423, 824 ; 1877, 10, 1562 ; 1882, 15, 2685, etc.

The distillate is at once extracted with ether and the extract, after having been separated from the water, is heated on the water bath until most of the ether has distilled. The residue, which contains unchanged phenol as well as the salicyaldehyde, is now vigorously shaken in a glass-stoppered bottle with two volumes of concentrated commercial sodium bisulphite solution. A thick paste of the bisulphite compound of the aldehyde is formed After this paste has stood for from half an hour to one hour the bisulphite compound is separated by filtration at the pump, pressed well on the filter funnel, and washed several times, first with alcohol and finally with ether, until completely free from adherent phenol. The crystals (small plates, iridescent like mother-of-pearl) are then decomposed with dilute sulphuric acid in a small round-bottomed flask which is fitted with an air condenser and gently warmed on the water bath. After the liquid thus produced has cooled, the aldehyde which separates is extracted with ether and the ethereal solution is dried with anhydrous sodium sulphate. The pure aldehyde which remains when the ether is evaporated distils at 196°. The yield amounts to 10–12 g.

From the residue after the steam distillation, filtered while hot and saturated with sodium chloride, p-*hydroxybenzaldehyde*, which is not volatile with steam, crystallises, often only after long standing. If the filtrate, obtained when the crystalline material is separated by filtration, be extracted with ether, a further amount of the *p*-compound is obtained. By recrystallisation from water, to which some aqueous sulphurous acid is added, both portions can be purified together. Melting point 116°. Yield 2–3 g.

If the sodium phenoxide separates at the beginning, the synthesis fails.

This reaction, which can also be used with substituted phenols, can be applied generally, but, because of side-reactions which accompany it, it is usually rather unprofitable. Its result very distinctly recalls that of Kolbe's salicylic acid synthesis (Chap. VI. 4, p. 240), and we might be tempted to picture its course as proceeding in the same way as has been demonstrated experimentally for Kolbe's synthesis, namely, by way of a double decomposition with elimination of a molecule of NaCl; part of the chloroform molecule would thus become attached to the oxygen of the phenol, the two remaining chlorine atoms would then be replaced by oxygen, and finally, in the phenyl formate so produced, a rearrangement involving the wandering of the formyl group (to the *o*- or *p*-position) would take place. This wandering corresponds to that which occurs in phenyl carbonate :

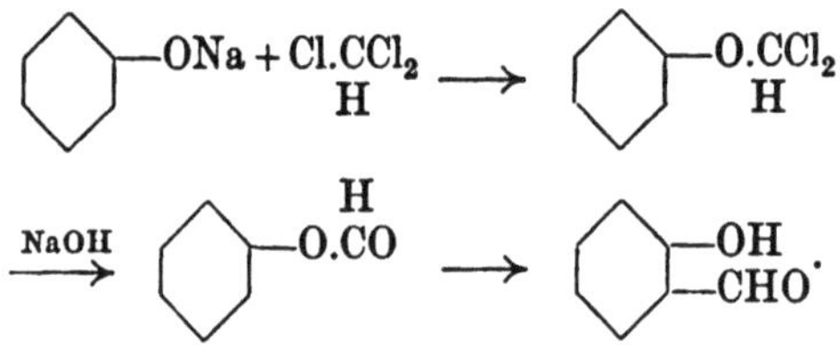

More probably, however, the process consists first in *an addition of chloroform to one of the double bonds of the ring* and then in a change which is readily understood from the following formulae :

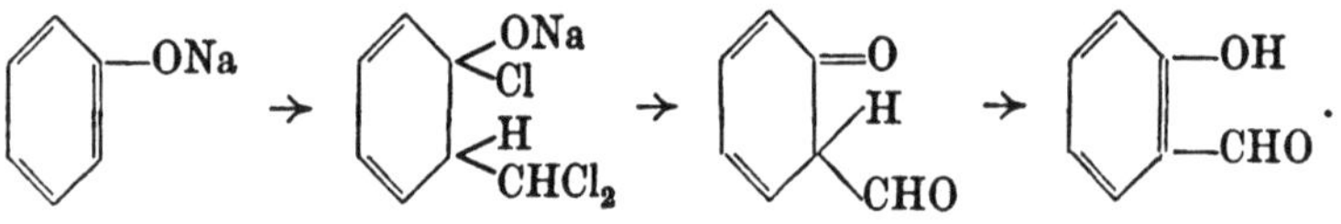

The isolation of addition compounds which can only have been formed in such a way (in the cases of *o*- and *p*-cresol ; Auwers, *Ber.*, 1902, **35**, 4207) gives decisive support to this view.

As will be seen, the primary product from *o*-cresol cannot reconstitute the aromatic ring because of the CH_3-group. Hence only NaCl is eliminated. The other two chlorine atoms are not hydrolytically removed and a quinol-like substance is formed, as the formula indicates :

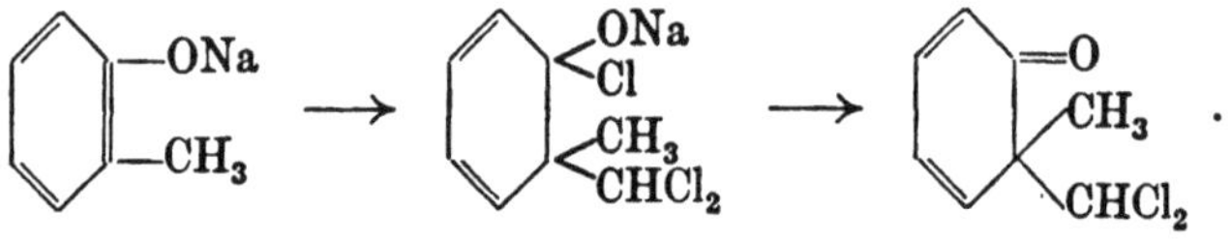

In the case of *p*-cresol the addition of chloroform takes place in the 1 : 4-positions :

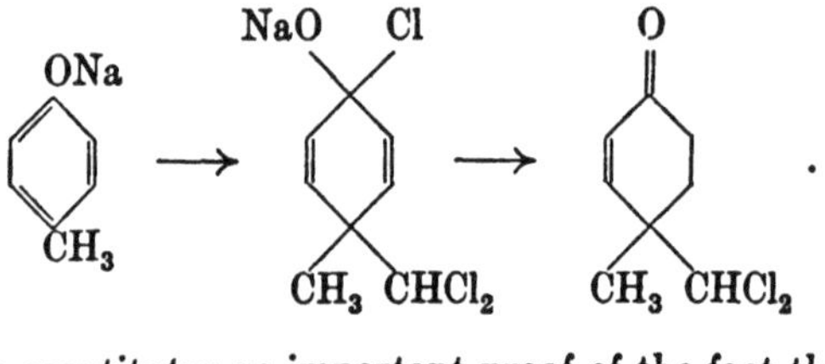

This reaction constitutes an important proof of the fact that in the benzene ring additions in the 1 : 4-positions occur and, at the least, it provides a valuable indication that the *p*-substitution, which so often accompanies *o*-substitution, is likewise preceded by 1 : 4-addition (cf. in this connection p. 101).

Experiment.—A few drops of salicylaldehyde are dissolved in sodium hydroxide solution; an intensely yellow sodium salt is formed.

A drop of ferric chloride solution produces an intense violet colour in an aqueous or alcoholic solution of the aldehyde.

In salicylaldehyde the phenolic odour predominates. The aldehyde is much less liable to autoxidation than is benzaldehyde.

With regard to this point compare the two aldehydes by exposing a few drops of each to the air on a watch-glass.

Guaiacol, by the action of chloroform and alkali, gives *vanillin* in poor yield along with the *o*-aldehyde.

From salicylaldehyde and sodium acetate by Perkin's synthesis *coumarin* is formed. (Formulate!)

CHAPTER VI

PHENOLS AND ENOLS. KETO-ENOL-TAUTOMERISM

1. CONVERSION OF A SULPHONIC ACID INTO A PHENOL. β-NAPHTHOL[1]

MATERIALS required: 70 g. (0·3 mole) of sodium naphthalene-β-sulphonate,[2] 210 g. of sodium hydroxide, 20 c.c. of water.

The sodium hydroxide is powdered and, along with the water, is heated with stirring in a copper or nickel crucible to 280°. The thermometer, which is at the same time the stirrer, is enclosed in a copper or nickel tube about 16 cm. long and 0·8 cm. wide. The bulb of the thermometer, which is maintained in position by a bored cork, dips into a layer of oil 1 cm. thick. (See Fig. 51.) *Goggles and gloves must be worn.* As soon as the temperature has reached 280° the sodium naphthalenesulphonate is added rather rapidly with stirring; heating is continued with a somewhat smaller flame and the temperature is maintained between 260° and 280°. After all the salt has been added, the flame is raised somewhat, causing the molten material to become more viscous as steam is evolved and bubbles are formed. Finally, at 310° the reaction proper sets in. After the temperature has been kept at 310°–320° for about five minutes, the material becomes more mobile and the reaction is over. The molten material is at once poured in a thin layer on to a thick copper plate the edges of which have been bent slightly upwards. The dark-coloured upper layer consists of the sodium salt of naphthol. After it has cooled, the material is powdered and dissolved in water, the naphthol is precipitated from the solution with concentrated hydrochloric acid and, after cooling, is extracted with ether. After the

FIG. 51.

[1] E. Fischer, *Anleitung zur Darstellung organischer Präparate*, 9th ed., Braunschweig, 1920. *Zeitschrift für Chemie*, 1867, 10, 299; *Compt. rend.*, 1867, 64, 749.

[2] Or the equivalent weight of naphthalene-β-sulphonic acid, prepared according to O. N. Witt. See p. 185.

ethereal solution has been dried with anhydrous sodium sulphate, the ether is removed and, finally, the naphthol which remains is distilled in a sausage flask. Boiling point 286°. Melting point 123°. Yield 30–35 g.

The reaction just described is carried out technically on a large scale in iron vessels provided with stirrers, for β-naphthol, as well as the numerous sulphonic acids which can be obtained from it by the action of sulphuric acid, is widely used for the preparation of azo-dyes. Moreover, by the action of ammonia under pressure, β-naphthylamine is produced from β-naphthol:

$$C_{10}H_7.OH + NH_3 = C_{10}H_7.NH_2 + H_2O\,.$$

β-Naphthylamine and also its sulphonic acids are likewise employed technically for the manufacture of azo-dyes. In the same way α-naphthol is made from naphthalene-α-sulphonic acid by fusion with sodium hydroxide although on a smaller scale than β-naphthol. α-Naphthylamine, on the other hand, is obtained by the reduction of α-nitronaphthalene (analogy to aniline). The fusion of alkali salts of arylsulphonic acids with alkali also serves technically for the production of *pure* phenol and of many phenol derivatives.

Because of the fact that the halogen is so very firmly attached to the benzene ring, the hydrolysis of chlorobenzene, a cheap substance, can only be carried out according to the following equation at very high temperatures and then with dilute solutions of alkali hydroxide (K. H. Meyer and F. Bergius).

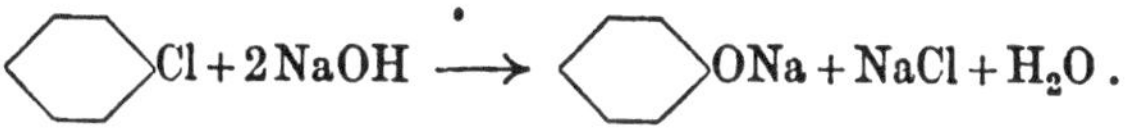

The hydrolysis proceeds much more easily when the halogen is mobilised by an *o*- or *p*-nitro-group. This question has already been discussed on p. 93.

A general method leads from the primary aromatic amines by way of the diazonium salts to the phenols.

By their properties and reactions the phenols are very clearly distinguished from the ordinary alcohols of the aliphatic series. The fundamental difference between the two classes of compounds is that the OH-group of the phenols is attached to a doubly bound C-atom. The phenols, therefore, behave like the similarly constituted *enols*, as may be deduced from the following formulae:

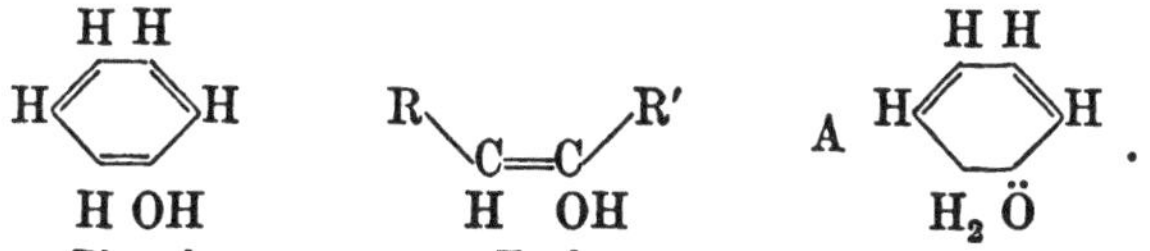

Phenol Enol

Simple ketones such as acetone and the like are not capable of existing in the " tautomeric " form $H_2C{=}\underset{OH}{C}{-}CH_3$. Whence comes it that an arrangement such as that shown in formula A does not obtain in phenol ? The reasons are doubtless the same as those which, in general, make a rearrangement of partially (in this case doubly) hydrogenated derivatives of benzene impossible. Reference to these reasons has already been made on several occasions (*e.g.* p. 169). The tendency to preserve the most saturated system of the aromatic ring, with its three double bonds, determines the formation of the stable arrangement containing the hydroxyl group and hence the peculiar character of phenols.

The phenols are *acidic* because the OH-group, as in the carboxylic acids, is attached to a doubly-bound atom.

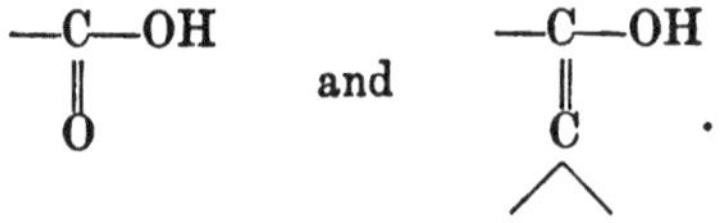

It is true that the acid character of phenol itself is not very pronounced, but the acidity increases as negative substituents such as NO_2 and halogen are introduced into the ring. The alkali salts of phenol are to a great extent hydrolysed in aqueous solution and they are completely decomposed by carbonic acid. Phenols and carboxylic acids can hence be separated by the action of carbon dioxide.

Experiment.—Carbon dioxide is passed into a moderately concentrated solution of β-naphthol in sodium hydroxide solution until the free naphthol is precipitated.

In other reactions also the OH-group of the phenols shows itself to be more reactive than that of the aliphatic alcohols. Phenols, but not alcohols, react easily with diazomethane. With other alkylating agents also, such as alkyl halides, and dialkyl sulphates, the phenols react even in aqueous alkaline solution whilst the alcohols do not react under such conditions. Benzoyl derivatives, most of which crystallise readily, are excellently adapted for the characterisation of phenols (Schotten-Baumann reaction).

Experiment.—In a test tube a small amount of crystallised phenol (0·5 g.) is dissolved in 5 c.c. of water, benzoyl chloride (0·5 c.c.) is added, and the mixture, after being made distinctly alkaline with concentrated sodium hydroxide solution, is gently warmed for a short time, with shaking, over a flame. If the reaction mixture is cooled under the tap with shaking and rubbing with a glass rod, the oil which separates solidifies to colourless crystals, which are filtered with suction, washed copiously with water, pressed on a porous plate, and recrystallised in a small test tube from a little alcohol. Melting point of the phenyl benzoate 68°–69°.

The experiment can be carried out in the same way with β-naphthol. Melting point of the benzoyl-compound 107°.

The necessary remarks on the importance of this much used reaction, which can also be applied to the amines, have already been made on pp. 113, 114.

The *naphthols* are in many respects still more reactive than phenol. This is shown most distinctly by the fact that the naphthyl ethers can be obtained by the method used for the preparation of esters of carboxylic acids, namely, directly by the action of hydrogen chloride on the phenol in the presence of the alcohol. The naphthols, moreover, react readily with zinc-ammonium chloride and with ammonium sulphite and ammonia to yield naphthylamines. The second of these two methods is a general one. It was investigated by H. Bucherer.

From all these facts it can be deduced that the phenols are much more closely related to the carboxylic acids than to the alcohols of the aliphatic series.

The influence of the OH-group on the reactivity of the benzene ring is of great importance. All substitution processes which are traced back to preliminary addition reactions take place *much more easily* with the phenols than with the hydrocarbons; as far as possible the groups which enter occupy the *o*- and *p*-positions. A number of reactions illustrating this statement is discussed below and in subsequent parts of this book. It may be mentioned here that by the action of bromine water, o-o-p-*tribromophenol* is at once precipitated from an aqueous solution of phenol. (Method for the quantitative determination of phenol.)

Experiment.—To an approximately 2 per cent phenol solution bromine water is added until the bromine ceases to be consumed. The colourless flocculent precipitate is separated at the pump and, after drying, is recrystallised from dilute alcohol. Melting point 95°.

According to K. H. Meyer, this surprising reactivity—which is also encountered amongst the enols—is explained by the fact that the OH-group "activates" the double bond adjacent to it and the two neighbouring double bonds of the conjugated system (Thiele's theory) are also involved in this activation. Phenol, therefore, can yield 1:2- and 1:4-substitution products with bromine, addition reactions first taking place:

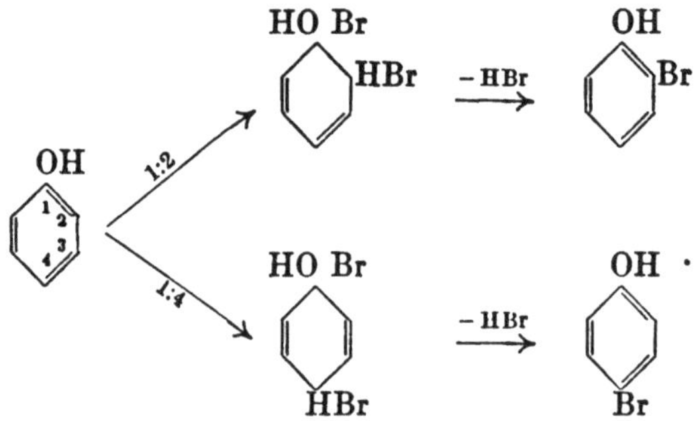

The substitution of the second and third bromine atoms proceeds in the same way until the three favoured positions (*o*-, *o*-, and *p*-) are occupied by bromine. If now a fourth molecule of bromine is allowed to act, in chloroform solution, on the tribromo-derivative, this molecule is attached at the 1:4-positions in fundamentally the same way as that assumed here :

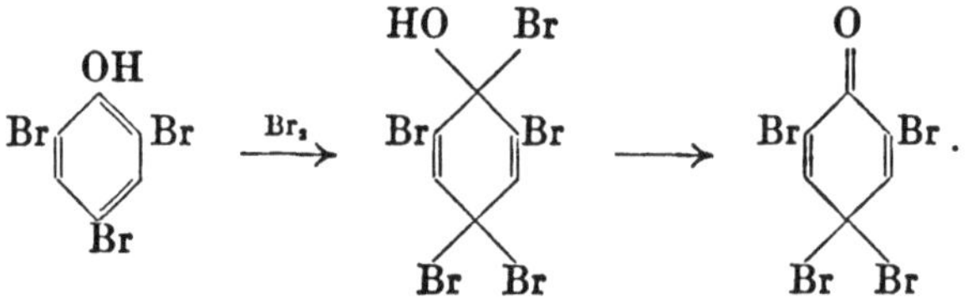

In this case, however, the elimination of HBr can no longer lead to the production of a true benzene derivative. Actually the so-called "*tribromophenol bromide*", the final product of the above reaction, is the ketobromide of a quinone and hence a derivative of dihydrobenzene.

The technical uses of phenol are important, particularly in the manufacture of *salicylic acid* (Chap. VI. 4, p. 240), and in that of the valuable synthetic resins of the "*bakelite*" type (condensation with formaldehyde). Under mild conditions phenol may be caused to combine with formaldehyde giving p-*hydroxybenzyl alcohol* :

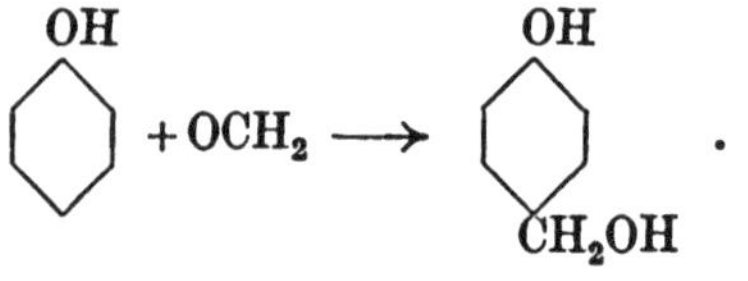

This alcohol, when warmed, loses water and polymerises.

The direct introduction of mercury into the benzene ring of the phenols must also be mentioned here. It is sufficient to heat with mercuric acetate (Balbiano, Pesci, Dimroth) in order to bring about this reaction.

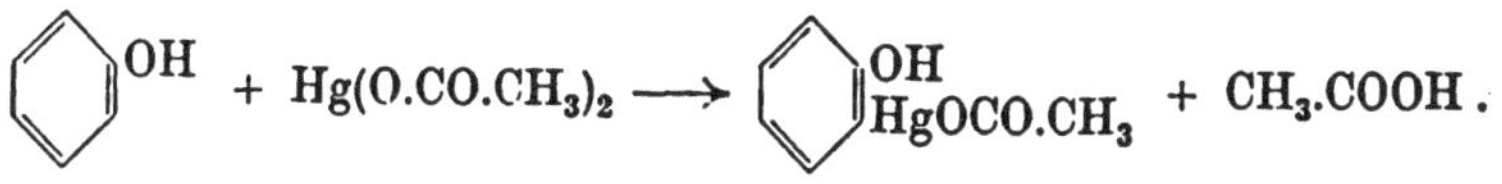

2. METHYLATION OF PHENOLS[1]

(*a*) **Anisole.**—Phenol (19 g.; 0·2 mole) is dissolved in 100 c.c. of 2 *N*-sodium hydroxide solution in a narrow-necked glass-stoppered bottle and about one-third of the dimethyl sulphate to be used (26 g. in all) is added in one lot. (Fume chamber.) The bottle is then vigorously shaken and methylation takes place with evolution of heat. After about five minutes, the second third of the sulphate is added and shaking is continued. Finally, after a short time the remainder of the methylating agent is added. During the periods of shaking the stopper of the bottle is removed from time to time. When the aqueous solution on which the resulting anisole floats is no longer acid, the contents of the bottle are poured into a small round-bottomed flask to which a reflux condenser is attached; the bottle is afterwards washed out with 20 c.c. of sodium hydroxide solution, which is also poured into the flask. To complete the reaction and to destroy any dimethyl sulphate which may still be present, the mixture is heated on the water bath for half an hour. Thereafter, when the contents of the flask have cooled, the aqueous layer is removed and the anisole, having been dried with calcium chloride, is finally distilled. Boiling point 155°. Yield 90 per cent of the theoretical. If difficulty is encountered in separating the anisole from the aqueous liquid, the methylated phenol is extracted with ether.

In an analogous way *phenetole* is prepared from phenol by the action of diethyl sulphate.

(*b*) **β-Naphthyl Methyl Ether.**—The procedure is the same as in the case of anisole, but the stoicheiometrical ratio is altered. Melting point 72°. The substance is crystalline. For a method by which it may be completely purified see Witt, *Ber.*, 1901, **34**, 3172.

The neutral esters of sulphuric acid and especially the dimethyl ester are very poisonous. All experiments in which they are used must therefore be carried out very cautiously in the fume chamber.

Methylations with dimethyl sulphate are always carried out in alkaline solution. They proceed very easily in the cases of carboxylic acids (method for the preparation of esters) and phenols, whilst in the

[1] Ullmann, *Annalen*, 1903, **327**, 114; 1905, **340**, 208.

case of aliphatic alcohols, *e.g.* the sugars, the alkylation takes place only with difficulty and preferably in alcoholic solution. It should be noticed that, in accordance with the equation

$$C_6H_5ONa + \begin{matrix}H_3C{-}O\\ H_3C{-}O\end{matrix}\!\!>SO_2 \longrightarrow C_6H_5O.CH_3 + \begin{matrix}NaO\\ H_3CO\end{matrix}\!\!>SO_2$$

only one of the methyl groups is transferred to the phenol. The salt of the methylsulphuric acid which is first formed gives up its methyl group for the same purpose only slowly and at the boiling point. Hence in practice this second methyl group is not utilised for preparative purposes. Esters of *arylsulphonic acids* (*e.g.* of toluenesulphonic acid) are also used for the alkylation of phenols.

How is dimethyl sulphate prepared?

An elegant methylating agent for phenols is furnished by *diazomethane*. A description of its use for this purpose is given later.

The ethers of the phenols are very stable substances in which the reactivity of the benzene ring, as compared with that of the phenols themselves, is appreciably reduced. The alkyl group is very firmly attached. It cannot be removed by the action of alkalis, and by that of mineral acids only at high temperature (in sealed tubes). The most useful agent for reconverting phenolic ethers into phenols is aluminium chloride, which acts in the manner shown in the following equation:

$$\begin{aligned} C_6H_5.O.CH_3 + AlCl_3 \longrightarrow\ & C_6H_5OAlCl_2 + Cl.CH_3 \\ & \downarrow 3H_2O \\ & C_6H_5.OH + Al(OH)_3 + 2\,HCl\,. \end{aligned}$$

All enolic ethers of the type $>C{=}\underset{\dot{O}R}{C}-$ can undergo a rearrangement, which, however, is only shown among phenolic ethers by *allyl* derivatives; when heated they yield allyl phenols (Claisen):

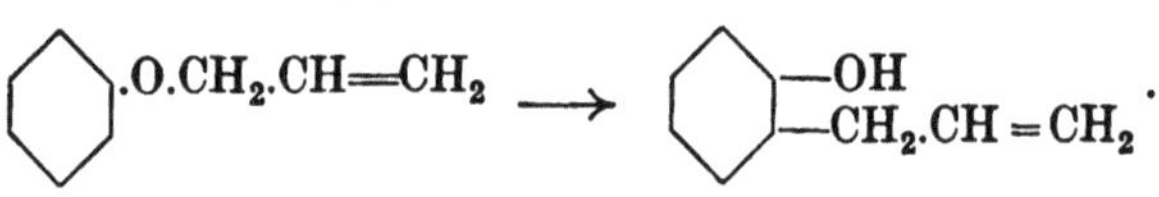

The recently discovered fission of phenolic ethers (and of aliphatic ethers) by metallic sodium is of special interest (Ziegler, Schorigin), *e.g.*

$$C_6H_5OCH_3 + 2\,Na \longrightarrow C_6H_5ONa + NaCH_3\,.$$

Of the substituted phenol ethers the amino-derivatives of anisole (*anisidine*) and phenetole (*phenetidine*) may be mentioned. They are prepared from the nitrophenols by alkylation and subsequent reduction of the nitro-group.

Alkaline reduction of *o*-nitro-anisole leads (as in the case of nitrobenzene) to the hydrazo-compound, which undergoes a benzidine transformation and is converted into the diphenyl base "dianisidine", an important intermediate for the production of blue azo-dyes (p. 178).

The well-known antipyretic "*phenaceline*" (I) and the sweetening agent "*dulcine*" (II) are derivatives of *p*-phenetidine:

Methylated phenols very often form part of the molecule of natural products, in particular, of alkaloids. In the elucidation of their constitution the quantitative determination of the methoxyl groups is of great importance. This determination is carried out by the excellent method of Zeisel, in which the methyl group is removed as methyl iodide by the action of concentrated hydriodic acid. This method (directions, p. 69) should be practised with the anisole already prepared.

3. ORTHO- AND PARA-NITROPHENOLS

Sodium nitrate (80 g.) or potassium nitrate (95 g.) is dissolved by heat in 200 g. of water in a round-bottomed flask, and before the solution has cooled completely, 100 g. of concentrated sulphuric acid are added with stirring. The liquid is cooled to 20° and then 50 g. of crystallised phenol, which has been liquefied by heating with 5 c.c. of water, is added drop by drop from a tap funnel, with frequent shaking; the temperature of the reaction mixture is kept between 20° and 25°. After the reaction mixture has stood for two hours, with frequent shaking, two volumes of water are added and the oil which separates is allowed to settle. As much as possible of the aqueous layer is poured off, and the oil, after being washed twice with water, is distilled with steam. *o*-Nitrophenol passes over. It is filtered at the pump from the aqueous distillate and dried on filter paper. If the material thus obtained is not pure, the steam distillation is repeated. Melting point 45°. Yield 30 g. The method of removing *o*-nitrophenol, which may solidify in the condenser, is given on pp. 26, 27.

From the residue after the steam distillation the non-volatile *p*-compound is isolated at the same time, at first in the form of its sodium salt. Sodium hydroxide solution (2*N*) is added until the reaction to Congo paper has just disappeared, 100 c.c. more of the hydroxide solution are then poured in, and the mixture is boiled with a little animal charcoal while steam is again passed through. The solution is now filtered through a folded paper and the filtrate is concentrated on a ring-burner until its volume is about 100 c.c. The sodium salt should crystallise when the concentrated solution is cooled. If a test with a small sample shows that the salt will not crystallise, 30 c.c. of sodium hydroxide solution (1:1) are added to the hot solution, which is then allowed to cool slowly. The salt is filtered off at the pump and washed with 2*N*-sodium hydroxide solution. From a hot solution of the salt, by precipitation with dilute hydrochloric acid, the *p*-nitrophenol is obtained as an oil which crystallises on cooling. If it is not sufficiently pure, *i.e.* if a sample cannot be recrystallised from hot, very dilute hydrochloric acid, it is purified by reconversion into sodium salt and reprecipitation. Melting point 114°. Yield 5–10 g.

The ease with which the phenols are nitrated has already been discussed. The process is not satisfactory, however, even when dilute nitric acid is used, because resinous by-products are formed as a result of oxidation and condensation. Nitration with nitrogen peroxide in non-aqueous solvents such as benzene and petrol ether gives better results (*Ber.*, 1921, **54**, 1776).

By further nitration with more concentrated acid *o*- and *p*-nitrophenols are converted into the same 2:4-dinitrophenol, and finally into *picric acid.* Polynitro-derivatives of benzene, such as picric acid and trinitrotoluene, can be caused to explode by detonation with mercury fulminate or lead azide. (The formulae of these two compounds should be written.) They are endothermic, *i.e.* the oxygen of the nitro-group can oxidise carbon and hydrogen within the molecule and heat is liberated. This intramolecular combustion is rather considerable in the case of picric acid, which is decomposed in accordance with the equation:

$$2C_6H_3O_7N_3 \longrightarrow 12CO + 2H_2O + 3N_2 + H_2.$$

***m*-Nitrophenol** cannot be prepared from phenol by direct nitration since the OH-group is a substituent of the first order, and therefore directs chiefly to the *o*- and *p*-positions. We are dependent on an indirect way—diazotisation of *m*-nitraniline and boiling the solution of the diazonium salt (p. 272).

In the pure state *m*- and *p*-nitrophenols are *colourless*, but the *o*-compound, on the other hand, is *yellow*. The salts of all three, however, are *intensely coloured*, the *o*- and *m*-compounds being orange-red and orange-yellow and the *p*-compound deep yellow. (Use of *p*-nitrophenol as an indicator.)

An attempt has been made to explain the strong colours of the salts of the nitrophenols as a rearrangement involving the formation of a quinonoid acid form (Hantzsch's *aci*-type) which absorbs light strongly.

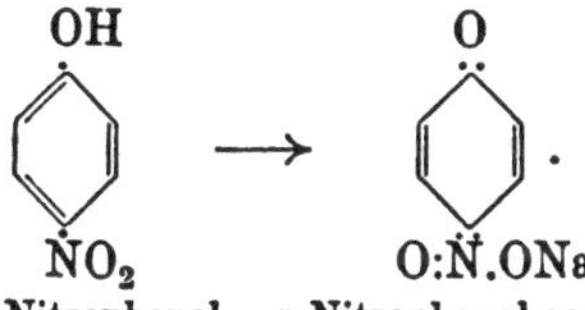

p-Nitrophenol *p*-Nitrophenol sodium

On various grounds, however, this explanation may be questioned. In particular it may be pointed out that *m*-nitrophenol behaves like the other two isomers and hence its alkali salts must also be quinonoid in form. But *m*-quinones are unknown throughout the whole range of the aromatic compounds. Moreover, there are numerous examples of substances which undergo a deepening in colour when they form salts but cannot change to a tautomeric quinone. Thus the disodium and dipotassium salts of yellowish-brown anthraquinol are deep blood-red in colour (p. 326).

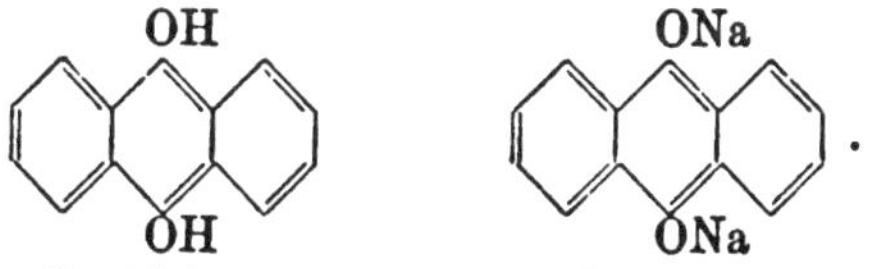

yellowish-brown blood-red

Finally, the alkali salts of phenol itself are more deeply coloured than is phenol. This fact cannot indeed be recognised subjectively, but investigation of the absorption of ultra-violet light demonstrates it. Thus it has been found that the absorption by sodium phenoxide much more nearly approaches the subjectively visible part of the spectrum than does that of the free phenol. The difference is so considerable that it provides also a satisfactory explanation of a subjectively perceptible deepening of colour from colourless to yellow. The colour of the salts of nitrophenols is therefore ascribed to the " bathychromic " (colour-deepening) effect of salt-formation.

Since nitro-groups in the *o*- and *p*-positions increase the mobility of halogens directly attached to the aromatic ring (p. 93), nitrophenols can also be obtained from chloronitrobenzenes. Thus *p*-chloronitrobenzene can be decomposed by alkalis in the autoclave, and 2:4-dinitrophenol, an important intermediate product for the manufacture of sulphur dyes, is produced from the corresponding chlorobenzene by an even milder treatment.

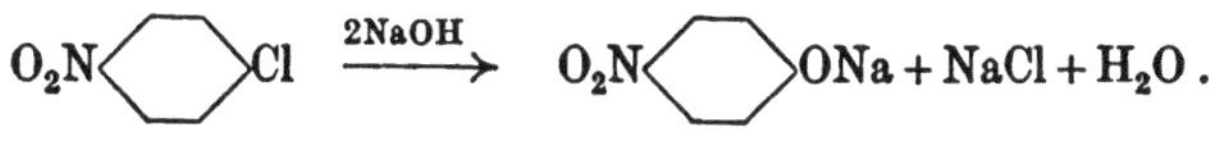

(If the sodium hydroxide is replaced by ammonia, *p*-nitraniline is formed.)

In trinitrochlorobenzene (picryl chloride) the chlorine is as mobile as in an acid chloride.

4. KOLBE'S SALICYLIC ACID SYNTHESIS[1]

In a porcelain (or preferably nickel) dish 13·5 g. of pure sodium hydroxide are dissolved in 20 c.c. of water and 31 g. (0·33 mole) of crystallised phenol are gradually added with stirring. The water is then evaporated by heating the constantly stirred solution, at first on a wire gauze and finally on a naked luminous flame, which is moved constantly to and fro under the dish. As soon as the lumps of material cease to agglomerate, it is powdered rapidly in a dry mortar and the fine powder is again heated in the nickel dish with efficient stirring until a perfectly dry dust is obtained.[2] This dust is poured into a tubulated retort (capacity 200 c.c.) which is dipped as deeply as possible into an oil bath. The bath is now heated to 110° and maintained at this temperature for one hour while dry carbon dioxide is passed over the sodium phenoxide (the end of the tube which delivers the gas being 1 cm. from the surface of the phenoxide). The temperature is now gradually raised, during the course of four hours, at the rate of 20° per hour to 190°, and finally, for one or two hours longer, it is maintained at 200°. During the whole of this time a moderately strong current of carbon dioxide is passed through the retort, the contents of which are stirred several times with a glass rod. The reaction product is cooled, transferred to a large beaker, and dissolved in water (the retort is also washed out several times with water). The salicylic acid is precipitated from the solution by addition of a large volume of concentrated hydrochloric acid. After the precipitate has become crystalline by cooling in ice it is collected at the pump, washed with a little water, and dried on porous plate. If a sample of the moist material can be directly recrystallised from hot water (with addition of decolorising charcoal) the whole may be purified in this way. But even in that case distillation of the crude product in superheated steam is recommended in order to become acquainted with this method. The *dry* material is heated in a *short-necked* flask in an oil bath at 170° and a moderately strong current of steam at 170°–180° is passed through (cf. p. 27). The flask should not be connected to the superheater until the oil bath and the steam have reached the temperature mentioned. Both the connecting

[1] *J. pr. Chem.*, 1874, [ii.] 10, 89; 1883, 27, 39; 1885, 31, 397.

[2] Completely dry phenoxide is essential for the success of the experiment. It is convenient to choose a time for this such that the dried salt remains over night in the dish, in a vacuum desiccator over sulphuric acid and solid potassium hydroxide, before the synthesis is started next morning.

tube and the condenser tube must be exceptionally wide. When the acid removed from the condenser tube is dissolved in the aqueous distillate from the receiver and the solution is cooled, the acid crystallises in long, completely colourless needles. Melting point 156°. Yield 10–12 g.

The first stage of Kolbe's synthesis is analogous to the well-known aliphatic synthesis of alkyl carbonates from alkoxides and carbon dioxide:

$$H_5C_2.ONa + CO_2 \longrightarrow H_5C_2{-}O{-}C\begin{matrix}\nearrow O \\ \searrow ONa\end{matrix} ;$$

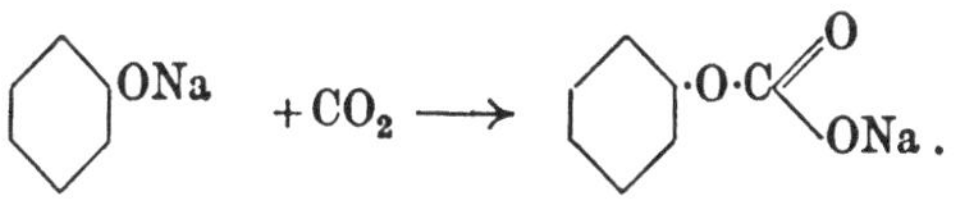

The sodium phenylcarbonate so produced then undergoes rearrangement on heating; the carboxy-sodium group wanders to the nucleus:

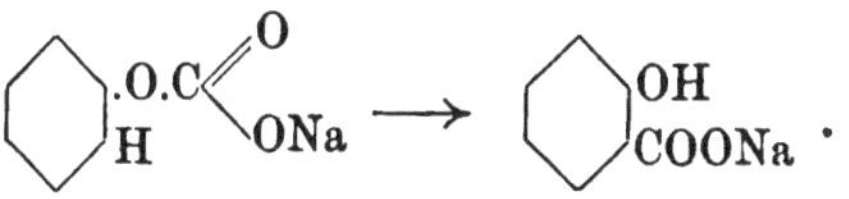

A relatively small amount of the *p*-compound is formed at the same time. It is remarkable that when *potassium* phenoxide is used the *p*-compound is the main product.

Since, in Kolbe's synthesis, as here described, the mono-sodium salicylate reacts to some extent with unchanged sodium phenoxide, producing the di-sodium salt, part of the phenol is liberated and excluded from the reaction. The reaction proceeds to completion if the sodium phenoxide is heated to about 150° for a long time, with carbon dioxide under pressure in the autoclave. This is the technical method of Schmitt.

In the case of polyhydric phenols the synthesis of the carboxylic acid can be carried out even in aqueous solution.

Ortho- and *para*-hydroxycarboxylic acids lose CO_2 when heated to high temperatures, and the ease with which this occurs increases with the number of OH-groups. (Preparation of pyrogallol from gallic acid.)

How is *m*-hydroxybenzoic acid prepared? Discuss the reduction of salicylic acid to *pimelic* acid (Einhorn).

Experiment.—A few drops of ferric chloride solution are added to an aqueous solution of salicylic acid. The colour reaction characteristic of phenols is obtained.

Salicylic acid is manufactured on a large scale. In the dye industry it serves for the production of valuable azo-dyes which exhibit great fastness. To some extent these dyes are applied to mordanted fibres. In addition, the acid and its derivatives are widely used in pharmacy.

Being a phenolcarboxylic acid it has a powerful disinfecting action (preservative). It has further proved itself an important anti-rheumatic and an analgetic. The derivative in which the phenolic hydroxyl group is acetylated (aspirin) has become especially popular. The first medicament of the series was the phenyl ester of salicylic acid, salol, which is produced as a by-product in the technical process. The preparation of salicylaldehyde has been described above (p. 226).

The alcohol *saligenin* $C_6H_4{<}^{CH_2OH\,(1)}_{OH\,(2)}$ is a constituent of the glucoside salicin obtained from the willow (*Salix*).

5. SYNTHESIS OF THE ESTER OF A β-KETONIC ACID

Acetoacetic Ester [1]

The quality of the ethyl acetate used is of great importance for the success of this experiment since the completely alcohol-free compound reacts only very slowly with sodium even on warming, whilst, on the other hand, material containing much alcohol reacts readily with sodium but gives varying and sometimes poor yields of ethyl acetoacetate.

The ethyl acetate already prepared (p. 130) still contains too much alcohol, and this is removed by allowing the liquid to stand for two hours over granulated calcium chloride. Shortly before it is used the decanted or filtered liquid is once more distilled, over a little calcium chloride, from the water bath. The condenser used is thoroughly dried before use and is attached to the receiver (a filter flask) by means of a cork. Clean sodium (26 g.; incrustations removed) is forced through a sodium press into absolute ether and the wire so obtained is transferred, without further treatment, to a flask (capacity 1 l.) containing 250 g. of the purified ethyl acetate. A reflux condenser, fixed in an oblique position, is at once attached to the flask. If the ethyl acetate has been correctly treated, vigorous boiling should not set in at once. Rather, gradual evolution of hydrogen should first begin, and then gentle boiling. After ten minutes the flask is placed on a gently heated oil bath, the temperature of which is so regulated that the ethyl acetate boils with moderate vigour only, until all the sodium has dissolved; about three hours are required. The yield is reduced if boiling is continued too long. The process is brought to an end as soon as possible and any *small* residue of sodium is neglected. The warm liquid is shaken

[1] The theoretical discussions concerning the sections 5 to 8 are collected on p. 248 *et seq.*

with a mixture of 70 c.c. of glacial acetic acid and 80 c.c. of water until it is just *acid.* An equal volume of cold saturated sodium chloride solution is then added and the lower aqueous layer is separated in a tap funnel from the upper layer, which consists of ethyl acetate and ethyl acetoacetate. The mixture of esters is shaken with a little cold saturated bicarbonate solution, which is then poured away, and the excess ethyl acetate is now removed by distillation from a flask connected to a downward condenser and fitted with a thermometer. A smoky flame is used without wire gauze. As soon as the temperature reaches 95° heating is stopped and the material which remains in the flask is distilled in a vacuum through a small water-cooled condenser (preferably) into the type of receiver shown in Fig. 16 or 17. If the pressure is below 16 mm. the flask is heated on the water bath, otherwise an oil or paraffin bath is used. After small amounts of ethyl acetate and water have passed over, the temperature soon becomes constant and the bulk of the ethyl acetoacetate distils within one degree. Boiling points at various pressures: 71°/12·5 mm., 74°/14 mm., 79°/18 mm., 88°/29 mm., 94°/45 mm., 97°/59 mm., 100°/80 mm. The yield amounts to 55 to 60 g. of ethyl acetoacetate. In the flask a substance which crystallises on cooling remains. It is *dehydracetic acid.* What is its formula? The various stages of the experiment should be carried out without any long interruption, because otherwise the yield suffers.

6. ACETYLACETONE [1]

Finely powdered sodamide,[2] which is kept in a closed container, is added gradually in small portions to a mixture of 120 c.c. of *ethyl acetate* (purified as for the preparation of ethyl acetoacetate) and 32 c.c. of dry *acetone.* During the addition of the sodamide the mixture is cooled in a freezing mixture. The flask is provided with a cork or rubber stopper carrying a calcium chloride tube. A vigorous evolution of ammonia at once begins. After all the sodamide has been added, the flask is allowed to stand, first in ice-water for two hours with frequent shaking, and then for twelve hours at room temperature. Ice and cold water (100 g. of each) are then successively added and, after the ethyl acetate which remains has been

[1] L. Claisen, *Ber.*, 1905, 38, 695.

[2] The powdering should be carried out as rapidly as possible, preferably in a metal mortar (*goggles to be worn*). The yield is determined by the quality of the sodamide.

separated, the aqueous layer is made just acid by addition of dilute acetic acid. From the solution thus obtained the acetylacetone is precipitated as copper salt by means of saturated aqueous copper acetate solution. Finely powdered copper acetate is dissolved in the necessary amount of boiling water. If basic copper acetate is present, a small amount of acetic acid is added. The solution is used while still lukewarm, before the salt crystallises again.

After some hours the blue-green compound of copper and acetylacetone is separated by filtration with suction, washed twice with water, transferred directly from the filter funnel to a separating funnel, and, after being covered with ether, decomposed by continuous shaking with 50 c.c. of 4 *N*-sulphuric acid. The ethereal solution is separated and the acid layer is extracted with ether; the extract is then combined with the ethereal solution, which is now dried over calcium chloride. After the ether has been removed by distillation the diketone is likewise distilled. The bulk of the material passes over at 125°–140° and, on repeating the distillation, at 135°–140°. The boiling point of the completely pure substance is 139°. Yield 15–20 g.

If the distillation is carried out under reduced pressure (about 50 mm. reduction), a purer and also more stable product is obtained.

Experiment.—A drop of ferric chloride solution is added to a few drops of an aqueous solution of acetylacetone. The reaction characteristic of enols takes place. If now the solution is cooled in ice and dilute bromine water is rather quickly added, the red colour of the iron enolate disappears for a short time and then returns rapidly.

Benzoylacetone. — $C_6H_5.CO.CH_2.CO.CH_3$ is prepared in an analogous way from acetophenone and ethyl acetate according to the procedure of Claisen, *Ber.*, 1905, **38**, 695. The yield may be as much as 75 per cent of the theoretical. The cheaper converse method—action of sodamide on ethyl benzoate and acetone—also succeeds in this case, although it fails when sodium or sodium ethoxide is used as condensing agent. In general the use of sodamide is to be preferred in the synthesis of 1:3-diketones.

7. DIETHYL MALONATE

Monochloroacetic acid (95 g., 1 mole) is dissolved in water in a large porcelain basin and the solution, gently warmed (to 50°), is neutralised with dry solid potassium carbonate, of which 75 g. are required. Pure, finely powdered sodium cyanide (55 g., or 70 g. of potassium cyanide) is then added, and the temperature is *very gradually* increased by heating the dish on a sand bath or an asbestos plate while the mixture is well stirred. (The whole operation must be carried out in the fume chamber.) After the cyanoacetic acid has been formed with vigorous boiling, the reaction mixture is stirred with a thermometer while it is being concentrated on a ring-burner until the temperature of the brownish viscous mass of salt is 135°. Stirring with a spatula is continued while the material is allowed to cool, because otherwise a hard mass is formed which can hardly be powdered. The solid is now rapidly and thoroughly broken up in a large mortar, transferred to a flask (capacity about 1 l.) fitted with a reflux condenser; with thorough shaking there are added first 50 c.c. of absolute alcohol and then gradually a cooled mixture of 200 c.c. of absolute alcohol and 150 c.c. of concentrated sulphuric acid. The pasty mass produced is heated for two hours with frequent shaking on the water bath (in the fume chamber), well cooled, and again shaken with 400 c.c. of water. Salt which does not dissolve is separated at the pump and is washed several times with ether, which is then used to extract the aqueous filtrate. This is then thoroughly extracted with two further lots of ether and the combined ether extracts are shaken with concentrated sodium carbonate solution until they are no longer acid. (*In view of the initially vigorous evolution of gas which takes place, the separating funnel used is at first not closed.*) The ethereal solution is dried over anhydrous sodium sulphate, the ether is removed, and the residue distilled. Boiling point 195°. Yield 90–100 g.

Diethyl Ethylmalonate.—Sodium (4·6 g.) is dissolved, in a small flask fitted with an efficient reflux condenser, in 75 c.c. of absolute alcohol and to the cooled solution 33 g. of diethyl malonate are added (precipitation of diethyl sodio-malonate). To the mixture thus obtained, ethyl bromide (25 g.) or ethyl iodide (35 g.) is added in small portions with shaking and the flask is then heated on the water bath until, after one to two hours, the contents are no longer alkaline. The alcohol is removed by distillation in a vacuum from the water bath at 40°–50° and the ester is extracted from the residue by shaking (twice or thrice) with ether. After the ether has been removed by distillation the crude ester is distilled. Boiling point 206°–208°. Yield about 30 g.

Ethylmalonic Acid.—Diethyl ethylmalonate (19 g.) is poured in small portions, with shaking, into a small round-bottomed flask containing a cooled solution of potassium hydroxide (15 g.) in water. The flask is fitted with a reflux condenser. The emulsion which is first formed soon sets to a solid mass of ethyl potassium ethylmalonate. When this is now heated moderately on the gently boiling water bath the hydrolysis sets in with vigorous evolution of heat. Heating is continued until the layer of oil has disappeared. The flask is then allowed to cool and the solid paste obtained is twice shaken with ether in order to remove any residual ester which may have escaped hydrolysis. (Close the flask with a rubber stopper.) The ether is simply decanted. The paste is next cooled in ice and made acid to Congo paper by addition of hydrochloric acid (26 c.c. of concentrated acid (d. 1·18) + 25 c.c. of water). The acid solution is shaken five times in a separating funnel with 25 c.c. portions of ether and the ethereal solution is dried over sodium sulphate. After the ether has been removed by distillation the residue is caused to crystallise by cooling and rubbing. Ethylmalonic acid can be recrystallised from benzene (a sample should be so treated), but for the purpose of conversion into butyric acid this is not necessary. Melting point 111°. Yield 12 g.

Butyric Acid from Ethylmalonic Acid.—A small distilling flask having a long side tube is fixed in an oblique position with the tube directed upwards. The ethylmalonic acid is placed in the flask, which is then corked and heated in an oil bath at 180° until, after half an hour, evolution of carbon dioxide has ceased. The residue is then distilled from the same flask in the usual way; the butyric acid passes over between 162° and 163°. Yield 80–90 per cent of the theoretical.

In an analogous way, by the action of the equivalent amount (exactly 1 mole) of *benzyl chloride* on diethyl sodiomalonate, *diethyl benzylmalonate*, and finally *hydrocinnamic acid*, are formed.

8. PHENYLNITROMETHANE[1]

(a) *aci*-Phenylnitroacetonitrile Sodium

$$\begin{array}{l} C_6H_5.C.CN \\ \quad\quad\;\; \| \\ \quad\quad\;\; NOONa \end{array}$$

In a round-bottomed flask (capacity 0·5 l.) 8 g. of sodium are dissolved in 120 c.c. of absolute alcohol. Into the solution, which

[1] W. Wislicenus and A. Endres, *Ber.*, 1902, 35, 1757.

is kept cool in water (precipitation of ethoxide is not detrimental) a mixture of 36 g. of benzyl cyanide (p. 126) and 32 g. of ethyl nitrate (p. 137) is poured in small portions. The salt formulated above separates gradually in the shape of almost colourless crystals. In order to allow the reaction to proceed to completion, the flask is left for one hour without cooling, but with exclusion of moisture; the precipitate is then collected at the pump and subsequently washed first with alcohol-ether (1:1) and then with ether alone. Yield 40–45 g.

With ferric chloride solution a sample of the salt, dissolved in alcohol, gives an intense olive-green colour.

(b) Hydrolysis to the Sodium Salt of *aci*-Phenylnitromethane

The sodium salt just prepared (about 40 g.) is heated to gentle boiling in an open round-bottomed flask on a conical air bath with 600 c.c. of 2*N*-sodium hydroxide solution. The concentration of the solution gradually increases and large amounts of ammonia are evolved. The hydrolysis is complete when this evolution ceases. The sodium salt of the *aci*-phenylnitromethane, which is sparingly soluble in the excess of alkali, frequently begins to separate in crystalline form even while the solution is hot. If this occurs before the reaction has ceased, hot water is added and boiling is continued until the evolution of ammonia is complete. Then the solution is allowed to cool and, with continuous shaking and *efficient ice-cooling*, is acidified with about 220 c.c. of rather concentrated hydrochloric acid (110 c.c. each of concentrated acid and water). Sufficient acid must be added to produce a distinct colour change on Congo paper and to cause complete precipitation of the *aci*-nitro-compound which separates in a flocculent condition. Large volumes of carbon dioxide are evolved. In order to allow time for the sensitive *aci*-compound to change into the stable phenylnitromethane the reaction-mixture is now left over night. Next day it is exhaustively extracted with ether and the extract, after washing with sodium carbonate solution, is evaporated without previous drying. After the ether has been removed the extracted material is distilled with steam and the distillate is again extracted with ether. The ethereal solution is dried over calcium chloride and the residue, after evaporation of the ether on the water bath, is distilled in a vacuum. The phenylnitromethane passes over as a light yellow oil at 118°–119° (16 mm.). Yield 14–18 g., *i.e.* on the average, 50 per cent of the theoretical.

On Keto-Enol Tautomerism

In the free state and in their reactions the simple aldehydes and ketones are in general known only in the aldo- and keto-forms. Erlenmeyer suggested the rule that the isomeric enol-structure, which might be formed at first in the production of acetaldehyde from glycol, for example, should in no case be capable of existence.

$$\begin{array}{l} CH_2\text{—}CH_2 \\ |\qquad\ \ | \\ OH\quad OH \end{array} \xrightarrow{-H_2O} \begin{array}{l} CH_2\text{=}CH \\ \qquad\ \ | \\ \qquad OH \end{array} .$$

Chiefly as a result of the work of Claisen, this rule has been proved to be erroneous. We know now that even simple aldehydes and ketones exhibit a detectable tendency to "enolise", by the wandering of an H-atom and the simultaneous shift of a double bond.

Thus it has been possible to show that in the *bromination* of acetone, a process which has been found to be unimolecular, not the normal keto-form, but the tautomeric enol-form reacts. The enol-form is present, in equilibrium with the keto-form, in amount too small to be measured. As soon as this amount has reacted a further quantity is formed and the process is repeated. That the reaction is unimolecular follows from the fact that it is the rate of rearrangement (I) which is measured, whilst the reaction of the enol with bromine (II) takes place with immeasurable rapidity (Lapworth).

We have, consequenity, the equations:

$$\underbrace{CH_3.CO.CH_3 \longrightarrow CH_3.\overset{}{\underset{OH}{C}}}_{I}\!\!=\!\!\underbrace{CH_2 \xrightarrow{Br_2} CH_3.\underset{OH}{CBr}}_{II}\text{—}CH_2Br$$

$$\xrightarrow{-HBr} CH_3.CO.CH_2Br .$$

It may be mentioned here that, if water is excluded, acetone and also acetaldehyde are converted by metallic sodium (with evolution of hydrogen), or better by sodamide, into very reactive "enolates", *e.g.*

$$CH_2\text{=}\underset{ONa}{C}\text{—}CH_3 .$$

Now the mobility of a hydrogen atom united to a C-atom which is adjacent to the C=O-group—for the change to the enol-form such mobility is necessary—is increased if still other *activating* groups, which are, in general, *unsaturated*, are united to the same C-atom. This is the state of affairs in *ethyl acetoacetate* in which the group —COOR exerts such activating influence.

The Mechanism of the Ethyl Acetoacetate Synthesis.—Before the tautomerism of ethyl acetoacetate is discussed we must consider the mechanism of its formation, which for decades has been the subject of lively discussion and was conclusively explained only in recent years (Scheibler). It has been found that even the C=O-group of the simple carboxylic esters, although in other respects inferior in activity to the true carbonyl group, can be enolised by alkali metals. Thus ethyl acetate is converted by potassium into the potassium salt of the tautomeric enol with evolution of hydrogen :

$$CH_3.\underset{\underset{O}{\|}}{C}{=}OR \xrightarrow{K} H_2C{=}\underset{OK}{C}{-}OR + H.$$

Now such enolates have an exceptionally reactive double bond which makes them capable of taking part in all kinds of addition reactions. The ethyl acetoacetate synthesis is the most important process of this type. The enolate is able to combine with a second molecule of the ester ; the latter is added as the two fragments $R_1{-}C{=}O$ and OR

$$H_2C{:}\underset{ONa}{C}.OR + CH_3.\underset{OR}{C}{:}O \longrightarrow CH_3.CO.CH_2.C\begin{cases}OR\\OR\\ONa\end{cases}.$$

The final product of the reaction, the *sodium derivative of ethyl acetoacetate,* is then formed by elimination of sodium ethoxide and salt formation with the ethyl acetoacetate, rearranged in the enol-form :

$$CH_3.CO.CH_2.C\begin{cases}OR\\OR\\ONa\end{cases} \longrightarrow CH_3.CO.CH_2.C\begin{cases}OR\\O\end{cases} + RONa$$

$$\longrightarrow CH_3.\underset{ONa}{C}{=}CH{-}C\begin{cases}OR\\O\end{cases} + ROH.$$

The condensation of esters with ketones proceeds in an exactly similar way ; the enol-form of the ketone (in the form of a salt) is the basis of the addition, *e.g.*

$$CH_3.\underset{\underset{ONa}{|}}{C}{:}CH_2 + C_6H_5.COOR \longrightarrow CH_3.\underset{ONa\;\;OR}{C}.CH_2.CO.C_6H_5$$

$$\longrightarrow CH_3{-}\underset{ONa}{C}{=}CH.CO.C_6H_5 + ROH.$$

Sodium benzoylacetone

In general, esters of β-keto-acids are formed from two molecules of ester, and β-diketones from ester and ketone. The use of formic esters leads, in both cases, to *hydroxymethylene* compounds :

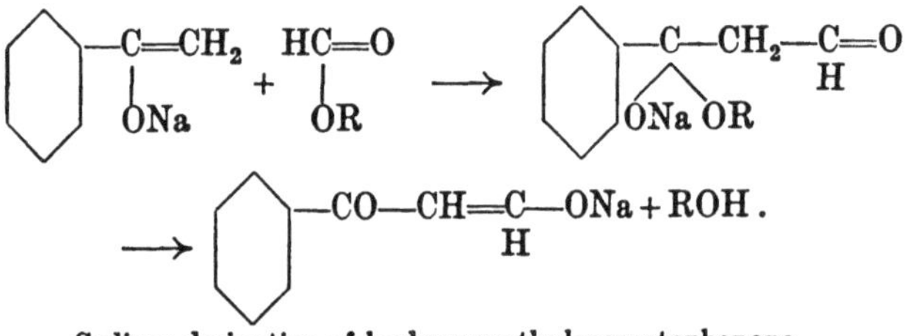

Sodium derivative of hydroxymethyleneacetophenone

The tendency to enolise is exceptionally pronounced in the formyl group if it is in the β-position.

In the case of the dicarboxylic esters of adipic and pimelic acids, *intramolecular condensation* gives cyclic esters of β-keto carboxylic acids (Dieckmann):

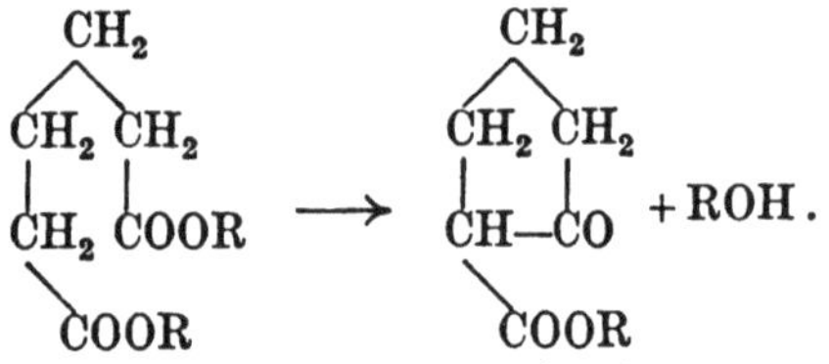

Ester of adipic acid　　Ester of cyclopentanone carboxylic acid

Esters of succinic acid condense to esters of succinylsuccinic acid (1:4-diketohexamethylene-2:5-dicarboxylic esters). Study Baeyer's work on the hydrogenated derivatives of benzene, based on this synthesis.

It is not only the esters of organic acids which combine, in the manner of the "ethyl acetoacetate synthesis", with the enolates of ketones and of esters; an analogous behaviour is shown by the esters of *nitrous* and *nitric acids*. The process which leads to the formation of *isonitroso-* and *acinitro-compounds* yields products fundamentally similar to those already described: just as with ethyl acetate the group $CO.CH_3$ enters, so here, the NO- and NO_2-groups are involved, and "enolise" exactly as does $>C{=}O$:

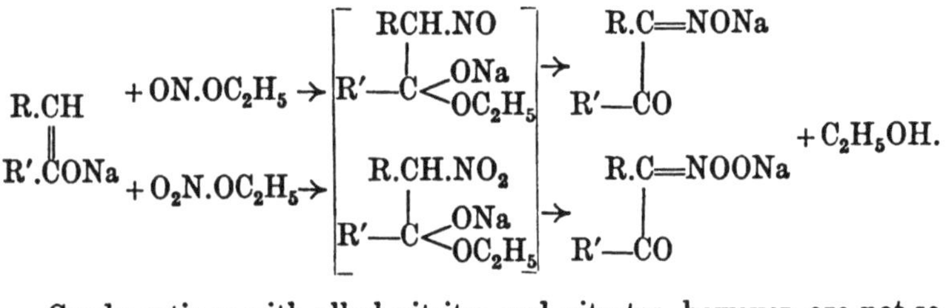

Condensations with alkyl nitrites and nitrates, however, are not so generally applicable as the true ethyl acetoacetate reaction, and the possibility is not excluded that they proceed in another way; compounds with " mobile " hydrogen might first be added to the inorganic part of the ester by means of an " aldol " condensation. The fact that *fluorene*, which contains no " active " double bond at all, combines with ethyl nitrate (as well as with ethyl oxalate) and sodium ethoxide in the same way, yielding *aci*-nitrofluorene, seems to support this second theory (W. Wislicenus).

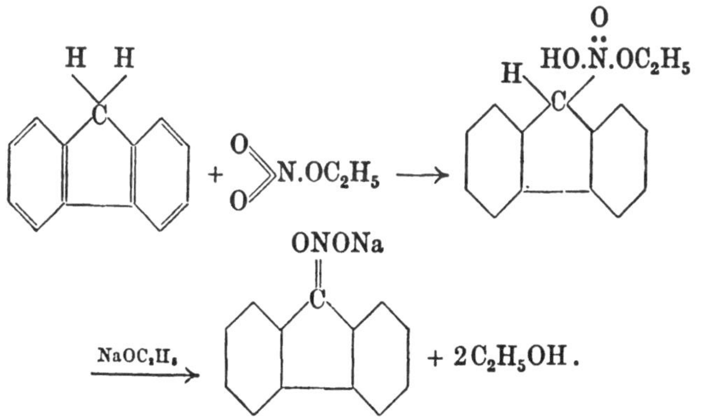

The synthesis of the sodium salt of *aci*-phenylnitroacetonitrile, described above, is an example of the preparative use of this reaction. The CH_2-group of the benzyl cyanide becomes reactive as a result of the proximity of C_6H_5 and CN. Here also it may be assumed that an *aci*-form $C_6H_5.CH{=}C{=}NH$ is the cause of this behaviour.

Constitution of the Esters of the β-Ketocarboxylic Acids and of the β-Diketones.—Ethyl acetoacetate is taken as example. It reacts like a ketone with phenylhydrazine, bisulphite, and other ketone reagents; on the other hand it shows an acid reaction, it dissolves in alkalis, and gives the colour reaction with ferric chloride characteristic of enols and also of phenols. From this double behaviour it was formerly concluded that it was either purely ketonic or purely enolic and that the reactions in the other form were to be attributed to a rearrangement caused by the reagents used. The true state of affairs was first disclosed by the quantitative investigation of the structural conditions (K. H. Meyer, L. Knorr, 1911). Ethyl acetoacetate, in the cold, takes up a limited amount of bromine, a reaction which, as was mentioned above in connection with acetone, involves the enol form only. Under suitable conditions, therefore, the amount of enol present in ethyl acetoacetate can be quantitatively determined with a standard solution of bromine. A solution of ethyl acetoacetate so titrated consumes, after a short time, a further quantity of bromine, *i.e.* fresh enol has been produced in the solution. It follows from this that in a solution of ethyl acetoacetate *keto-* and *enol-forms* are present *in equilibrium*. Under the experimental conditions chosen for the titration with bromine this state of equilibrium is attained so slowly that the accuracy of the method is not noticeably affected.

Experiment.—About 0·5 c.c. of ethyl acetoacetate is dissolved with shaking in the necessary amount of water, a few drops of ferric chloride solution are added, and to the cold solution dilute (1:10) bromine water is added, drop by drop, but rather quickly from a tap funnel, until the red colour of the ferric enolate has disappeared. The enol has now been completely used up by the bromine, but since, in order to restore the equilibrium, more enol is formed, the colour reappears after a short time and can at once be destroyed again by the addition of a few drops of bromine. The procedure can be repeated until the whole of the ethyl acetoacetate is converted into ethyl bromoacetoacetate. By means of this experiment the keto-enol rearrangement can be subjectively perceived.

The ratio in which the keto- and enol-forms are present at equilibrium is greatly dependent on the nature of the solvent. The following table gives figures for ethyl acetoacetate :

Solvent.	*Percentage of Enol.*
Water	0·4
Ethyl Alcohol	12·0
Glacial Acetic Acid	5·7
Benzene	16·2
Petrol Ether	46·4

The following simple general formula expresses the important connection which exists between the proportions of the tautomeric compounds at equilibrium and their solubilities in the solvent concerned (van't Hoff, Dimroth) :

$$\frac{C_a}{C_b} = \frac{L_a}{L_b} G.$$

C_a and C_b are the concentrations and L_a and L_b the solubilities of the isomers a and b, and G is a constant independent of the solvent. In the case of ethyl acetoacetate we thus deduce, by reference to the table, that in water the keto-ester should be the more soluble, and in petrol ether the enol-ester ; this is actually the case.

Liquid ethyl acetoacetate consists of 92·5 per cent of the keto-form and 7·5 per cent of the enol-form. The freshly distilled substance contains considerably more enol since the enol-ester, because of its lower boiling point, distils first and then more is formed in the distillation flask.

Experiment.—Ethyl acetoacetate (2·5 g.) is dissolved in 20 c.c. of N-alkali hydroxide solution, the solution is cooled in ice to 0° and

20 c.c. of cooled *N*-hydrochloric acid are added in one lot, with shaking. A turbid milky solution is formed which, however, becomes clear in a few seconds. The enol, which is less soluble in water than the keto-form, at first separates, but changes very rapidly and almost completely into the more soluble keto-form, as the conditions of the equilibrium in water require.

K. H. Meyer's "bromine method"[1] makes it possible to determine the enol content in almost all solutions of tautomeric substances.

It has also been possible, in various ways which cannot be detailed here, to prepare both the keto- and enol-forms of ethyl acetoacetate in the pure state (Knorr, K. H. Meyer). Their physical constants are altogether different. The refractive index, for example, is 1·4225 (D_{10°) for the keto-form and 1·4480 for the enol-form. From determinations of the refractive indices of equilibrium mixtures the content of both forms can be calculated by interpolation (Knorr, 1911), and these results have been confirmed spectroscopically (Hantzsch, 1910).

Whether or no both forms of a tautomeric substance are capable of isolation in the free state depends chiefly on the velocity of rearrangement of the more labile form. The isolation of keto- and enol-forms in a permanent crystalline state was first carried out with unsymmetrical dibenzoylacetone (Claisen, 1896):

$$(C_6H_5.CO)_2{:}CH.CO.CH_3 \quad \text{and} \quad (C_6H_5CO)_2{:}C{=}\underset{\displaystyle OH}{C}{-}CH_3 .$$

For such cases in which the two forms have been brought to light solely as a result of improved experimental methods, the term "tautomerism" has been modified to "desmotropism". In desmotropic substances the tautomeric relations are therefore particularly clear and well defined; numerous examples have become known, which now also include aceto-acetic ester. The case of *acetylacetone* is quite similar, but here the formation of the enol-form is much more favoured. The liquid substance contains up to 80 per cent of enol.

In *benzoylacetylacetone* the tendency to enolise is so pronounced that this substance exists as enol only. The keto-form is unknown.

$$C_6H_5.CO.\underset{\displaystyle H_3C{-}CO}{C}{=}\underset{\displaystyle OH}{C}{-}CH_3 .$$

Nor can there be any question of real tautomerism in the case of *phenol*. In its chemical properties phenol resembles the aliphatic enols in all respects. We need only recall the agreement in the acid character, the production of colour with ferric chloride, and the reactions

[1] *Annalen*, 1911, 380, 212.

with halogens, nitrous acid, and aromatic diazo-compounds (coupling), caused by the "activity" of the double bond and proceeding in the same way in phenols and aliphatic enols. The "enol nature" of phenol provides valuable support for the conception of the constitution of benzene as expressed in the Kekulé-Thiele formula, since it is an expression of the tendency of the ring to maintain the "aromatic" state of lowest energy. In this connection the hypothetical keto-form of phenol (*A*)—not yet obtained—would be of interest in comparison with the aliphatic ketone (*B*).

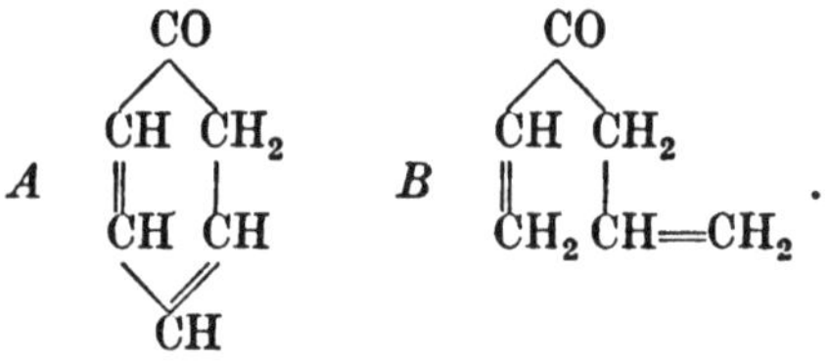

The tautomerism of the *aliphatic nitro-compounds* is very closely allied to that of the ketones and aldehydes. Here also there are two forms—the neutral and the acid, or so-called *aci*-form (Hantzsch):

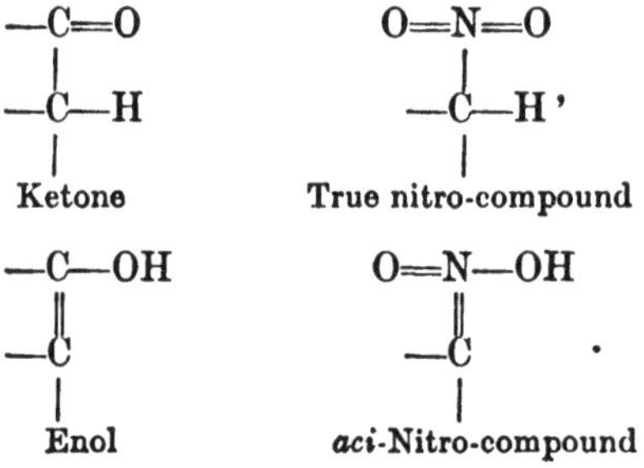

In respect of their properties, conditions of rearrangement, and reactions, we simply refer to what was said about the keto-enol change. Here, also, the bromine method enables the points of equilibrium to be determined quantitatively. The oldest and most important example of desmotropy in nitro-compounds was found in *phenylnitromethane*, which exists as a stable neutral nitro-compound (an oil) and as a labile crystalline *aci*-nitro-compound (Hantzsch).

$$C_6H_5.CH_2.NO_2 \text{ and } C_6H_5.CH{=}NOOH.$$

Experiment.—Phenylnitromethane (about 2–3 g.) is shaken in a wide test tube with 15 c.c. of 2*N*-sodium hydroxide solution. Since it is sparingly soluble in water, the neutral nitro-compound undergoes rearrangement, *i.e.* dissolves quite slowly. (In alcoholic solution the salt formation proceeds very rapidly.) The process of dissolution can be hastened by heating. When this is complete the solution is cooled in a small beaker and a few small pieces of ice are added, followed by 20 c.c. of $2N$-sulphuric acid in one lot. The free *aci*-phenylnitromethane separates in colourless crystalline flakes which are at once separated by filtration at the pump, washed with water, and pressed on porous plate. Part of the material may be recrystallised, without delay, from light petroleum (to which a few granules of calcium chloride are added). A small sample is dissolved in a little alcohol and a drop of ferric chloride solution is added. To a second larger portion a few drops of cold alcoholic bromine solution are added with cooling; the colour of the bromine is destroyed. The same tests, applied to the phenylnitromethane in its original condition, prepared as described above, give negative results.

The remainder of the *aci*-nitro-compound, dissolved in alcohol, is allowed to stand over night. The solution then neither takes up bromine nor gives the colour reaction with ferric chloride. If a few particles of the substance are left exposed on a watch-glass over night they are transformed into an oil.

The *aci*-form of phenylnitromethane is evidently only capable of transitory existence because of its low velocity of rearrangement: it cannot exist in equilibrium.

The Use of Ethyl Acetoacetate and Ethyl Malonate for Synthetic Purposes

Free diethyl malonate has the constitution which corresponds to the usual formula; there is no evidence for the existence of an enol-form $ROOC{-}CH{=}C{<}^{OR}_{OH}$. Nevertheless, by the action of sodium on ethereal solutions of the ester, the so-called *diethyl sodio-malonate* is produced with evolution of hydrogen; it also results even from the ester and ethoxide and is the enolate of the above tautomeric form. In the reactions which are here discussed, ethyl sodio-acetoacetate, which may be chosen as an example for the following reactions, behaves exactly as diethyl sodio-malonate.

When an *alkyl halide* is added to a solution of ethyl sodio-acetoacetate a derivative having the alkyl group attached to carbon is formed, not the substance which would be expected with this group united to oxygen. It must be assumed, then, that the alkyl halide is first added to the reactive double bond and that the sodium halide is afterwards eliminated; it is not merely a case of double decomposition.

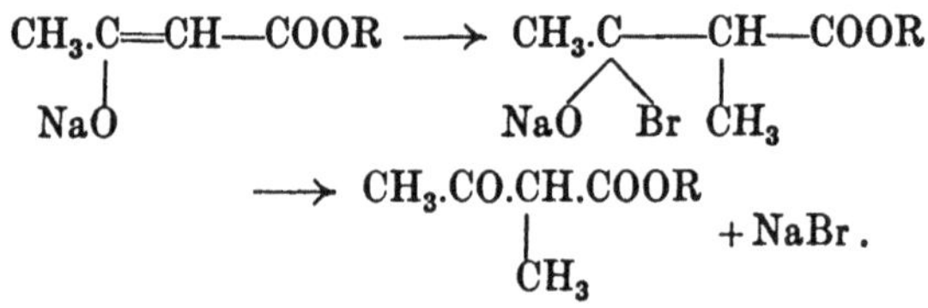

With acid chlorides the reaction is similar.

On the other hand, the action of acid chlorides on ethyl acetoacetate in *pyridine* leads to the formation of O-acyl-derivatives, whilst the O-alkyl-derivatives can only be obtained indirectly from the acetals by elimination of alcohol (Claisen).

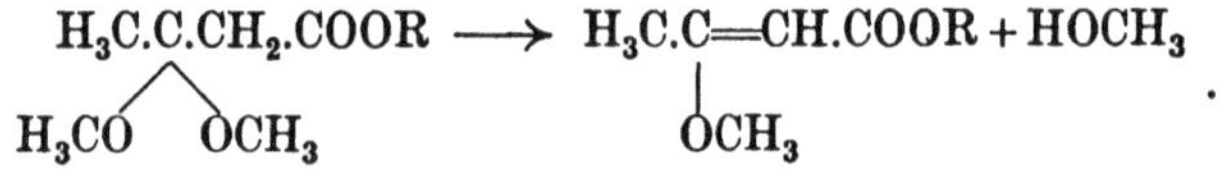

O-alkyl and O-acyl compounds do not undergo rearrangement to form C-derivatives under the conditions in which the latter are prepared, as above described. This change takes place, however, as the result of the catalytic action of solid potassium carbonate in indifferent solvents (Claisen), *e.g.*

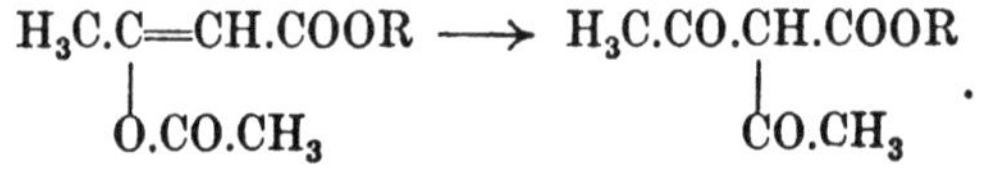

Since, now, the simple C-alkylated or C-acylated derivatives of ethyl acetoacetate and diethyl malonate are again capable of yielding enolates, a second alkylation or acylation on the same carbon atom can be carried out. In the choice of the groups to be introduced at both stages the greatest variety is possible. The synthesis can be performed with all substances which contain reactive halogen and is hence not limited to halogenated hydrocarbons and acid chlorides. If dihalogenated paraffins are chosen, the reaction can also be used for the synthesis of simple carbon rings (W. H. Perkin, jun.), *e.g.*

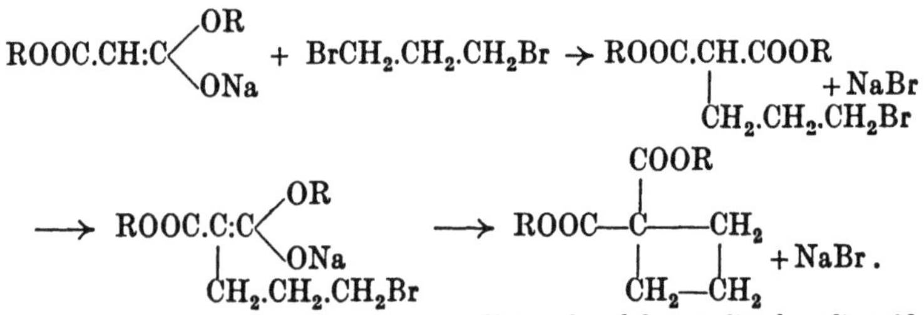

Ester of cyclobutanedicarboxylic acid

Another important advantage of the ethyl acetoacetate and

diethyl malonate syntheses is that such ring-compounds may be readily broken down to simpler compounds.

From the fact that malonic acid loses carbon dioxide on fusion and is converted into acetic acid, we conclude that a carbon atom has not the power of holding two carboxyl groups firmly. Now this also applies to all *substituted* malonic acids which can be readily obtained by hydrolysis from the esters. This constitutes a desirable simplification of the final product.

Example : When *isopropyl bromide* is used *isovaleric acid* is obtainable.

The product of the reaction may be still further broken down by eliminating the second carboxyl group. (Preparation of cyclobutane from the dicarboxylic ester formulated above.)

In ethyl acetoacetate the methylene group is united to $—CO.CH_3$ and —COOR. Free acetoacetic acid is even much less stable than malonic acid and, on merely warming in solution, decomposes in fundamentally similar fashion, into acetone and carbon dioxide. Since all synthetic derivatives of ethyl acetoacetate behave in the same way, so that the acetoacetic acids, obtained by hydrolysis of their esters with aqueous mineral acids, decompose spontaneously with loss of carbon dioxide when heated, numerous derivatives of acetone are made available by this synthesis, which is called *ketonic* hydrolysis, *e.g.*

$$\begin{array}{l}H_3C.C{=}CH.COOR + H_2CCl.COOR \longrightarrow H_3C.CO.CH.COOR \\ \quad\;\, | \qquad\qquad\qquad\qquad\qquad\qquad\qquad\qquad\qquad\quad | \\ \quad ONa \qquad\qquad\qquad\qquad\qquad\qquad\qquad\qquad\quad CH_2.COOR \\ \qquad \longrightarrow \underset{\text{Laevulinic acid}}{CH_3.CO.CH_2.CH_2.COOH} + CO_2 + 2\,ROH\,.\end{array}$$

Concentrated alkali hydroxide decomposes the acetoacetic acid produced by hydrolysis of the ester in a different manner. The cleavage does not take place between the carboxyl group and the rest of the molecule, but between the latter and the $—CO.CH_3$-group, so that two molecules of acetic acid are produced. This *acidic* hydrolysis introduces a new variation into the synthesis as a whole. The practical importance of this acid hydrolysis may be illustrated by the same example, the condensation product of ethyl acetoacetate with ethyl chloroacetate.

$$\begin{array}{l}H_3C.CO.CH.COOR \longrightarrow H_3C.COOH + H_2C.COOH \\ \qquad\qquad\;\; | \qquad\qquad\qquad\qquad\qquad\qquad\quad | \qquad\qquad + 2\,ROH\,. \\ \qquad\quad CH_2.COOR \qquad\qquad\qquad\qquad\; \underset{\text{Succinic acid}}{H_2C.COOH}\end{array}$$

Another type of synthesis in which ethyl acetoacetate is used consists in the coupling of two molecules of the ester to *ethyl diacetylsuccinate* by the action of iodine on the sodium compound :

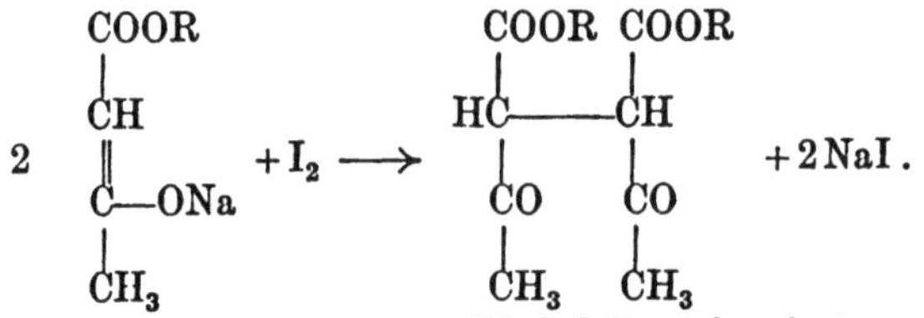

Diethyl diacetylsuccinate

Here, also, as in the case of alkylation, it is found that the union takes place between carbon and carbon. Information about the interesting isomerism of the esters of diacetylsuccinic acid can be found in the literature (L. Knorr).

CHAPTER VII

THE DIAZO-COMPOUNDS

GENERAL

THE most important reaction of the nitrogen-hydrogen compounds, *i.e.* of ammonia and all its derivatives in which hydrogen remains directly united to the nitrogen, is, doubtless, that with *nitrous* acid. The great variety of phenomena which this reaction exhibits, should be considered from a general point of view. Together with salt formation, which indeed takes place normally with ammonia itself and with the aliphatic amines, the reactive N=O double bond of nitrous acid is of outstanding importance.

The simplest form of the reaction occurs in the case of the secondary amines.

Dimethylamine forms, first, dimethylammonium nitrite. When the temperature is raised, intramolecular change involving the double bond of the nitrous acid takes place, water is eliminated, and *acylation*, *i.e.* formation of *nitrosamine*, occurs.

$$(CH_3)_2NH + HONO \longrightarrow (CH_3)_2N\begin{cases} H \\ O\text{—}N\text{=}O \\ H \end{cases}$$

$$\longrightarrow (CH_3)_2N\text{—}\underset{\displaystyle OH}{\underset{|}{N}}\text{—}OH \longrightarrow (CH_3)_2N.NO + H_2O\,.$$

This reaction is completely analogous to other acylation processes, such as the formation of the acetyl derivative from dimethylamine acetate. The only difference is, that since the C=O double bond of acetic acid is less reactive than the N=O double bond of nitrous acid, a higher temperature is required in the former case.

$$(CH_3)_2NH \;+\; O\text{=}\overset{\displaystyle OH}{\overset{|}{C}}\text{—}CH_3 \longrightarrow (CH_3)_2N\text{—}\underset{\displaystyle OH}{\underset{|}{\overset{\displaystyle OH}{\overset{|}{C}}}}\text{—}CH_3$$

$$\longrightarrow (CH_3)_2N\text{—}\underset{\displaystyle O}{\underset{\|}{C}}\text{—}CH_3 \;+\; H_2O\,.$$

Dimethylacetamide

The great velocity of esterification of nitrous acid, exceeding that of all other acids, is doubtless to be traced to the same cause (p. 136).

$$CH_3.CH_2OH + O{:}N.OH \longrightarrow \begin{matrix} CH_3.CH_2{-}O \\ HO \end{matrix}\!\!>\!N{-}OH$$

$$\xrightarrow{-H_2O} CH_3.CH_2{-}O{-}N{=}O\,.$$

Ethyl nitrite

If we apply these considerations to the behaviour of ammonia it is clear that the acyl-product which is formed on heating must decompose into nitrogen and water.

$$H_3N + O{:}N.OH \longrightarrow \left[\begin{matrix} HO \\ H_2N \end{matrix}\!\!>\!N{-}OH\right] \longrightarrow N{:}N + 2H_2O\,.$$

Fundamentally the same holds for primary aliphatic amines:

$$H_3C.NH_2 + O{:}N.OH \longrightarrow [H_3C.NH{-}\overset{OH}{N}{-}OH] \longrightarrow [H_3C.N{:}NOH]$$

$$\longrightarrow N{\equiv}N + H_3C.OH\,.$$

The second intermediate product, in brackets, is known in the form of salts; we recognise that under the conditions of its formation it must decompose into nitrogen and alcohol.

The simple primary amines of the aliphatic series, then, do not form diazo-compounds because the reaction which would lead to their formation only occurs at a temperature at which they are destroyed. The reactivity of the NH_2-group can, however, be increased by a neighbouring carbonyl group. Thus we come to the case of the esters of the *a-amino-carboxylic acids* and of the *a-amino-ketones*. The ethyl ester of glycine can be diazotised even in the cold; the diazo-compound which does not decompose under these conditions undergoes stabilisation by elimination of water and change into ethyl diazoacetate:

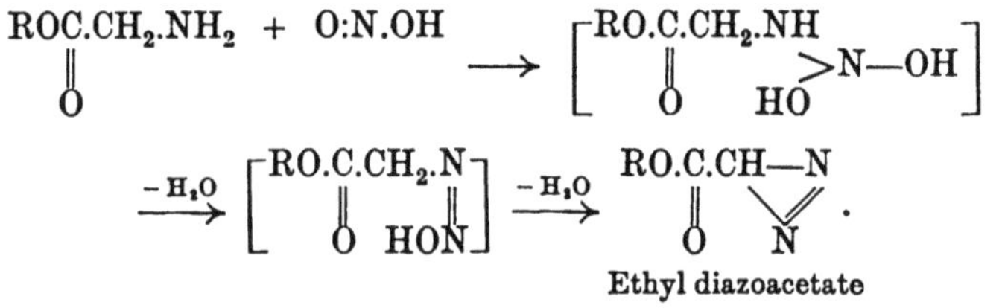

Ethyl diazoacetate

The unexpected feature of the reaction of the primary *aromatic* amines with nitrous acid is that the diazo-compound, which is doubtless formed at low temperatures according to the scheme so far used, undergoes rearrangement by the acid present in the solution, and forms a base, the salt of which, a *diazonium* salt, is obtained as the product of the reaction.

$$C_6H_5.NH_2 + O{:}NOH \longrightarrow C_6H_5N{:}NOH$$
$$\xrightarrow{HCl} C_6H_5.\underset{Cl}{N}{:}N + H_2O\,.$$

Here we encounter a specific property of the aromatic compounds. In the aliphatic series diazonium salts are unknown because here substances of the aniline type $-C{=}\underset{NH_2}{C}-$ cannot exist.

The possibility is not excluded that the ethyl ester of glycine is so easily diazotised because of a tautomeric change:

$$\begin{array}{ll} ROC-CH_2 \\ \;\;\| \quad\;\; | \\ \;\;O \quad NH_2 \end{array} \longrightarrow \begin{array}{ll} ROC{=}CH \\ \;\;| \quad\;\; | \\ OH\;NH_2 \end{array},$$

But even then the fundamental difference between the two series remains, viz. the lack of basic properties in the aliphatic diazo-compounds.

For the present it must be considered inexplicable that an aromatic nucleus—but no alkyl group—can confer strong basic properties on the nitrogen of a diazo-group united to it. As we shall learn later (p. 347), several aromatic nuclei together have a similar effect on a carbon atom to which they are all directly united (carbonium salts). It may be recalled that the basicity of the amines themselves is greatly reduced by the aromatic ring whilst alkyl groups increase it.

For the study of the diazo-compounds we recommend the excellent work *Die Diazoverbindungen*, 1921, by A. Hantzsch, revised by Reddelien, Leipzig, 1921.

A. ALIPHATIC DIAZO-COMPOUNDS

1. DIAZOMETHANE [1]

The apparatus in which the substance is prepared is essentially the same as that indicated for the preparation of acetaldehyde (p. 197), but the flask must be smaller, and, in order to avoid explosions, the calcium chloride tube between the condensers is omitted. The observation and regulation of the rate at which the chloroform is run in is facilitated by choosing a dropping funnel provided below its tap with a bulb, into which the liquid drops or flows visibly from a small delivery tube.[2] Both the tube which delivers nitrogen and the delivery tube of the dropping funnel (a straight calcium chloride tube fused on to its lower end) dip under the surface of the material in the flask.

[1] Staudinger and Kupfer, *Ber.*, 1912, 45, 505.

[2] Such dropping funnels are unsuitable for extractions with ether.

Internal pressure is overcome in the way indicated for acetaldehyde.

The flask (capacity 0·5 l.) contains a hot solution of 75 g. of potassium hydroxide in 180 c.c. of absolute alcohol to which a solution of 15 g. of hydrazine hydrate in 50 c.c. of absolute alcohol is added. From the dropping funnel a mixture of 40 g. of chloroform with 50 c.c. of absolute alcohol is allowed to run slowly at such a rate that the vigorous reaction which ensues does not become too violent. Three wash-bottles in series, containing respectively 50, 40, and 30 c.c. of absolute ether, are strongly cooled in a freezing mixture and serve as receivers. During the experiment a gentle current of nitrogen is passed through the liquid.

If the ether in the third wash-bottle acquires a strong yellow colour, a fourth is added.

When all the chloroform has been run in, the diazomethane which remains in the apparatus is driven into the receivers by passing nitrogen for ten to fifteen minutes longer.

If the diazomethane is not to be used at once after the experiment, the contents of the receivers are combined in a small narrow-necked glass bottle closed in the manner described for drying ether over sodium (p. 107 footnote) and kept in a cool place. The diazomethane solution can be kept for several days, but undergoes continuous, albeit slow decomposition with evolution of nitrogen. The vessel in which the liquid is kept must, therefore, not be tightly closed.

Diazomethane is a yellow gas, boiling point − 24°. When required for preparative purposes it is used in solution only. In the free state it is explosive. In addition to ether, the alcohols, benzene, and petrol ether may be used as indifferent solvents. Acetone also may be used for short periods.

Determination of the Diazomethane Content of Solutions (according to Marshall and Acree, *Ber.*, 1910, 43, 2324).—An aliquot portion of the solution (about one-twentieth) diluted with absolute ether is run with shaking into an ice-cooled 0·2 *N* ethereal solution of benzoic acid. This latter solution, which must contain an excess of acid with respect to the diazomethane, is prepared by dissolving 1·22 g. of purest benzoic acid in absolute ether in a 50 c.c. measuring flask. The presence of an excess of acid is proved by the continued evolution of nitrogen, until the last portions of diazomethane have been added, when the solution should remain colourless. The excess of benzoic acid is titrated with 0·1 *N*-sodium hydroxide solution.

The yield of diazomethane amounts to 2–3 g.

In most scientific institutes this substance is a very welcome reagent since it enables valuable acids and phenols to be methylated conveniently and smoothly. Alcoholic OH-groups are practically never attacked by diazomethane, nor are amines affected.

Experiments.—To an ice-cooled solution of 2–3 g. of a phenol (phenol, cresol, β-naphthol, salicylaldehyde, quinol) in a little ether, acetone, or methyl alcohol, the diazomethane solution prepared as described above is added in small portions until evolution of gas no longer takes place and the solution is coloured faintly yellow.

In *coloured* solutions an excess of diazomethane is recognised by pouring a few drops into a small test tube and dipping in a glass rod which has been moistened with glacial acetic acid. An immediate evolution of gas indicates the presence of diazomethane.

After evaporation of the solvent the products of reaction are purified by distillation, or, if they are solid, by crystallisation. Use one of the phenols available in the laboratory and report on the nature of the methyl ether obtained. Treat carboxylic acids (*p*-toluic, phenylacetic, cinnamic, oxalic, terephthalic, salicylic) in the same way.

Some phenols react slowly with diazomethane. In such cases the substance, mixed with more than the calculated amount of diazomethane solution, is allowed to stand for several days in a vessel provided with a capillary tube in the stopper.

Diazomethane, the simplest aliphatic diazo-compound, was first prepared by von Pechmann [1] in the following way:

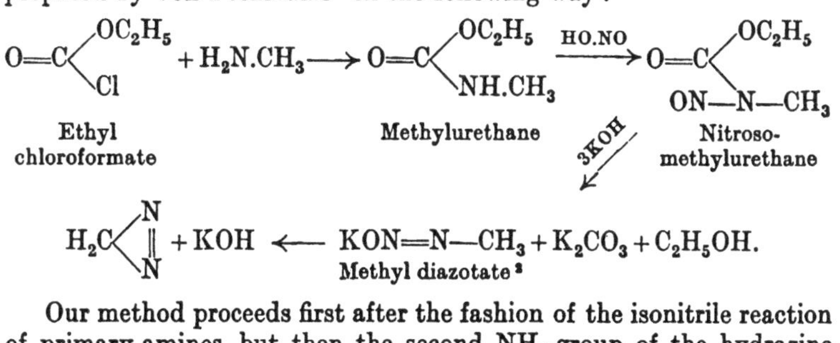

Our method proceeds first after the fashion of the isonitrile reaction of primary amines, but then the second NH_2-group of the hydrazine also takes part in the reaction:

[1] *Ber.*, 1895, **28**, 855.

[2] Hantzsch and Lehmann, *Ber.*, 1902, **35**, 897. According to Thiele (*Annalen*, 1910, **376**, 253) methyl isodiazotate (probably a stereoisomeride) is also formed.

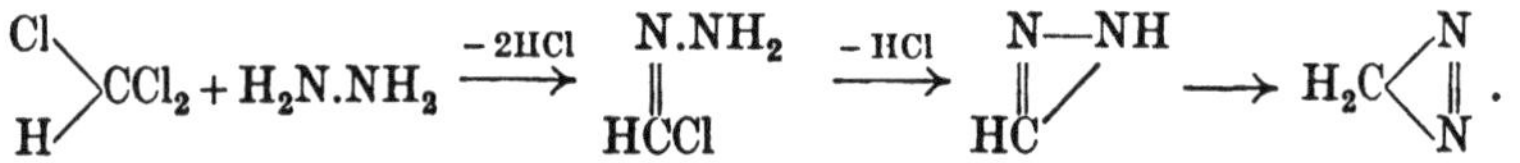

At room temperature diazomethane is gaseous; it is very reactive and poisonous. In ethereal solution it can be preserved for a long time. Information about its numerous reactions (with hydrogen cyanide, acetylene, quinone, etc.) can be found in the special literature of the subject. As has already been mentioned, it is an important methylating agent, especially for phenols. With them it reacts in such a way that the two nitrogen atoms are eliminated as elementary nitrogen, and the two valencies which thus become free are occupied by H and OR.

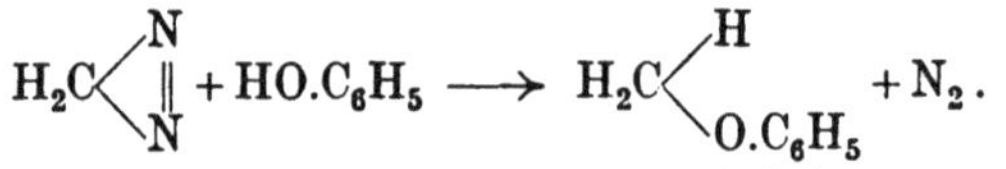

Anisole

The aliphatic diazo-compounds can also be prepared by careful dehydrogenation of the hydrazones (with HgO) (Curtius, Staudinger) and, conversely, they are converted into the latter by hydrogenation:

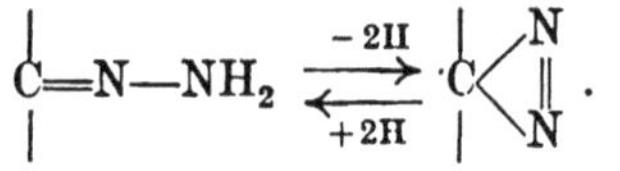

For this and various other reasons another, open chain, formula has also been ascribed to them (Angeli, Thiele), namely >C=N≡N.

The azo-compound corresponding to azobenzene, *azomethane* $CH_3.N{=}N.CH_3$, is a colourless explosive gas (at low temperatures a pale yellow liquid) obtained by dehydrogenation of the corresponding hydrazine (hydrazomethane) (Thiele, *Ber.*, 1909, **42**, 2575).

For further reactions of the aliphatic diazo-compounds see ethyl diazoacetate (below).

2. ETHYL DIAZOACETATE

(*a*) **Glycine Ethyl Ester Hydrochloride,**[1] $H_5C_2OOC.CH_2NH_2.HCl$.—Chloroacetic acid (94 g.; 1 mole) dissolved in 30 c.c. of water is run, at 15°, with shaking, into a litre of concentrated ammonia (d. 0·913). The flask is stoppered and left for twenty-four hours. Then the liquid is heated in a basin on a ring burner (fume chamber!) until the large excess of ammonia has evaporated and its odour is hardly perceptible. Concentrated hydrochloric acid (100 c.c.) is next added to make the solution distinctly acid to Congo paper and heating is continued over an open flame, with *continuous stirring*, until a sample

[1] Hantzsch and Silberrad, *Ber.*, 1900, **33**, 70.

of the pale yellow material, which is semi-solid while hot, becomes completely hard on cooling. At this stage overheating must be avoided by lowering the flame and stirring very vigorously.

While the hot mass is cooling it is thoroughly ground in a porcelain mortar and before the subsequent esterification the water which is still present is removed by transferring the powdered mixture of ammonium chloride and glycine hydrochloride to a short-necked, round-bottomed flask which is suspended in a boiling water bath and evacuated with a pump. After four hours the mass is again powdered and the heating *in vacuo* is continued for three hours, this time in an oil bath at 115°. The salt mixture,[1] which is now in the form of a dry dust, is at once transferred to a double-necked round-bottomed flask connected by means of ground glass connections to a reflux condenser and a gas delivery tube (Fig. 43, p. 91) and boiled with 350 c.c. of absolute alcohol, whilst a strong current of dry hydrogen chloride is passed into the boiling mixture until thick fumes escape from the condenser. (Because of the "bumping" which occurs during boiling the ring of the water bath on which the flask stands is wrapped in a towel.) The hydrogen chloride apparatus is now disconnected and boiling is continued for one hour. By filtration at the pump the ammonium chloride is separated from the hot liquid, and when the filtrate cools the hydrochloride of the glycine ester separates in crystalline form. After twelve hours the crystals are collected by filtration at the pump and are recrystallised from the smallest possible amount of alcohol—frequently some ammonium chloride remains undissolved, and hence care must be taken not to use too much alcohol. In this way the salt is obtained completely pure. Melting point 144°. The thoroughly dried crude product, of which the yield amounts to 50–60 g., can be used for the preparation of ethyl diazoacetate. The yield can be increased by concentrating the mother liquor or by adding ether.

In addition to this, the simplest method of synthesising α-amino-acids (a method which is less satisfactory for the preparation of higher members of the series), there are two other processes, both starting from aldehydes. Strecker obtained the nitrile of the amino acid, Chap. V. 7, p. 220, by addition of ammonium cyanide to the next lower aldehyde, and Erlenmeyer jun. condensed *hippuric acid* with the aldehyde containing two carbon atoms less than the required amino-acid.

[1] A few grammes of the salt mixture are laid aside for the preparation of hippuric acid as described below.

In this latter process an *αβ*-unsaturated *α*-benzoylamino-acid is formed and is converted into the *α*-amino-acid by hydrogenation and subsequent removal of the benzoyl group by hydrolysis.

$$R.CO(H) + H_2C(NH.COC_6H_5).COOH \rightarrow R.CH{=}C(NH.COC_6H_5).COOH \rightarrow R.CH_2.CH(NH.COC_6H_5).COOH$$
$$\longrightarrow R.CH_2.CH(NH_2).COOH + C_6H_5.COOH\,.$$

Recall the importance of the amino-acids as units from which the proteins are built up. What amino-acids have hitherto been isolated from proteins by digestion and by acid hydrolysis ?

By the method of E. Fischer it is possible to carry out a separation, in some degree quantitative, of the individual *α*-amino-acids. They are esterified as described above, and the esters are then separated from one another by fractional distillation of the mixture *in vacuo*.

It is supposed that in the protein molecule the individual amino-acids are united through amide linkages. For the synthesis of the so-called peptides E. Fischer condensed a molecule of the chloride of an amino-acid with a molecule of amino-acid, *e.g.*

$$H_2N.CH_2.COCl + H_2N.CH(CH_3).COOH \longrightarrow H_2N.CH_2.CO.NH.CH(CH_3).COOH\,.$$

Glycylalanine

This procedure was also used in building up polypeptides of high molecular weight.

Amino-acids in solution are quantitatively determined as follows :

1. Method of van Slyke. The nitrogen, evolved when nitrite is added to an acid solution of the amino-acid, is measured.

2. Sörensen's method. The amino-acid is condensed with formaldehyde and the substance produced is now a sufficiently strong acid to be titrated.

3. Method of Willstätter and Waldschmidt-Leitz. Direct titration in alcoholic solution with $0.1N$-alkali using phenolphthalein as indicator.

Experiment. Hippuric Acid.—A few grammes of the mixture of glycine hydrochloride and ammonium chloride just prepared are boiled with 10–14 c.c. of alcohol. The mixture is freed from the undissolved ammonium chloride by filtration and the filtrate is evaporated to dryness on the water bath (remove alcohol completely !) The residue, dissolved in a little water, is benzoylated in a small glass-stoppered bottle by the Schotten - Baumann method (p. 232). An excess (about 2–3 moles) of benzoyl chloride is gradually added in small portions, with continuous vigorous shaking, to the solution, which is kept alkaline throughout and is as concentrated as possible. When the odour of the chloride is no longer perceptible concentrated hydrochloric acid is added until the solution is acid to Congo paper. After some hours the crystalline sludge which has formed is filtered at the pump and the reaction product, when dried, is freed from benzoic acid by washing with ether. The hippuric acid is then recrystallised from hot water. Melting point 187°.

Hippuric acid is a normal product of metabolism. It is produced in the kidney by the enzymic union of benzoic acid and glycine (Schmiedeberg and Bunge, 1877). In birds benzoic acid is rendered innocuous by combination with *ornithine* (αδ-diamino-valeric acid) to form the dibenzoyl-derivative, so-called *ornithuric acid* (Jaffé).

(*b*) **Ethyl Diazoacetate.**[1]—Glycine ethyl ester hydrochloride (47 g.; 0·33 mole), freed from excess of hydrochloric acid by thorough drying, is dissolved in a separating funnel (capacity 0·75 l.) in just sufficient water and the cold saturated aqueous solution of 26 g. of commercial sodium nitrite is then run in, followed by 4 *N*-sulphuric acid, added drop by drop (through a dropping tube) with vigorous shaking. The ethyl diazoacetate which separates as a turbidity is removed each time by shaking with supernatant ether. After 20 c.c. of the sulphuric acid have been dropped in, not too slowly, the aqueous solution is run off, so that the ester, when it becomes concentrated, is not exposed too long to the action of the free acid. Then fresh ether is added and more sulphuric acid is dropped in. Separation of the aqueous solution and renewal of the ether are repeated five or six times.

Finally, when no further turbidity is produced and the excess of nitrous acid dissolves in the ether with a green colour, the addition of acid is stopped and the combined ether extracts are shaken, first with a little sodium carbonate solution (until a permanent red colour is obtained) and then, twice, with a little water. The ethereal solution is now dried over a little calcium chloride for about half an hour with frequent shaking, and most of the ether is removed by distillation at ordinary pressure from a distilling flask (having its side tube high up in the neck) heated on a water bath, at first at a temperature not exceeding 40°. The last third of the ether, however, is distilled

[1] Curtius, *J. pr. Chem.*, 1888, 38, 396; W. Fränkel, *Z. physikal. Chem.*, 1907, 60, 202.

in vacuo at not more than 25°.[1] Water (20 c.c.) and solid barium hydroxide (2 g.) are added to the residue, which is then purified in the same flask by distillation *in steam under reduced pressure.* This distillation is carried out like an ordinary steam distillation. The steam is generated in a round-bottomed flask not quite half-filled with water and closed with a two-holed rubber stopper through which pass a capillary and a suitably bent delivery tube. The flow through the capillary, which should not be too narrow, can be regulated by a screw-clip. The delivery tube, attached by means of a rubber stopper to the distilling flask fixed in an oblique position, passes nearly to the bottom of this flask and is bent slightly downwards at its lower end. The sufficiently long side tube of the flask, bent downwards at an obtuse angle, is directly attached to a second distilling flask which serves as receiver, and it is so fixed that its lower end reaches to about the middle of the bulb of this receiver. The second distilling flask is completely immersed in an efficient freezing mixture with its side tube bent upwards and attached to the pump.

The steam generator is now warmed to 40° in a water bath while the distillation flask is kept at 30°–35° by means of a basin of warm water. At a pressure of 20–30 mm. the ethyl diazoacetate distils with water in forty-five to sixty minutes. The contents of the receiver are twice extracted with ether, and after the combined extracts have been dried with calcium chloride, the ether is removed by distillation, at first from a water bath at 40° and finally *in vacuo.* Lastly the ethyl diazoacetate is purified by distillation in a vacuum. A short condenser and a receiver cooled in a freezing mixture are used. Boiling point 45°/12 mm. Yield 20–25 g. (theoretical 38 g.). The pure product can be kept for a long time, but should not be stored in a tightly closed vessel.

Ethyl diazoacetate, discovered in the year 1888 by T. Curtius, was the precursor of *hydrazine* and *hydrazoic acid.*

Hydrazine was first obtained by hydrolytic decomposition of " bis-diazoacetic acid ". The latter, a tetrazine derivative, is obtained from ethyl diazoacetate in the form of alkali salt by the (catalytic) action of strong alkali. Two molecules of the acid simply unite and, at the same time, the ester group is hydrolysed :

[1] The method of purification described below yields a very pure product such as is necessary for the determination of the concentration of H·-ions by the method of Bredig and Fränkel (pp. 269, 270). The yield, however, is somewhat smaller than when the distillation *in vacuo* is at once carried out at this point.

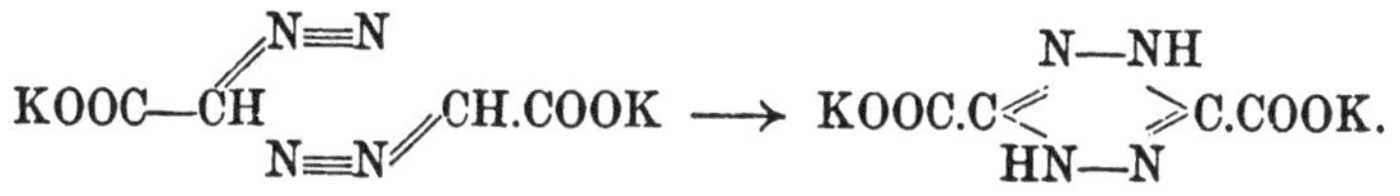

When bis-diazoacetic acid is boiled with acid it is decomposed into hydrazine and oxalic acid. For details see *Ber.*, 1908, **41**, 3161.

Experiment.—Ethyl diazoacetate (5 g.) is added in small portions with shaking to a warmed solution of 8 g. of sodium hydroxide in 12 c.c. of water. At each addition a vigorous reaction takes place. After cooling the yellow mass is shaken with twice its volume of alcohol, the liquid is poured off, and the process is repeated. The sodium salt is isolated by filtration at the pump and can be re-crystallised from water. To obtain the free *bis-diazoacetic acid* six parts of 30 per cent sulphuric acid (1 part by weight of H_2SO_4 to 2·2 parts by weight of water) are poured over the salt and the resultant solution is left for twelve hours. The acid is then collected at the pump. Melting point 152° (decomp.).

In order to *decompose* the acid it is boiled with fifteen parts of 4 *N*-hydrochloric acid until the yellow colour has disappeared, the hydrochloric acid is buffered with 10 g. of crystalline sodium acetate, and the *oxalic acid* which has been produced is precipitated with calcium chloride. The filtrate from the calcium oxalate yields the *hydrazine* as *benzalazine* by shaking it with benzaldehyde, added drop by drop. Melting point of the substance crystallised from alcohol, 93°.

Ethyl hydraziacetate produced by the hydrogenation of ethyl diazoacetate is likewise hydrolysed by acid yielding *hydrazine*, glyoxylic acid, and alcohol. The hydraziacetate is the hydrazone of ethyl glyoxylate ROOC.C(H)=N.NH_2 (cf. in this connection p. 264).

Like those of all the simple aliphatic diazo-compounds the manifold reactions of ethyl diazoacetate are determined by the lability of the nitrogen. The elimination of the latter is catalytically accelerated by aqueous acids, and, indeed, the velocity of decomposition is directly proportional to the hydrogen ion concentration, so that a means is provided by which this concentration can be measured for acids of unknown strength. (Bredig, Fränkel, *Z. physikal. Chem.*, 1907, **60**, 202).

After the nitrogen has been eliminated an unsaturated radicle ROOC.CH< remains and undergoes various further changes. In the presence of water, the components of the latter are added on to give esters of hydroxyacetic acid. Addition of HCl gives esters of chloroacetic acid, that of iodine esters of diiodoacetic acid; carboxylic acids, alcohols, aldehydes, amines, NO_2, and triphenylmethyl are added in a similar way. If, as in thermal decomposition, an addendum is lacking, two radicles unite to give fumaric esters.

$$2\,ROOC.CH< \longrightarrow ROOC.CH=CH.COOR\,.$$

Experiments.—In order to learn, at least qualitatively, the influence of the hydrogen ion concentration on the velocity of decomposition, about 0·5 c.c. of ethyl diazoacetate is dissolved in a little 50 per cent alcohol and the solution is divided into two portions in small beakers to which respectively 1 c.c. of 0·1 *N*-hydrochloric acid and 1 c.c. of 0·1 *N*-acetic acid (prepared in a measuring cylinder from glacial acetic acid) are added.

Further, to an ethereal solution of the ester a little ethereal solution of iodine is added. The colour of the solution does not disappear until after some time and nitrogen is evolved.

In addition to the reactions mentioned, ethyl diazoacetate takes part in many condensations with acetylene and ethylene derivatives, when the nitrogen is retained in the molecule. Thus, with esters of fumaric acid, for example, *pyrazoline tricarboxylic esters* are produced :

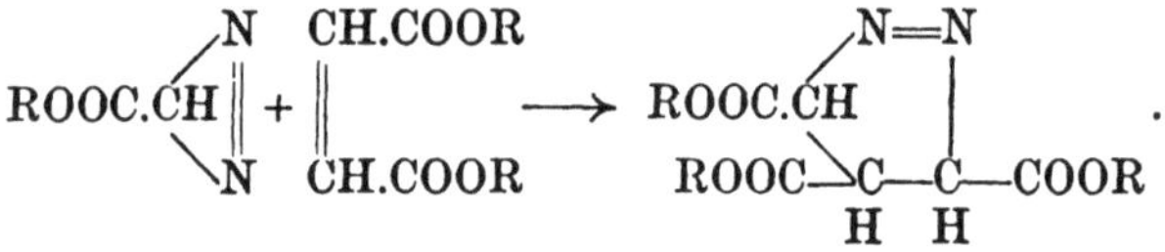

The decomposition of these latter esters at high temperatures constitutes the important synthesis of *cyclo-propane derivatives* (Buchner) :

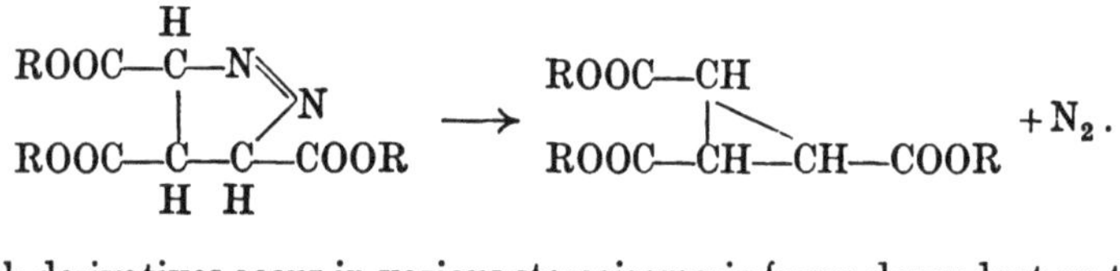

Such derivatives occur in various stereoisomeric forms dependent on the plane of the ring.

Benzene reacts in the same way, at high temperature, with ethyl diazoacetate (Buchner)

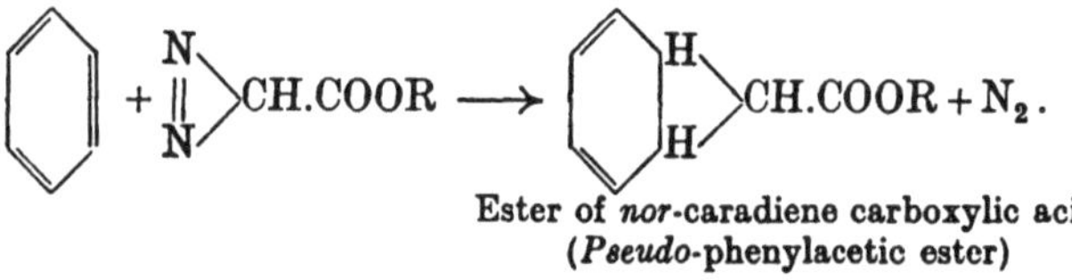

Ester of *nor*-caradiene carboxylic acid
(*Pseudo*-phenylacetic ester)

When the cyclopropane ring of this interesting reaction product is broken as a result of isomerisation two substances may be formed:

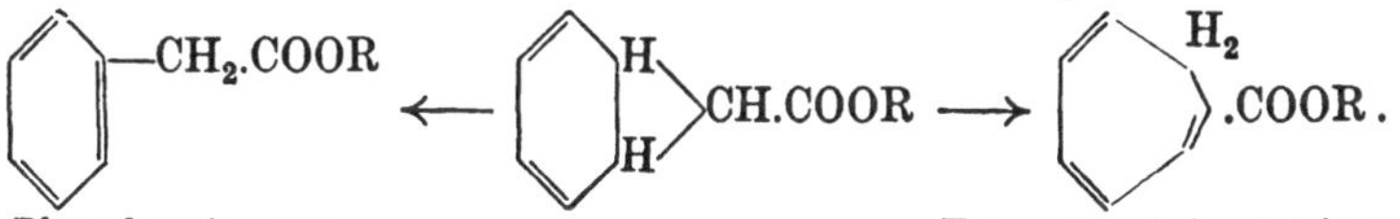

Phenylacetic ester | Ester of *cyclo*-heptatriene carboxylic acid

Ethyl diazoacetate condenses with diethyl malonate to yield a derivative of 4-*hydroxypyrazole* (Bertho and Nüssel, *Annalen*, 1927, **457**, 278):

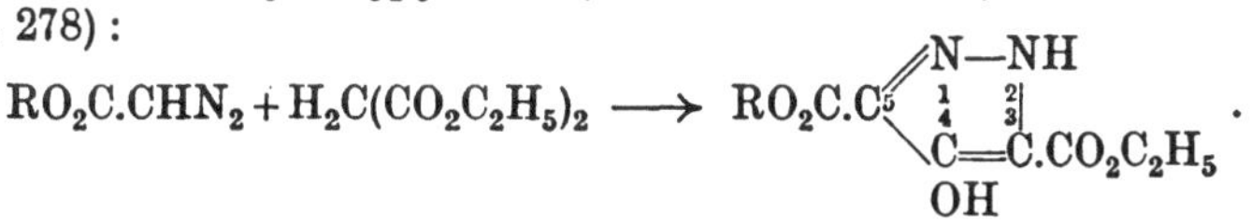

Further details of the reactions of the aliphatic diazo-compounds are to be found in: H. Wieland, *Die Hydrazine*, Stuttgart, 1913, pp. 97 *et seq.*

B. AROMATIC DIAZO-COMPOUNDS

3. DIAZOTISATION OF ANILINE. PHENOL FROM ANILINE. ISOMERISM OF THE DIAZO-COMPOUNDS

(*a*) **Preparation of a Diazonium Salt in Solution.**—Concentrated sulphuric acid (10 c.c.) is poured with good stirring into 50 c.c. of water in a jar or beaker (capacity 0·5 l.) and to the hot dilute acid 9·3 g. of freshly distilled aniline are added. The aniline sulphate solution is now cooled in ice (*not* in a freezing mixture) and, after 100 g. of ice have been added gradually in small portions, with the result that some of the sparingly soluble salt has been precipitated, the solution of 7·5 g. of sodium nitrite in 30 c.c. of water is run in gradually from a dropping funnel. During the addition of the nitrite solution the contents of the beaker are vigorously stirred. When most of the nitrite has been added a test with potassium iodide-starch paper [1] is made in order to discover whether or no an excess of nitrous acid is present. In this connection it must be remembered that towards the end of the reaction the concentration of the reactants decreases greatly, so that the diazotisation becomes slow. A few minutes must, therefore, be allowed to elapse (after the last

[1] A small piece of starch the size of a pea is finely powdered, dropped, with good stirring, into 200 c.c. of boiling water, and kept at the boil for a short time. After the liquid has cooled a fragment of potassium iodide as large as a lentil is dissolved in a little water and added. Long strips of filter paper about 3 cm. wide are soaked in the liquid and then dried in an acid-free atmosphere on a stretched string. After drying, the strips are cut up and kept in a closed container.

addition of nitrite) before the test is made. When, finally, after five minutes, free nitrous acid in small amount can still be detected, the diazotisation is at an end. The aniline sulphate must, of course, have dissolved completely.

A sample of the reaction mixture should not become turbid when sodium acetate solution is added. But if now a few drops of a solution of an aniline salt are poured in, there is precipitated yellow *diazoaminobenzene*, which can be re-dissolved with concentrated hydrochloric acid after a few small pieces of ice have been added. Further, when a few particles of β-naphthol or of R-acid are dissolved in a small excess of $2\,N$-sodium hydroxide solution and a sample of the diazo-solution is added, an intense red colour is produced. The development of this colour, which results from " coupling ", is an infallible test for a diazonium salt and hence, also, for the corresponding primary aromatic amine.

(*b*) **Preparation of Phenol from the Diazonium Salt Solution by Boiling.**—Nitrogen is slowly evolved from the solution, even on standing at room temperature. The decomposition is, however, caused to proceed at a slightly raised temperature (40°–50°), in a round-bottomed flask on a gently boiling water bath. After the evolution of gas has slackened, the resulting phenol is directly distilled with steam. When all the phenol has passed over (test a sample with bromine water !) the distillate is saturated with common salt and extracted several times with ether. The extract is dried with calcium chloride and after the usual treatment the phenol is distilled from a small flask. Boiling point 183°. Yield 6–7 g. The product should soon solidify.

For preparative diazotisations it is important to use a sufficient excess of acid and to keep the temperature down. Two moles of acid are required for each mole of amine, one for salt formation and one for liberating the nitrous acid from the nitrite. As a rule 2·5–3·0 moles are used. The excess is required to prevent condensation of the diazonium salt with unchanged base to diazoamino-compound ; such condensations take place in a faintly acid medium. The test for unchanged amine, accordingly, consists in buffering the free mineral acid with sodium acetate, and so providing a solution faintly acid with acetic acid, under which conditions the diazoamino-compound is formed. The latter is decomposed by mineral acids into diazonium salt and amine salt, *e.g.*

$$C_6H_5.N{=}N{-}NH.C_6H_5 \xrightarrow{2HCl} C_6H_5.\underset{Cl}{N}{\equiv}N + HCl.NH_2.C_6H_5 .$$

There are, however, a few diazonium salts which couple with the parent base even in acid solution, *e.g.* *m*-phenylenediamine (Bismarck brown).

In diazotisation the addition of nitrite is controlled by means of potassium iodide-starch paper. When the reaction is complete the paper should become blue, but the excess of nitrite should be as small as possible. It should be remembered that diazotisation is not an ionic reaction, but requires time; especially towards the end a few minutes must elapse before the test is made.

Sparingly soluble salts of primary aromatic amines are diazotized in suspension, with vigorous stirring. Very weak bases such as halogenated anilines and nitroanilines require a large excess of acid for salt formation; they are first dissolved in just sufficient hot concentrated hydrochloric acid, which is then simultaneously cooled in ice and diluted. In this way the salts, which are mostly sparingly soluble, are precipitated in a finely divided condition. Dissolution in concentrated sulphuric acid and direct diazotisation of the sulphate, precipitated as a fine powder by means of ice, is often to be recommended. The free amines, however, should never be diazotised in acid *suspension* because they react far too slowly. Salt formation should first be ensured.

The stability of the diazonium salts varies; there are some, for instance, in the anthraquinone series, which can be crystallised from hot water.

On account of their great lability the diazonium salts of the simple primary amines cannot be isolated from aqueous solution. On the other hand, they crystallise from alcohol when ether is added. Since the metallic salts of nitrous acid are insoluble in alcohol, its esters are used instead for diazotisation in alcohol. These esters are hydrolysed by acid with extraordinary rapidity and therefore behave almost like salts (see p. 136).

(*c*) **Solid Phenyldiazonium Chloride.**—Aniline hydrochloride (3·5 g.) is dissolved in absolute alcohol (20 c.c.) and to the solution 0·5 c.c. of alcoholic hydrogen chloride and, with ice cooling, 3 g. of ethyl nitrite or 4 g. of amyl nitrite are successively added. After the mixture has stood for ten minutes the diazonium salt is completely precipitated by gradual addition of ether; it is collected at the pump and washed, first with alcohol-ether (1:1) and then with ether alone. The salt is kept moist with ether and only a small sample is dried on filter paper. The dried material is caused to explode by a blow or in a flame. Even the material which is moist with ether must not be touched with a spatula or other hard object.

The diazonium salt on the filter is dissolved in ice-water and is used for the preparation of *phenyldiazonium perbromide* and *phenyl azide* (see below).

When nitrous gases (from arsenious oxide and nitric acid, d. 1·35) are passed into a well-cooled suspension of aniline nitrate in water, and alcohol and ether are then added gradually, crystalline *phenyldiazonium nitrate* is obtained. At most only 2 g. of aniline are used, and of the diazonium salt only as much as covers well the tip of a knife-blade is dried on porous plate, after collecting at the pump and washing with alcohol-ether (1:1).

The nitrate detonates on heating on a spatula or on being struck with a hammer. The chloride is less liable to explode, but it too must not be preserved in the dry state. In general, work with dry diazonium salts must be carried out cautiously.

The diazonium salts are colourless. Their aqueous solutions react neutral. If the acid is removed from the salts by means of alkali, very unstable diazonium hydroxides result, which can only be demonstrated in the solution during quite a short time. These hydroxides change by addition of alkali and elimination of water into salts of acid diazohydroxides, the so-called *diazotates.*

$$\begin{array}{c} C_6H_5.N{\equiv}N \\ \quad\;\, | \\ \quad\;\, Cl \end{array} \xrightarrow{\text{NaOH}} \begin{array}{c} C_6H_5.N{\equiv}N \\ \quad\;\, | \\ \quad\;\, OH \end{array} \xrightarrow{\text{NaOH}} \left[\begin{array}{c} \quad\;\, H \\ \quad\;\, | \\ C_6H_5.N{=}NONa \\ \quad\;\, | \\ \quad\;\, OH \end{array}\right]$$

Diazonium salt — Diazonium hydroxide

$$\xrightarrow{-H_2O} C_6H_5.N{=}NONa\,.$$

Sodium phenyldiazotate

If, now, the diazotate solution so obtained is re-acidified, the diazonium salt is re-formed:

$$C_6H_5.N{=}NONa \xrightarrow{2HCl} \left[\begin{array}{c} \quad\;\, H \\ \quad\;\, | \\ C_6H_5.N{=}NOH \\ \quad\;\, | \\ \quad\;\, Cl \end{array}\right] + NaCl$$

$$\xrightarrow{-H_2O} \begin{array}{c} C_6H_5.N{\equiv}N \\ \quad\;\, | \\ \quad\;\, Cl \end{array}.$$

An important reversible relationship therefore exists between the *diazonium* and *diazohydroxide* types.

The diazohydroxide, which is formed from the diazonium hydroxide by isomerisation, is said to be the *pseudo-base* of the latter (Hantzsch), because although it is itself not a base (but rather an acid) it nevertheless combines with acid to form the diazonium salt.

Before the diazo-compounds are further treated from a preparative point of view another rearrangement which members of this interesting class undergo will be discussed. By energetic treatment with strong alkali, phenyldiazotate is converted into the salt of an isomeric acid; the *isodiazotate* type (Schraube and Schmidt) is produced. For many years the constitution of the *iso*-diazotates was the subject of a historic controversy between Bamberger and Hantzsch. Most chemists now regard the question as decided in favour of the latter's view that the isomerism is spatial and depends on the different arrangement of the C_6H_5- and OH-groups with respect to the fixed plane of the doubly bound nitrogen. The same conception had already provided an explanation of the isomerism of asymmetrically substituted oximes (pp. 335, 336). It coincides, in principle, with the theory of the *cis-trans*-isomerism of ethylene derivatives (fumaric and maleic acids).

According to this theory the labile normal diazotates are regarded as the *syn-* (=*cis-*) compounds, the stable *iso*-diazotates as the *anti-* (=*trans-*) compounds.

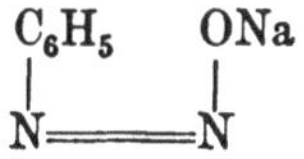

Normal diazotate

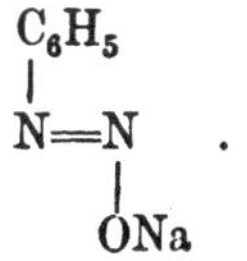

Iso-diazotate

Whilst the conversion of the simple phenyl-*syn*-diazotate into its isomeride requires strong alkali, the *syn*-forms of other diazotates are so labile that, almost instantaneously after formation from the diazonium salts, they change into the *anti*-form and hence can never be obtained in solution. An important example of this change is described under (*d*), namely, *p*-nitrophenyldiazotate, which, on coupling with β-naphthol, yields the much-used dye " para-red ".

Phenyldiazonium Perbromide.—To a fresh ice-cold solution of one of the solid diazonium salts, prepared as described above, or to the diazo-solution from 2 g. of aniline, there is added the solution of 1·5 c.c. of bromine in 15 c.c. of potassium bromide solution (25 per cent), with ice cooling, until precipitation of dark-coloured oil ceases. The aqueous solution is then decanted; when the residual perbromide is washed a few times with ice-water it crystallises.

In order to convert it into *phenyl azide* (" diazobenzeneimide ") the perbromide is covered with water and ice and about 10 c.c. of concentrated ammonia solution are added, with cooling. A vigorous reaction takes place, resulting in the formation of the pungent smelling phenyl azide, which is purified by steam distillation. It can also be distilled in a vacuum without decomposition. Since it explodes when rapidly heated it must be handled cautiously.

The bromides of organic bases form with bromine insoluble perbromides, in the present case the addition compound $C_6H_5.\overset{\cdot}{N}\equiv N$. $Br.Br_2$

The mechanism of the reaction with ammonia is as follows. The perbromide bromine is converted into hypobromite and at the same time the diazonium salt undergoes rearrangement to *syn*-diazohydroxide, which immediately couples with NH_3 to give *phenyltriazene* (" diazobenzeneamide "). The latter is then dehydrogenated by the hypobromite yielding *phenyl azide* (Dimroth):

$$\underset{Br.Br_2}{C_6H_5.N{\equiv}N} \xrightarrow[2H_2O]{3NH_3} C_6H_5.N{=}NOH + 2NH_4Br + NH_4OBr\,,$$

$$C_6H_5.N{:}NOH + NH_3 \longrightarrow C_6H_5.N{:}N.NH_2 + H_2O\,,$$

$$C_6H_5.N{:}N.NH_2 + NH_4OBr \longrightarrow C_6H_5.N\!\left\langle\begin{matrix}N\\ \|\\ N\end{matrix}\right. + NH_4Br + H_2O\,.$$

By very careful hydrogenation (with stannous chloride in ethereal hydrogen chloride) phenyl azide has been converted into the exceedingly sensitive phenyltriazene (Dimroth), which, as has been shown, can be reconverted into the former by dehydrogenation. As in the case of the aliphatic diazo-compounds, an *open* chain structural formula has lately also been assigned to hydrazoic acid and its esters, so that the changes just mentioned may be formulated as follows:

$$C_6H_5.N{=}N{\equiv}N \underset{-2H}{\overset{+2H}{\rightleftarrows}} C_6H_5.N{=}N{-}NH_2\,.$$

Phenylazide is best prepared from phenylhydrazine (p. 289).

The aryl azides are very reactive compounds. By the action of acids, for example, they lose the two terminal N-atoms as N_2; the residual $C_6H_5N<$ combines with water to form aryl hydroxylamine, but this at once undergoes rearrangement and becomes an aminophenol.

When the benzene ring contains a negative substituent the azido-group is eliminated by alkalis in the same way as is halogen, so that an alkali azide and a phenol are produced.

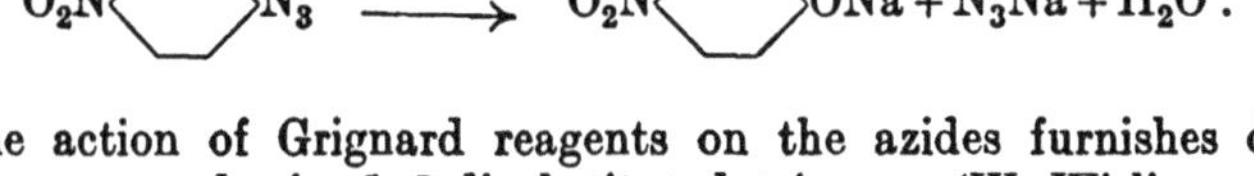

The action of Grignard reagents on the azides furnishes diazoamino-compounds, *i.e.* 1:3-disubstituted *triazenes* (W. Wislicenus and Dimroth), *e.g.*

$$C_6H_5.N_3 + CH_3MgBr \longrightarrow C_6H_5.N{=}N{-}NH.CH_3\,.$$

Diethyl malonate reacts with azides to form triazolone derivatives, noteworthy because of their interesting tautomeric relationships (Dimroth).

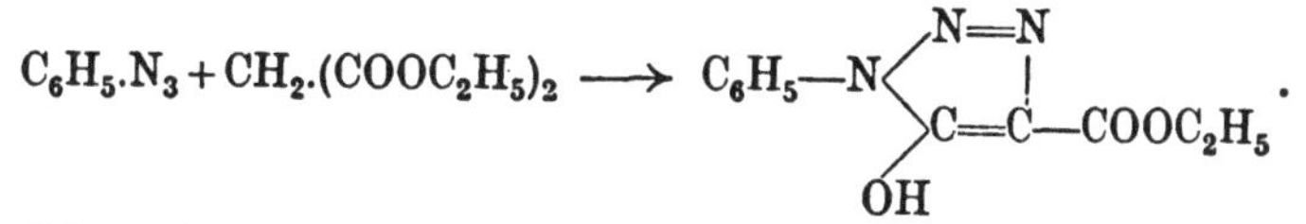

This condensation is completely analogous to that of ethyl diazo-acetate, mentioned on p. 271, and, in general, azides and aliphatic diazo-compounds are strikingly similar in the manner in which they react with unsaturated substances like acetylenes, olefine derivatives, and hydrogen cyanide to yield heterocyclic compounds.

(*d*) **Sodium *p*-Nitrophenyl-*anti*-Diazotate.**[1]—Nitraniline (14 g.; 0·1 mole) is dissolved by heating in 60 c.c. of hydrochloric acid (30 c.c. of concentrated acid and 30 c.c. of water) and the solution is poured into a small filter jar containing 80 g. of ice. The solution is now diazotised at 5°–10° with the solution of 8 g. of sodium nitrite in 20 c.c. of water added in one lot with vigorous stirring. After it has been ascertained that the diazotisation is complete the solution is poured, with stirring, into 400 c.c. of approximately 4 *N*-sodium hydroxide solution which has been warmed to 40°–50°. As the mixture cools the *anti*-diazotate is deposited in the form of beautiful golden plates. After several hours the salt is collected at the pump and washed with saturated brine. When dried on porous plate the substance can be preserved for any desired time, and can be freed from adherent sodium chloride by dissolution in alcohol at 60°, filtration of the solution, and evaporation of the solvent. Yield upwards of 18 g.

Verify that the aqueous solution of the diazo salt does not couple with the sodium salt of β-naphthol nor with a solution of R-salt.[2]

But if the dilute solution is acidified with hydrochloric acid and filtered from undissolved flocculent material an azo-dye results on renewed coupling. The combination with the phenol of the *syn*-diazotate which is first formed proceeds more rapidly than its re-arrangement to the *anti*-form.

4. IODOBENZENE. BENZENE FROM ANILINE

Aniline (14 g.), dissolved in 35 c.c. of concentrated hydrochloric acid which has been diluted with two volumes of water, is diazotised with ice cooling, according to the method described for the preceding preparation. A solution of 11 g. of sodium nitrite in 50 c.c. of water is used.

From 50 c.c. of the diazo-solution *benzene* is prepared and the rest is used for the preparation of *iodobenzene*.

[1] Schraube and Schmidt, *Ber.*, 1894, 27, 518. [2] On R-acid, see p. 292.

(*a*) **Iodobenzene.**—To the portion of the solution set aside for the purpose in a round-bottomed flask (capacity 0·5 l.) a solution of 20 g. of potassium iodide in 20 c.c. of water is added, and the mixture, kept cool in water, is allowed to stand for some hours. The flask, fitted with an air condenser, is then heated on the water bath (which is kept boiling moderately) until evolution of nitrogen has ceased. The contents are now made strongly alkaline with concentrated sodium hydroxide solution in order to fix the phenol formed as a by-product, and the iodobenzene is distilled with steam; it is separated in a funnel from the aqueous portion of the distillate, dried with a few lumps of calcium chloride, and finally distilled. If the separation in the funnel is sharp no ether need be used. Boiling point 189°–190°. Yield 18–20 g.

(*b*) **Benzene.**—While the major part of the diazonium solution is being mixed with KI the remainder is kept cool in ice. An alkali stannite solution is prepared by mixing a solution of 25 g. of sodium hydroxide in 30 c.c. of water with the turbid solution of 20 g. of stannous chloride in 100 c.c. of water; the stannite is gradually added to the diazo solution, with efficient ice cooling, and in small portions. Each portion is added only after the evolution of nitrogen, due to the preceding portion, has ceased. When the reaction is complete the benzene formed is distilled through a downward condenser. The hydrocarbon passes over before much water distils and is collected in a test tube. After being dried with a little calcium chloride the benzene is distilled from a small distilling flask which has a condenser jacket slipped over the side tube and passes over almost completely at 81°. Yield 3–4 g.

Although practically important in other cases, the replacement of an NH_2-group by hydrogen *via* the diazo-compound naturally has none in the example here given. Thus *m*-nitrotoluene (and from it *m*-toluidine) is obtained from *p*-toluidine by nitration of the acetylated base and replacement of the NH_2-group (after elimination of the acetyl) by H, as shown above.

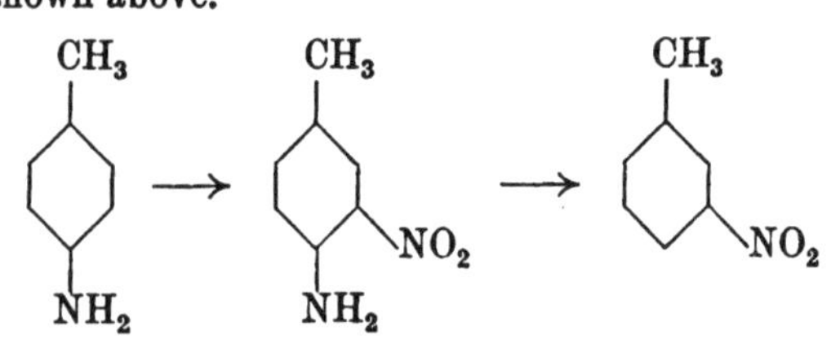

The course of the reaction is very remarkable. We must suppose that the diazotate takes up hydrogen at the N=N-double bond, that then H_2O or NaOH is eliminated; finally the unstable *phenyldiimine* would decompose into benzene and nitrogen.

$$C_6H_5.N{=}NONa \xrightarrow{2H} C_6H_5.NH{-}NHONa \longrightarrow C_6H_5.N{=}NH + NaOH \longrightarrow C_6H_6 + N_2.$$

Instead of by alkali stannite the diazo-group can be exchanged for hydrogen by prolonged boiling of the diazonium salt in *alcohol*, which is thereby oxidised to aldehyde.

A secondary reaction yields at the same time *phenol ether* by the replacement of the diazonium group by alkoxyl. This is clearly analogous to the conversion of diazonium salts to phenols.

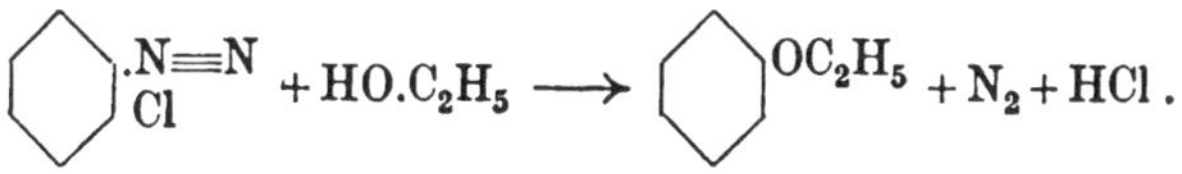

Derivatives of Iodobenzene

It is remarkable that the *aromatic iodocompounds* can be converted *via* addition products with chlorine into substances containing multivalent iodine (V. Meyer, Willgerodt).

Phenyl Iodochloride.—Iodobenzene (3 g.) is dissolved in chloroform (15 c.c.), and chlorine from a cylinder is passed into the solution (cooled in ice) until absorption of the gas ceases. The beautiful light yellow crystals produced are collected at the pump, washed with chloroform, and dried in the air on filter paper.

The phenyl iodochloride $C_6H_5I{<}^{Cl}_{Cl}$ so produced behaves as the salt of a weak di-acid base $C_6H_5I{<}^{OH}_{OH}$, of which other salts such as the diacetate are also known. The base itself is known only in the form of its anhydride $C_6H_5I{=}O$, *iodosobenzene.*

Iodosobenzene.—Phenyl iodochloride (2 g.) is thoroughly ground in a mortar with 3 *N*-sodium hydroxide solution (10 c.c.). After the mixture has stood over night the resultant iodosobenzene is collected at the pump, washed with water, and dried on porous plate. The substance is not crystalline.

A colourless salt is precipitated from the alkaline filtrate (without the washings) by passing in sulphur dioxide, which serves to reduce the iodic acid produced. After the mixture has stood for some time, the salt, *diphenyliodonium iodide*, is collected at the pump and recrystallised from hot water.

Iodoxybenzene.—The major portion of the iodosobenzene so prepared is made into a paste with a little water and submitted to steam distillation in a round-bottomed flask. The distillation is continued until all the material has dissolved and the iodobenzene produced has passed over (use a condenser and receiver). The residue after the distillation (if still turbid) is filtered while hot and the filtrate is concentrated on the water bath until a sample, poured into a test tube, gives a copious deposit of crystals. After the concentrated solution has cooled the crystalline material is collected at the pump, etc.

The formation of *iodosobenzene* is readily understood, that of the peculiar *iodonium salt* less readily. In general the iodonium bases are formed from iodoso- and iodoxy-compounds in the presence of alkali, preferably silver oxide; the two organic iodine compounds combine with elimination of iodate.

$$C_6H_5.IO + \genfrac{}{}{0pt}{}{O}{O}\!\!>\!I{-}C_6H_5 \xrightarrow{NaOH} \genfrac{}{}{0pt}{}{C_6H_5.\dot{I}.C_6H_5}{OH} + NaIO_3 .$$

Iodoxybenzene is formed, together with iodobenzene, from iodosobenzene by intermolecular dismutation:

$$C_6H_5.IO + OI.C_6H_5 \longrightarrow C_6H_5.I\!\!<\!\genfrac{}{}{0pt}{}{O}{O} + C_6H_5I ,$$

just as nitrous acid undergoes dismutation to nitric acid and nitric oxide. This reaction takes place to a slight extent even in the cold; hence the appearance of the iodonium base as a by-product in the preparation of iodosobenzene.

The iodoso- and, in particular, the iodoxy-compounds decompose with a puff when heated, since the oxygen which they contain is present in a state of tension. They liberate an equivalent amount of iodine from acidified potassium iodide solution and are thereby reconverted into iodobenzene.

The basic function of the iodine in the iodonium bases, which correspond exactly to the ammonium, sulphonium, and oxonium bases, is most interesting. A molecule of diphenyliodonium iodide has the same atoms as two molecules of iodobenzene.

Experiment.—A small sample of the salt in a test tube is carefully heated over a small flame. Fusion sets in at some point and then spreads without further application of heat while ebullition also occurs. The *iodobenzene* formed is recognised by its odour and by conversion into phenyl iodochloride.

Whereas in most dissociations (N_2O_4, NH_4Cl, PCl_5) energy must be supplied, it is liberated in the present case. In contrast to the others, this dissociation is not reversible, since iodobenzene, of course, does not polymerise to form diphenyliodonium iodide. As is still the case to-day with the diazonium compounds, so it was for long thought that the aromatic compounds containing multivalent iodine, which were simultaneously and independently discovered by V. Meyer and by C. Willgerodt, formed a class of substances peculiar to the aromatic series. But in 1909 Thiele found analogues to all these compounds amongst the olefines, with chloroiodoethylene CHCl=CHI as their simplest parent substance. Even methyl iodide, at low temperatures, can combine with chlorine, but the product decomposes again readily, yielding methyl chloride and iodine chloride (replacement of iodine by chlorine). The derivatives of multivalent iodine are not stable, unless the iodine is attached to a doubly bound carbon atom.

5. *p*-TOLUNITRILE FROM *p*-TOLUIDINE (SANDMEYER'S REACTION [1])

A solution of 55 g. of potassium cyanide in 100 c.c. of water is added gradually with continued warming to copper sulphate solution, prepared by heating 50 g. of the salt with 200 c.c. of water in a two-litre flask on the water bath. (Evolution of cyanogen! Fume chamber!)

While the cuprous cyanide solution is warmed gently (to 60°–70°) on the water bath, a solution of *p*-tolyldiazonium chloride is prepared as follows: Heat 20 g. of *p*-toluidine with a mixture of 50 g. of concentrated hydrochloric acid and 150 c.c. of water until dissolution is complete. Immerse the solution in ice-water and stir vigorously with a glass rod so that the toluidine hydrochloride separates as far as possible in a microcrystalline form. Then cool the mixture in ice and diazotise with a solution of 16 g. of sodium nitrite in 80 c.c. of water, added until the nitrous acid test with potassium iodide-starch paper persists. The diazonium chloride solution so obtained is poured during the course of about ten minutes into the warm cuprous cyanide solution, which is meanwhile shaken frequently. After the diazo-solution has been added the reaction mixture is heated under an air condenser on the water bath for a further quarter of an hour and then the toluic nitrile is separated by distillation with steam (fume chamber, HCN!). The nitrile (which passes over as a yellowish oil) is extracted from the distillate with ether, the *p*-cresol produced as a by-product is removed by shaking the ethereal extract twice

[1] *Ber.*, 1884, 17, 2650; 1885, 18, 1490; 1889, 22, 2178.

with 2*N*-sodium hydroxide solution, the ether is evaporated, and, by shaking the warm residue with a solution of 4 g. of stannous chloride in 10 c.c. of concentrated hydrochloric acid, the azotoluene which gives the material its yellow colour is removed.[1] The liquid is diluted with water and the toluic nitrile which soon solidifies is collected at the pump and dried on porous plate. If the material remains partly oily it is dissolved in ether and shaken again with sodium hydroxide solution in order to remove stannous chloride. The ethereal solution is then dried and the nitrile is distilled. Boiling point 218°. Melting point 38°. Yield 12–14 g.

Benzonitrile.—In the same apparatus *benzonitrile* can be prepared in a corresponding yield from the diazonium chloride solution obtained from 18·6 g. of aniline. Liquid boiling at 186°.

***p*-Toluic Acid.**—If the hydrolysis of a nitrile to an acid has not already been carried out (benzyl cyanide → phenylacetic acid, p. 129) this process should be learned here.

In a small round-bottomed flask mix 20 c.c. of concentrated sulphuric acid with 10 c.c. of water and add 5·5 g. of toluic nitrile gradually in small portions. Then heat the mixture to boiling under reflux condenser for about one hour on the sand bath or on wire gauze. After cooling, dilute with water, collect the crystalline acid at the pump, remove any amide which may be present by dissolving the crude material in dilute alkali solution and filtering, then precipitate the toluic acid from the filtrate with hydrochloric acid. A purer product is obtained by hydrolysis for five hours at 150° (in the oil bath). For purification dissolve the material (not previously dried) in the minimum quantity of boiling alcohol, add water to the boiling solution until turbidity just fails to persist, and continue the boiling for a few minutes longer with a little animal charcoal, but do not add the latter while the solution is boiling. Filter and allow the filtrate to cool. The acid which crystallises melts at 177°. Yield 4 g.

The best preparative method of obtaining *terephthalic acid* is to oxidise the sodium salt of *p*-toluic acid with permanganate at the temperature of the water bath. In the same way toluene can be converted into benzoic acid, and an important technical example of this reaction is the oxidation of *o*-tolylsulphonamide to *saccharin*.

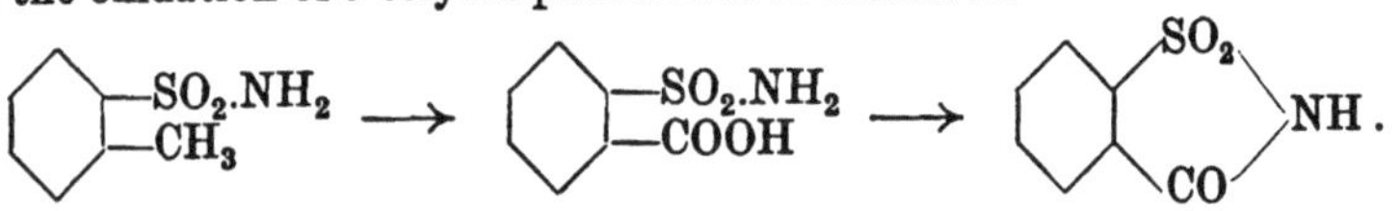

[1] Henle, *Organ. chem. Praktikum*, 3rd. Ed., p. 149.

By permanganate oxidation long side chains are broken down to carboxyl groups attached to the ring.

Sandmeyer's Reaction.—The ready formation of iodobenzene in the above method is due to the spontaneous decomposition of the diazonium iodide into iodobenzene and nitrogen

$$C_6H_5.\underset{I}{N}{\equiv}N \longrightarrow C_6H_5I + N_2 .$$

In the bromide and chloride, however, after the nitrogen has been eliminated, the halogen wanders only to a slight extent to the "gap" so left, and the formation of phenol predominates.

In 1884, Sandmeyer, however, made the important discovery that in the presence of the corresponding *cuprous salt* chlorine and bromine are also directed to the nucleus. This catalytic action has not yet been explained. Perhaps a double salt is formed, or else a complex salt in which the halogen is more firmly bound than in the simple halide. According to Gattermann, the cuprous salt may be replaced by *copper powder*. In general, the decomposition of labile diazo-compounds, by elimination of elementary nitrogen, is accelerated by copper.

The replacement of the amino-group by halogen is a process of great importance. There is no other preparative method of obtaining the aromatic iodo-compounds. The introduction of chlorine and bromine in this way is important because single homogeneous halogen derivatives can be obtained from the amine, whereas we know that such is not always the case when the parent substance is directly chlorinated or brominated. Thus when toluene is brominated (in the ring) both *o*- and *p*-bromotoluene are formed, and can be separated only with difficulty. But by means of Sandmeyer's reaction the two toluidines yield exclusively *o*- or *p*-bromotoluene; *m*-bromotoluene, moreover, can only be obtained from *m*-toluidine.

Sandmeyer's synthesis of aromatic nitriles is far more elegant than the removal of water from the ammonium salts of carboxylic acids, which latter reaction is also applicable to benzene derivatives. In particular, the former synthesis permits of the preparation of carboxylic acids *via* the nitriles, and so provides a complete substitute for Kolbe's synthesis (alkyl halide and potassium cyanide), which is inapplicable to aromatic compounds. The simplest example is the conversion of aniline into benzoic acid. The converse transformation is Hofmann's degradation (benzamide → aniline, see p. 143).

6. ARSANILIC ACID FROM *p*-NITRANILINE[1]

***p*-Nitrophenylarsinic Acid.**—Diazotise 13·8 g. of *p*-nitraniline in the manner described for the preparation of the *anti*-diazotate (p. 277). Dilute to one litre with water and ice, add 4 *N*-sodium

[1] H. Bart, *Annalen*, **1922**, **429**, 95.

hydroxide solution, with stirring, until the free acid is so far neutralised that Congo paper just fails to turn blue, and then pour the diazonium salt solution in a thin stream into a large filter jar containing 800 c.c. of 5 per cent solution of hydrogen disodium arsenite.[1] Stir the mixture with a glass rod. A vigorous evolution of nitrogen takes place and the reaction is complete almost instantaneously. Now concentrate the liquid to about 400 c.c. in a porcelain basin, and precipitate resinous by-products from the dark-coloured solution by making feebly acid with hydrochloric acid. When the precipitation is finished filter the liquid (now lighter in colour) through a folded filter and concentrate the filtrate, which is now acid to Congo paper, until crystallisation begins. On cooling 8–10 g. of faintly yellow needles of *p*-nitrophenylarsinic acid separate. If the solution is still too strongly coloured after filtration, boil with animal charcoal before concentrating.

The product should be readily soluble in cold sodium carbonate solution. Otherwise arsenic trioxide, which can thus be removed, is present.

Reduction.—Place 10 g. of iron powder (*ferrum reductum*), 100 c.c. of water, and 2 c.c. of concentrated hydrochloric acid in a 250 c.c. flask and attach to it the extraction apparatus shown in Fig. 28 (p. 34). Into the thimble put 6·5 g. of nitrophenylarsinic acid.

Heat the contents of the flask to boiling so that about every two seconds a (yellow) drop of solution falls back into the flask. The extraction should be complete in about half an hour. Continue the boiling for a quarter of an hour longer, add 25 c.c. of 5 *N*-sodium hydroxide solution, boil for five minutes more, and pour off from the bulk of the iron sludge into a Büchner funnel. Then boil out the sludge twice with 100 c.c. portions of hot dilute sodium hydroxide solution (about 0·2 *N*). Now concentrate the combined filtrates to 75 c.c., add concentrated hydrochloric acid until the liquid is just acid to Congo paper, and neutralise the excess of mineral acid with sodium acetate solution. The arsanilic acid separates on prolonged keeping. Recrystallise it from 40–50 c.c. of hot water, adding a little animal charcoal if necessary. Yield 3–4 g.

Experiment.—Demonstrate the presence of the primary NH_2-group by dissolving a small amount of the acid in a little sodium hydroxide solution, adding approximately one equivalent of sodium

[1] Prepared by dissolving 23·5 g. of powdered arsenious oxide in 240 c.c. of previously titrated 2*N*-sodium hydroxide solution and diluting to 800 c.c.

nitrite, cooling the solution by addition of ice and acidifying with hydrochloric acid. The diazonium solution when mixed with an alkaline solution of β-naphthol produces the corresponding red azo-dye. (Formulate !)

The introduction of the arsinic acid group into the aromatic nucleus is of great interest in connection with the therapeutic application of arsenic compounds in combating certain infectious diseases (*atoxyl* = sodium arsanilate ; *salvarsan*).

Arsanilic acid was first synthesised, in very poor yield, by melting arsenic acid with aniline :

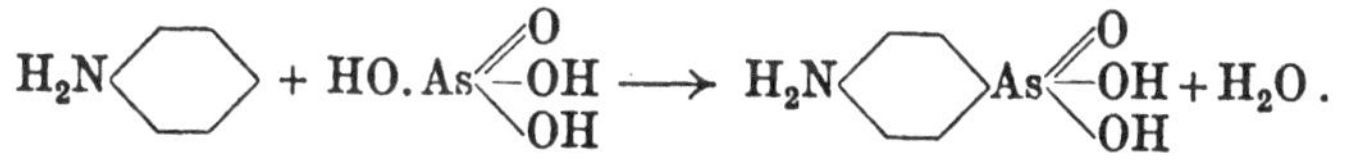

Compare this process with sulphonation and nitration, and note especially the difference between nitrogen and arsenic, which are introduced respectively as the neutral nitro-group and the corresponding hydrate, viz. the dibasic arsinic acid group.

The reduction of the arsinic acids to *arsenobenzenes* corresponds to that of the nitro-compounds to azo-compounds :

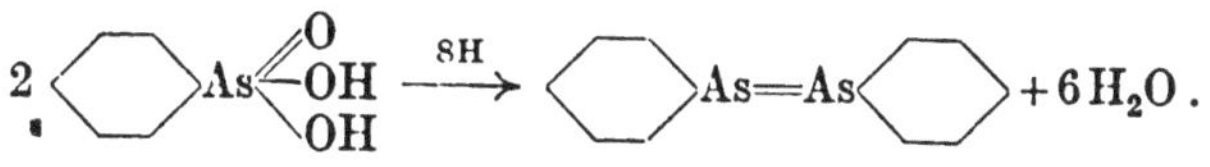

If the *p*-hydroxy-compound, obtained by boiling diazotised arsanilic acid with water, is nitrated, and the nitro-group so introduced is reduced to the amino-group, further reduction yields the corresponding arseno-compound *salvarsan*. (Formulate these reactions !)

The introduction of the arsinic acid group into the benzene ring, which generally takes place in Bart's reaction as above described, probably involves the formation of an intermediate product having a structure analogous to that of diazosulphonate (p. 287, footnote). This intermediate product does not undergo rearrangement to the stable *anti*-form so rapidly as does the diazosulphonate, but decomposes with evolution of nitrogen.

$$C_6H_5.\underset{Cl}{N}{\equiv}N + NaAs\begin{cases}=O\\-ONa\\-OH\end{cases} \longrightarrow C_6H_5.N{=}N.As\begin{cases}=O\\-ONa\\-OH\end{cases} + NaCl$$

$$\longrightarrow C_6H_5.As\begin{cases}=O\\-ONa\\-OH\end{cases} + N_2 .$$

7. PHENYLHYDRAZINE [1]

Dissolve 47 g. (0·5 mole) of aniline in a mixture of concentrated hydrochloric acid and water (100 c.c. of each) and diazotise the well-

[1] E. Fischer, *Annalen*, 1877, 190, 78.

cooled solution, in the manner repeatedly described, with a solution of 38 g. of sodium nitrite in 100 c.c. of water. Before diazotising the aniline, prepare as concentrated a solution as possible in water of 158 g. (1·25 mole) of neutral anhydrous sodium sulphite or of 315 g. of the hydrated (7 H_2O) salt. The sulphite content of this solution is equivalent to the amount of hydrochloric acid taken, and represents a 25 per cent excess in respect of the aniline.

The cheapest way of preparing the sulphite solution is to neutralise commercial bisulphite solution, *the content of which must have been determined by titration*, with the necessary amount of sodium hydroxide solution. Of good 40 per cent bisulphite solution 325 g. are required, and are neutralised with 110 g. of 50 per cent sodium hydroxide solution. The success of the experiment depends on the correct preparation of the sulphite solution. Pour the freshly prepared diazonium chloride solution rapidly into the cold sulphite solution contained in a round-bottomed flask (capacity 2 l.). A sample of the orange-red solution formed should not become turbid when boiled in a test tube. If, nevertheless, turbidity does develop, add more sulphite. Now add gradually 100 c.c. of concentrated hydrochloric acid in small portions with shaking. The colour of the solution is thus changed to yellow. Now heat on the water bath, add a few cubic centimetres of glacial acetic acid, and make the solution lighter in colour by dropping in a little zinc dust. Filter while hot, add without delay 300 c.c. of concentrated hydrochloric acid to the filtrate, and allow it to cool slowly.

Filter the crystals of phenylhydrazine hydrochloride at the pump, press the salt as dry as possible on the funnel, wash with hydrochloric acid (1:3), and then decompose in a separating funnel containing 150 c.c. of 4 *N*-sodium hydroxide solution and ether. Extract twice with ether, dry the ethereal solution of the base with anhydrous potassium carbonate, and finally distil the phenylhydrazine *in vacuo*, using an Anschütz-Thiele adapter (Fig. 17, p. 21). Boiling point 120°/12 mm. Yield about 30 g.

After being cooled for some time in water the preparation should solidify completely and it should dissolve in dilute acetic acid without turbidity. Melting point 23°. Distillation under atmospheric pressure is always accompanied by decomposition (N_2, NH_3, $NH_2.C_6H_5$, and C_6H_6) and does not yield pure phenylhydrazine.

The method of V. Meyer, reduction of the diazonium chlorides to arylhydrazines with strongly acid stannous chloride solution, is less elegant. The difference in the actions of stannous salt in acid and in alkaline solution should be noted.

Emil Fischer's classical method here described proceeds *via* the phenyl *anti*-diazosulphonate[1] which had already been prepared by Strecker and Römer, and frequently separates in beautiful orange-yellow crystals at the beginning of the reaction.

$$\begin{matrix} C_6H_5.N{\equiv}N \\ \quad Cl \end{matrix} + Na_2SO_3 \longrightarrow \begin{matrix} C_6H_5.N \\ \| \\ N.SO_3Na \end{matrix} + NaCl\,.$$

The sulphurous acid liberated in the second phase of the process by the addition of hydrochloric acid hydrogenates the azo-double bond, probably *via* an addition product A, of which one SO_3H-group is easily removed by hydrolysis with the formation of the sodium salt of phenylhydrazine sulphonic acid.

$$\text{A}\quad \begin{matrix} C_6H_5.N{-}NHSO_3Na \\ \dot{S}O_3H \end{matrix} \xrightarrow{H_2O} C_6H_5.NH{-}NH.SO_3Na + H_2SO_4\,.$$

The hydrogen evolved by the zinc dust serves to complete the hydrogenation. Finally, the more firmly bound sulphonic group is also removed, as sulphuric acid, by the hot concentrated hydrochloric acid.

Phenylhydrazine is prepared on a large scale by this method. The base is an indispensable reagent for characterising aldehydes and ketones (phenylhydrazones) and for many syntheses, particularly for the technical production of *antipyrine* and *pyramidone*. Study the course of these syntheses. The salts of phenylhydrazine contain only one molecule of acid.

Experiment.—Add three drops of glacial acetic acid to a mixture of five drops of phenylhydrazine and 5 c.c. of water. Then add two drops of benzaldehyde (on a glass rod) and shake. At first a milkiness appears, but very soon a flocculent precipitate of *benzylidene-phenylhydrazone* is formed. In this way very small quantities of benzaldehyde may be detected.

In sugar chemistry phenylhydrazine has become of outstanding importance for the separation, characterisation, and transformation of the various kinds of sugars. The fundamental results of this domain could

[1] The diazonium sulphite $\begin{matrix} C_6H_5.N{\equiv}N \\ \dot{S}O_3Na \end{matrix}$ which is doubtless formed first, undergoes spontaneous rearrangement to the diazotate-form; the same holds for the arsenites (p. 285) and cyanides.

$$\begin{matrix} C_6H_5.N{\equiv}N \\ \dot{C}N \end{matrix} \longrightarrow C_6H_5.N{=}N.CN\,.$$

hardly have been obtained without this reagent. When one molecule of a sugar reacts with one molecule of phenylhydrazine a *normal hydrazone* is formed, *e.g.*

$$\underset{\text{Dextrose}}{CH_2.OH.(CH.OH)_4.CHO} + C_6H_5.NH.NH_2 = CH_2.OH.(CH.OH)_4.CH{:}N.NH.C_6H_5 + H_2O\,.$$

If, however, an excess of phenylhydrazine is used, the base has an oxidising effect on the sugar, *i.e.* it removes hydrogen so that, for example, the CH.OH-group adjacent to the aldehyde group in the above example is dehydrogenated to a ketonic group which again reacts with the hydrazine. The substances formed, the *osazones*, have already been mentioned on p. 214. In the above case we obtain :

$$\begin{array}{l} CH_2.OH.(CH.OH)_3.C{-}CH{=}N.NH.C_6H_5 \\ \qquad\qquad\qquad\qquad\quad \| \\ \qquad\qquad\qquad\qquad\quad N{-}NH.C_6H_5 \end{array}\,.$$

Like all hydrazones, osazones lose phenylhydrazine when heated with hydrochloric acid. Naturally, it is not the sugar originally taken which is thus recovered, but an oxidation product, a so-called *osone.* In the example chosen the osone is :

$$CH_2.OH.(CH.OH)_3.CO.CHO\,.$$

When this is reduced it is not the ketonic group which is affected (which would regenerate the original sugar), but rather the aldehyde group, and the compound

$$CH_2.OH.(CH.OH)_3.CO.CH_2.OH$$

is obtained.

An aldose has been converted into a ketose, *d*-glucose into *d*-fructose.

Experiment.—Heat a solution of 2 g. of phenylhydrazine in dilute acetic acid (1·5 c.c. of acid, 15 c.c. of water) with 1 g. of *d*-glucose dissolved in 5 c.c. of water on the water bath at 80°. After about twenty minutes the osazone begins to separate in fine small yellow needles. After the lapse of an hour collect the crystals at the pump, wash with water, and dry in the air. Melting point 205°.

Phenylhydrazine can donate hydrogen, but, in certain circumstances, can also accept it; it can, therefore, both reduce and oxidise. In the former case *benzene* and *nitrogen* are formed by way of the already mentioned phenyldiimine (action of copper sulphate, ferric chloride, Fehling's solution, ammoniacal silver nitrate solution) ; in acid solution diazonium salts can be re-formed by careful oxidation.

Experiment. Benzene from Phenylhydrazine.—Allow 5 g. of

phenylhydrazine, dissolved in a mixture of 5 c.c. of glacial acetic acid and 10 c.c. of water, to run slowly into an ordinary distilling flask in which a solution of 25 g. of copper sulphate in 75 c.c. of water is heated to the boiling point. A vigorous evolution of nitrogen takes place and the benzene at once begins to distil with the steam. Collect and purify (as described on p. 278). Yield 2–3 g.

Like hydrazobenzene, phenylhydrazine decomposes when over-heated; one molecule hydrogenates a second.

$$2C_6H_5.NH.NH_2 \longrightarrow C_6H_5.NH_2 + NH_3 + (C_6H_5.N{=}NH) \longrightarrow C_6H_6 + N_2.$$

As in the case of hydrazobenzene, the decomposition is catalytically accelerated by finely divided platinum.

Test the behaviour of phenylhydrazine towards Fehling's solution and towards ammoniacal silver solution.

If sodium nitrite solution is run drop by drop into an aqueous solution of a phenylhydrazine salt the yellow, poisonous *a-nitrosophenyl-hydrazine* is formed; it can be converted, with elimination of water, into *phenyl azide.*

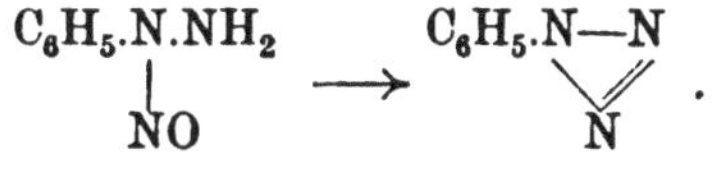

For details about the azides see p. 276.

Experiment. E. Fischer's Indole Synthesis.—Mix 2 g. of phenyl-hydrazine with 2 c.c. of acetone in a test tube. Water is eliminated and a turbidity appears. Suspend the tube in the boiling water bath for forty-five minutes, then add 6 g. of dry zinc chloride and heat the mixture for a few minutes with stirring in an oil bath at 180°. Now wash the dark-coloured melt into a small round-bottomed flask with four volumes of dilute hydrochloric acid and separate the resultant *a-methylindole* by distillation with steam. The substance collects as an oil which soon solidifies. After drying crystallise it from a little petrol ether. Melting point 59°.

Pine-Wood Reaction.—Dip a small piece of pine-wood in concentrated hydrochloric acid and then hold it in the vapour from boiling water containing a little methylindole. An intense red colour is produced.

The mechanism of this elegant, surprising, and widely applicable synthesis of indole derivatives was only explained recently (R. Robinson). It must be assumed that the keto-phenylhydrazones, in tautomeric hydrazo-form, undergo a species of benzidine rearrangement which, like the latter, can often occur even in dilute acid solution, *e.g.* with the phenylhydrazone of acetone.

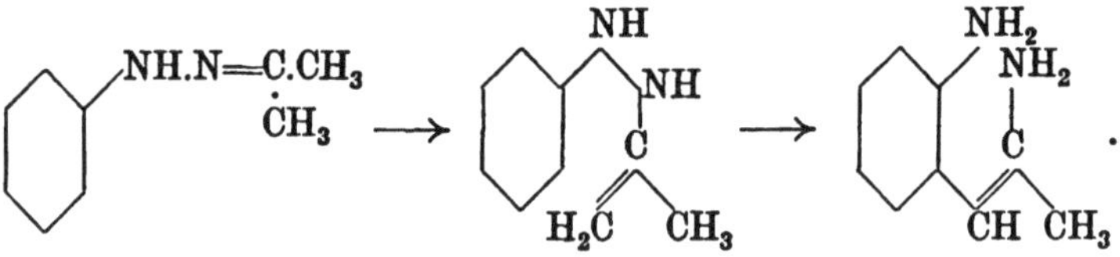

The last-formulated hypothetical diamine is transformed into an indole derivative by loss of ammonia, after the fashion of well-known analogues (1:4-diaminobutane to pyrrolidine).

8. PREPARATION OF AZO-DYES

(*a*) **Helianthine.**—Dissolve 20 g. of sulphanilic acid in 50 c.c. of 2*N*-sodium hydroxide solution and add a solution of 8 g. of sodium nitrite in 100 c.c. of water. Cool in ice and pour into 50 c.c. of 2*N*-hydrochloric acid. Next mix the solution of sodium diazobenzenesulphonate, so produced, with a previously prepared solution of 12 g. of dimethylaniline in 100 c.c. of *N*-hydrochloric acid, and make the mixture distinctly alkaline with sodium hydroxide. Very soon the sodium salt of the dye separates in beautiful orange-brown crystalline leaflets. Leave for several hours, then filter as dry as possible at the pump and, if desired, recrystallise the product, which is already fairly pure, from a little water. The yield is nearly quantitative.

Alternatively, dissolve 12 g. of dimethylaniline in a suspension of 20 g. of sulphanilic acid in 100 c.c. of water, cool in ice, and add the nitrite solution slowly. The sodium salt of the dye then separates directly.

As a variation diazotised anthranilic acid may be converted into " methyl red " by coupling with dimethylaniline.

The azo-dye obtained is the indicator *methyl orange* which is much used in alkalimetry. The yellow dilute solution of helianthine is turned red by acids.

The coupling proceeds according to the equation:

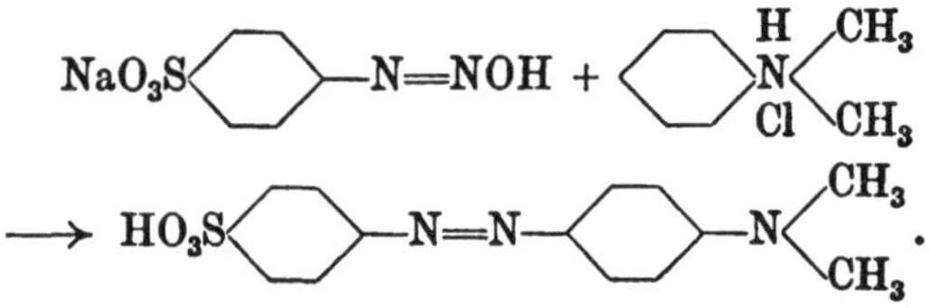

The yellow sodium salt is derived from this "azo-" form, whilst a red quinonoid salt is formed by the action of acids.

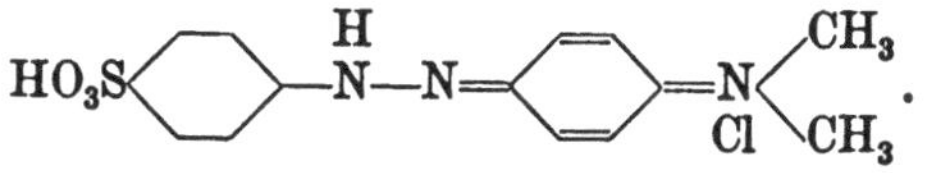

Probably the free acid, which is also red, has the constitution of an intramolecular quinonoid salt.

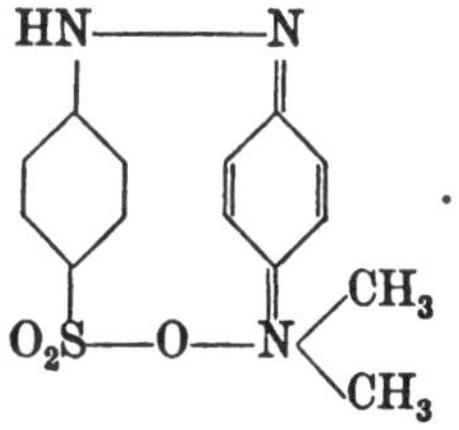

In view of the exceptional technical importance of the innumerable azo-dyes which are produced in this way, a general method, in use for their analysis, deserves mention. All azo-dyes are split on reduction by stannous chloride or also by sodium hydrosulphite; four hydrogen atoms are taken up and two molecules of primary amine are formed (in the case of polyazo-dyes correspondingly more). In contrast to the simple azo-dyes amino- and hydroxy-hydrazo-compounds are so unstable towards reducing agents that they are immediately decomposed at the hydrazine linkage. As a glance at the formula shows, helianthine, when reduced in this manner, yields sulphanilic acid and *p*-dimethylphenylenediamine. It is evident, then, that the nitrogen introduced by diazotisation ultimately appears in the "azo-component" (here dimethylaniline) as an additional NH_2-group, whilst the diazotised amine (the sulphanilic acid) is recovered in its original form.

Experiment.—Dissolve 3 g. of helianthine in the minimum amount of hot water and add hot stannous chloride solution (8 g. in 20 c.c. of concentrated hydrochloric acid) until decolorisation takes place. On cooling and rubbing with a glass rod *sulphanilic acid* crystallises. After some time collect it at the pump. Add excess of concentrated alkali hydroxide solution to the filtrate and extract it with ether. Dry the ethereal solution with a small piece of potassium hydroxide and evaporate the ether, when the *diamine* formed along with the sulphanilic acid remains; it can be recognised by the colour reaction described on p. 309 (Wurster's red). It becomes crystalline on cooling. Its *acetyl* derivative, which is obtained by warming the crude base on the water bath for a short time with 0·5 c.c. of acetic anhydride (in a test tube), is also suitable for characterising the diamine. The solution is diluted with water and, since the acetyl compound in virtue of the $—\mathrm{N}{<}^{\mathrm{CH_3}}_{\mathrm{CH_3}}$ group is still basic, the acetic acid is neutralised with sodium carbonate. Colourless crystals which can be recrystallised from water. Melting point 130°.

The method has also preparative significance since, ultimately, it is a way of introducing an amino-group. Compare the preparation of aminonaphthols from dyes of which the azo-components are α- or β-naphthol.

The development of the intense colour of the azo-dyes provides a delicate and distinctive test for primary aromatic amines. Since the naphthalene derivatives are more deeply coloured than those of benzene, it is customary to use not phenol but β-naphthol or the so-called R-acid (" acid for Red "), *i.e.* β-naphthol-3:6-disulphonic acid:

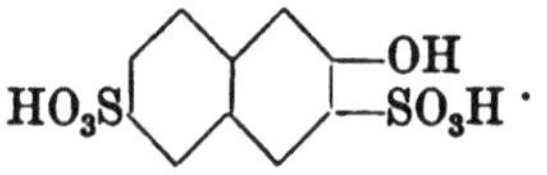

(*b*) **Congo Red.**[1]—Dissolve 4·6 g. of benzidine in a hot mixture of 12 c.c. of concentrated hydrochloric acid and 100 c.c. of water, add 150 c.c. more water, cool the solution to 2°–3°, and diazotise with 3·6 g. of sodium nitrite in 20 c.c. of water, added within the lapse of one minute. Leave the " tetrazo "-solution for five minutes and then pour it with stirring into a solution of 16 g. of sodium naphthionate and 20 g. of crystallised sodium acetate in 250 c.c. of water. When a sample of the liquid, on warming with hydrochloric acid, no longer evolves nitrogen, dissolve the blue-black precipitate of the dye-acid by warming with sodium carbonate and so produce the red sodium salt; filter and salt out the product from the filtrate with common salt (not too much). Collect the precipitate at the pump and wash with brine. The blue acid can be precipitated from the solution of the sodium salt with hydrochloric acid.

Congo red is a stronger acid than methyl orange; the change of colour requires a higher hydrogen ion concentration. Congo paper therefore serves to distinguish between organic acids and mineral acids.

Congo red is the prototype of the " substantive " cotton dyes of the benzidine group. The tinctorially valuable property of dyeing cotton directly (*i.e.* without a mordant) is doubtless to be attributed to the intimate adsorption of the colloidal particles of the dye on the fibre.

[1] After Möhlau-Bucherer, *Farbenchemisches Praktikum*, p. 156.

(*c*) *β*-**Naphthol Orange.**—Pour a mixture of 10 g. of sulphanilic acid with nitrite (4 g.), prepared as described under (*a*), with ice cooling and stirring, into 50 c.c. of 4 *N*-hydrochloric acid. Mix the sludge of *p*-diazobenzenesulphonic acid rather quickly, with stirring, with an alkaline solution of *β*-naphthol (8 g. in 100 c.c. of 2 *N*-sodium hydroxide solution) at room temperature. After a short time crystallisation of the sodium salt of the dye sets in (orange-yellow platelets). Complete the process by addition of saturated brine. Collect the product at the pump and wash with cold water. Yield 15–16 g.

The Coupling of Aniline

As we have already seen in the case of aniline, primary aromatic amines do not couple normally; a triazene derivative, *diazoamino-benzene*, is produced by linkage through the NH_2-group just as in the case of aliphatic amines, *e.g.* dimethylamine:

$$C_6H_5.N{:}NOH + NH_2.C_6H_5[HN.(CH_3)_2] \rightarrow C_6H_5.N{:}N.NH.C_6H_5[.N.(CH_3)_2].$$

Compounds of this type are soon reconverted by acids into diazonium- and amine-salt. In the case of the diaryltriazenes this reconversion is already caused by a quite feebly acid reaction, *e.g.* that of aniline salts in the presence of excess of base; but under these conditions the diazonium salt can combine with the amine present in excess to form an azo-dye (Rosenhauer). The equation

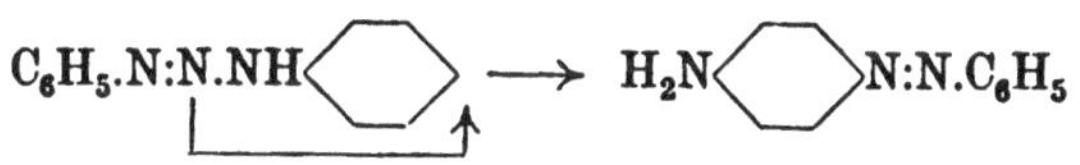

therefore gives information about the result only, but not about the *mechanism* of the reaction.

Diazoaminobenzene and *p*-Aminoazobenzene.—Diazotise 9·3 g. of aniline to the extent of one half under the usual conditions with half the amount of nitrite (3·8 g.) and add to the solution, with stirring, a solution of 25 g. of sodium acetate in 100 c.c. of water. After the liquid has become clear, filter at the pump, wash the yellowish-brown precipitate with water, and dry thoroughly, first on porous plate and then in a vacuum. Then recrystallise from petrol ether (boiling point 50°–60°). Melting point 98°. Heat a sample with dilute hydrochloric acid in a test tube.

Further, add 1 g. of *dry* finely powdered aniline hydrochloride to 5 g. of aniline and heat the mixture in a test tube on the water bath at 30°, with 2 g. of dry diazoaminobenzene. Continue heating the frequently stirred mixture for half an hour. Then raise the temperature to 45° and heat again for half an hour. When now a sample no longer evolves nitrogen on heating with hydrochloric acid, dissolve the aniline by adding 24 c.c. of 10 per cent hydrochloric acid (6 c.c. of concentrated acid and 18 c.c. of water). Recrystallise the *aminoazobenzene* hydrochloride, which remains undissolved, from 100 parts of hot water to which a little hydrochloric acid has been added. In order to obtain the orange-yellow base decompose the salt with sodium carbonate.

On the Theory of Dyes.—In carbon compounds absorption in the visible part of the spectrum, *i.e.* subjective colour, is conditioned by the presence of a so-called *chromophoric* group in the molecule. The nitroso-group is strongly chromophoric, the nitro-group much less so, whilst the azo-group is quite considerably chromophoric, but only in aromatic systems. Azomethane is colourless.

The intensely red azobenzene, however, is no more a dye than is nitrosobenzene. In order to make it into a dye, another group is necessary which, in virtue of its chemical nature, confers affinity towards the fibre. The most important of these groups, which are called *auxochromes*, are OH and NH_2. We have encountered their auxochromic influence in the simple cases of *o*-nitrophenol and the nitranilines.

Wool and silk are protein-like substances and hence are amphoteric. Accordingly, they can combine with acids as well as with bases. For this reason wool and silk can be dyed directly by dyes in virtue of their auxochromic groups.

It is otherwise with cotton, which is almost chemically pure cellulose, and hence is chemically indifferent in a tinctorial sense. Here combination with the dye results from the use of mordants which are adsorbed colloidally on the fibre before dyeing. The mordant can then enter into chemical union with the dye as a complex compound. For an important group of acid dyes (p. 326) the mordants are chiefly metallic hydroxides, namely, those of chromium, aluminium, iron, antimony, tin, etc., whilst for basic dyes tannin is the usual mordant.

In all cases, however, besides the *chemical* combination, the *physical* action due to surface adsorption plays a decisive part.

It is this action alone which enables a relatively small number of dyes, the so-called substantive cotton dyes, to be absorbed directly by the *unmordanted* vegetable fibre. The most important of these dyes are the bis-azo dyes, such as Congo-red and related compounds, which are derived from doubly diazotised benzidine. In aqueous solution they are present as sols and are colloidally adsorbed by the fibres as irreversible gels.

On the Coupling Reaction of the Diazo-Compounds

In its simplest form this reaction, by means of which the extremely numerous technical azo-dyes are manufactured, consists in condensation of aromatic diazo-compounds with phenols or aromatic amines to form azo-compounds. From the labile diazo-system the very stable azo-complex is produced. The azo-dyes, therefore, are, without exception, derivatives of azobenzene or else of azonaphthalene, etc.

It is a rule that combination with phenols takes place only in alkaline or neutral solution, whereas the aromatic amines couple in feebly acid solution (usually acetic acid).

The simplest azo-dye—devoid of technical importance, however—is formed from *phenyl diazotate* and *phenol*; the diazo-group takes up the *p*-position—in the case of β-naphthol the adjacent α- or *o*-position.

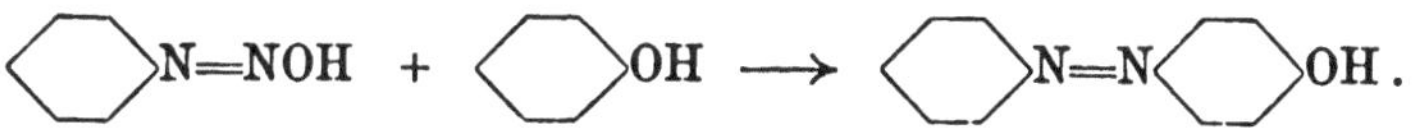

p-Hydroxyazobenzene

The coupling of aniline was already encountered above in connection with the testing of the diazonium solution for free aniline; after the addition of sodium acetate the presence of the latter was revealed by the appearance of the insoluble *diazoaminobenzene*.

It cannot be decided with certainty whether in this case the diazonium salt itself couples or whether, more probably, we must assume a partial hydrolysis to diazohydroxide and acid in the feebly acid solution. Our deductions will be based on the second explanation.

As follows from the preparation of helianthine, dimethylaniline condenses in exactly the same way as phenol: p-*dimethylamino-azobenzene* is formed.

The azo-dyes derived from phenols are called acid dyes, those derived from amines basic dyes. But since, in industry, the starting materials, not only *diazo-components* (diazotised amines), but also *azo-components* (coupled phenols or amines), are nearly always *sulphonic acids*, this distinction is pointless. The vast majority of the azo-dyes actually belong to the class of acid dyes. The object of the sulphonic group is to make the dye soluble in the dye bath.

The most important technical azo-dyes are derived from naphthalene, and indeed there is hardly any branch of organic chemistry which has been so thoroughly investigated in all its details as that of the intermediate products which are here involved. The number of possible combinations is almost without limit, and we can only refer to the specialist works, particularly F. Mayer, *Chemie der organ. Farbstoffe*, Berlin, 1924; Möhlau-Bucherer, *Farbenchem. Prakt.*, Berlin, 1926; Fierz-David, *Farbenchemie*, Berlin, 1924; Ristenpart, *Chem. Technol. der organ. Farbstoffe*, Leipzig, 1925.

The *mechanism of the coupling reaction* has been very exhaustively studied. Summarising first what has already been mentioned, it must be noted that the reaction is not confined to the aromatic series, for diazo-compounds condense also with enols and with the very closely related *aliphatic* aci-*nitro-compounds*. The final products of these reactions are not azo-compounds, but the *isomeric hydrazones* formed from them by rearrangement.

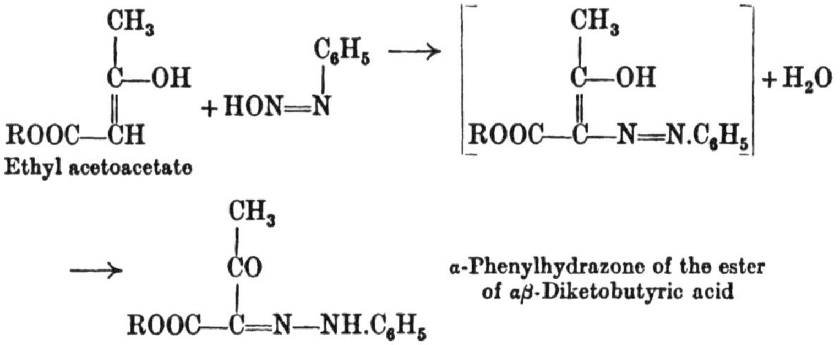

Ethyl acetoacetate

α-Phenylhydrazone of the ester of αβ-Diketobutyric acid

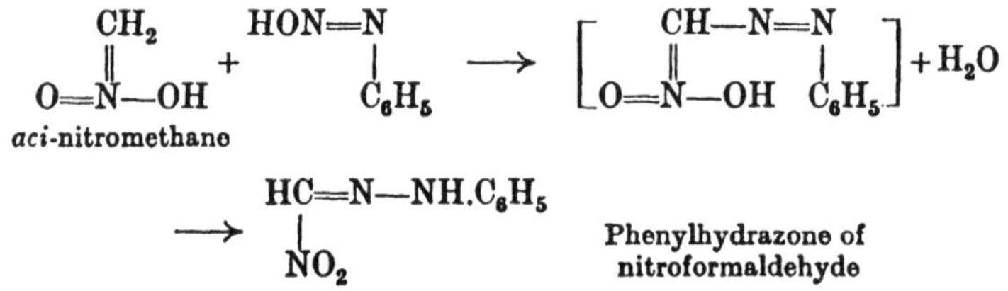

aci-nitromethane

Phenylhydrazone of nitroformaldehyde

Even the doubly unsaturated hydrocarbons, such as *butadiene*, can be coupled with suitable diazo-compounds. Finally, not only phenols, but also *phenol ethers*, such as *anisole*, are capable of coupling (K. H. Meyer[1]).

1. The simplest theory, developed by K. H. Meyer, traces the ability to couple to the activity of the double bond which adds itself to the diazohydroxide. By elimination of water the azo-compound or phenylhydrazone is formed.

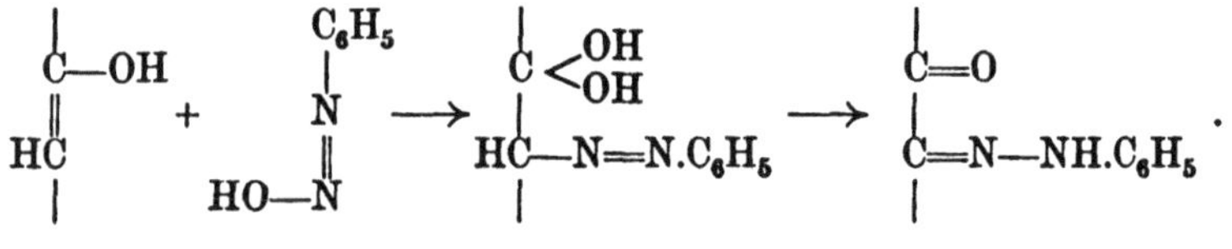

When the coupling takes place in the *p*-position the addition correspondingly occurs in the 1:4-positions.

2. Under certain conditions the first products of the reaction, in

[1] *Annalen*, 1913, 398, 66; *Ber.*, 1914, 47, 1741.

the case of nitrophenols, are *diazophenol ethers* (diazohydroxy-compounds), which then very easily undergo rearrangement

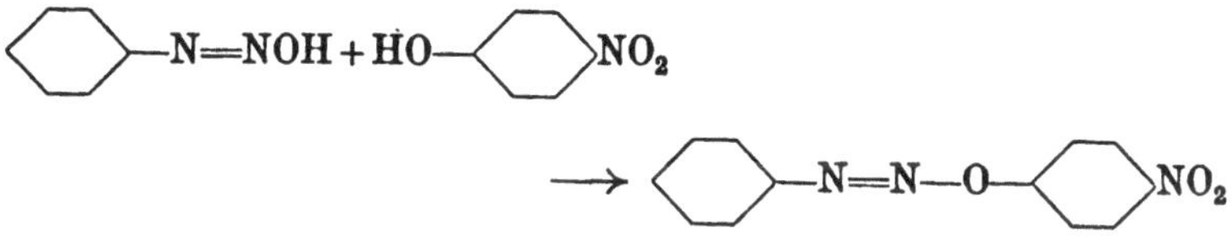

into the isomeric hydroxyazo-compounds (Dimroth [1]) just as diazoaminobenzene changes into *p*-aminoazobenzene. In the case of the tertiary amines we have also support for the view that the basic group first adds itself on to the diazo-complex. Thus, dibutylaniline couples with diazotised sulphanilic acid with elimination of a butyl group to form monobutylamino-azobenzenesulphonic acid (Karrer [2]). Apparently in this reaction a butyl radicle is displaced and the diazoamino-compound, which then immediately undergoes rearrangement in the way described on p. 293, is formed:

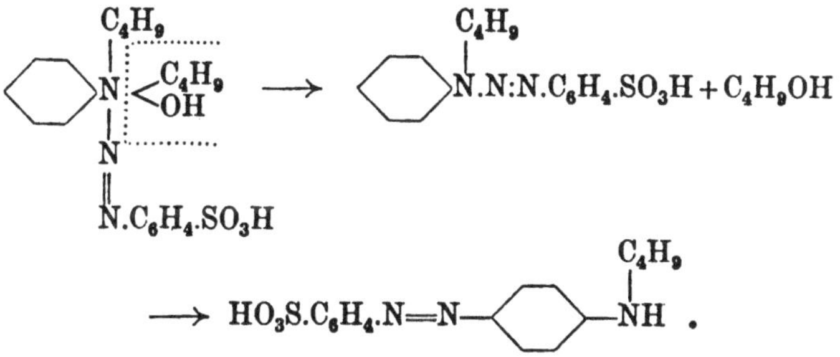

The coupling of the phenol ethers can be traced to the formation of an analogous primary oxonium compound:

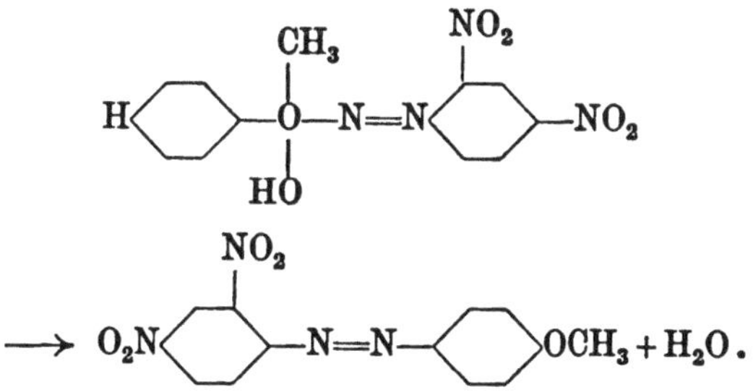

For the reaction with butadiene and other hydrocarbons, however, such an explanation fails.

[1] *Ber.*, 1907, 40, 2404, 4460; 1908, 41, 4012.
[2] *Ber.*, 1915, 48, 1398; Auwers, *Ber.*, 1914, 47, 1275.

3. The compound which couples adds itself to the double bond of the diazo-hydroxide either in virtue of (*a*) the mobile hydrogen atom or (*b*) the double bond :

(*a*) 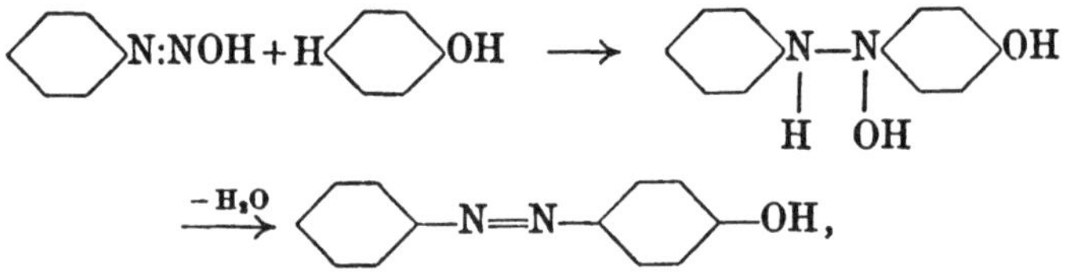

(*b*) 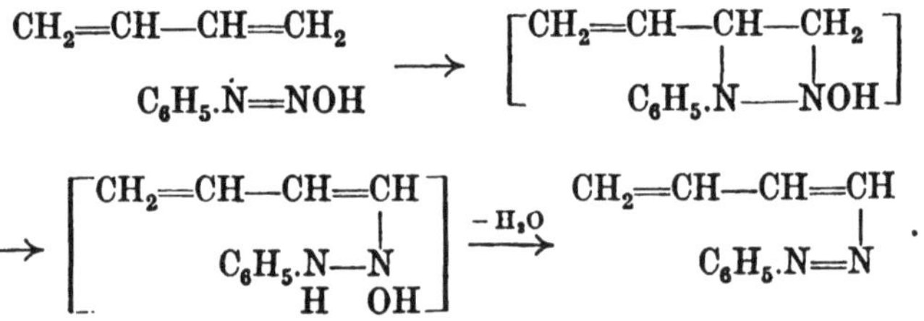

The method of reaction of the diazobenzene hydroxide is extraordinarily similar to that of *nitrous acid*; practically all substances which are capable of coupling also react with this acid. Comparison of the formulae shows that in the diazo-hydroxide the bivalent radicle $C_6H_5N<$ takes the place of the oxygen atom in nitrous acid. What has been said of the course of the coupling reaction can be applied almost without restriction to the analogous reactions of nitrous acid (cf. in this connection also p. 306). Whereas in coupling it is the phenylazo-radicle that is introduced, in the nitrous acid reactions it is the nitroso-group (nitrosophenol, nitrosodimethylaniline); whilst the products of coupling, *e.g.* in the case of enols, undergo rearrangement to phenylhydrazones, those from the action of nitrous acid change into oximes, *e.g.*

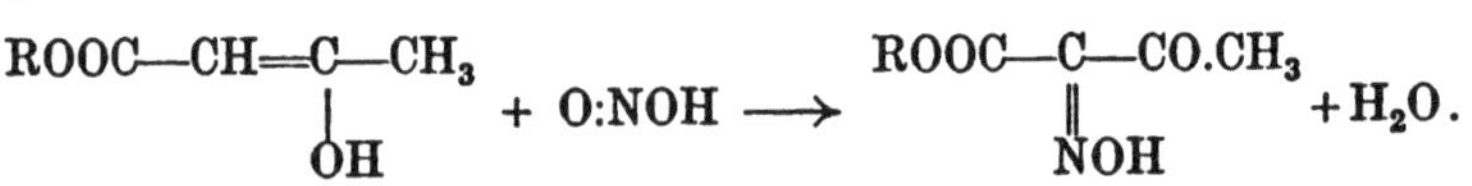

Ethyl acetoacetate — *Iso*-nitroso-derivative of ethyl acetoacetate

Accordingly, diazobenzene and nitrous acid are related to each other in the same way as are phenylhydrazine and hydroxylamine; the latter two substances, in fact, yield the same reaction products with carbonyl compounds as the former two do with the corresponding methylene derivatives.

It is consequently intelligible that certain quinones condense with phenylhydrazine to the same end products as are formed by the coupling of the corresponding phenols with diazobenzene hydroxide (Zincke), *e.g.*,

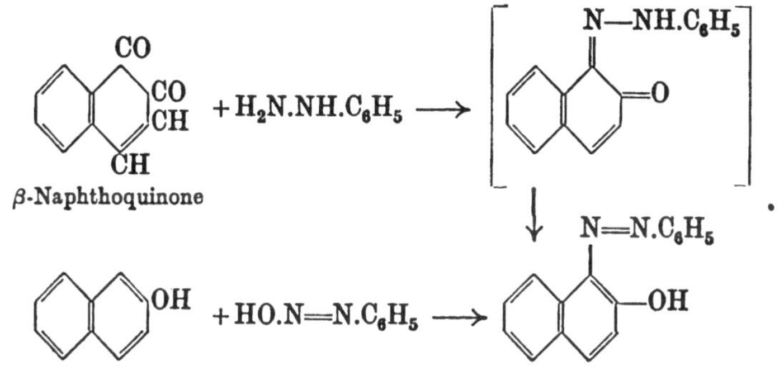

For reasons already frequently mentioned—cf., for example, pp. 93, 169, 186—the partially hydrogenated aromatic nucleus is not stable; hence the bracketed intermediate product tends to change into the "benzenoid" form, while hydrogen wanders from the nitrogen to the oxygen atom.

CHAPTER VIII

QUINONOID COMPOUNDS

1. QUINONE FROM ANILINE[1]

FROM a tap funnel a solution of 30 g. of sodium dichromate in 75 c.c. of water is gradually run, with stirring and ice cooling (Fig. 47, p. 136), into a filter jar containing a solution of 23 g. aniline (¼ mole) in a mixture of 100 c.c. of pure concentrated sulphuric acid and 500 c.c. of water. The temperature should not rise above 10°. The reaction mixture is then left in a cool place over night and next morning 40 g. of dichromate in 120 c.c. of water are added as before. The dark-brown solution, after standing for a further six hours, is filtered at the pump through a large Büchner funnel and the solid is washed with a little water. Then the filtrate is extracted twice, each time with 0·5 l. of ether; the ethereal solution is transferred to a distilling flask suitable for steam distillation and the ether is promptly distilled. The distilled ether, divided into two portions, is used to extract the reaction mixture again and the solvent is again removed from the extracts by distillation.[2] The residual crude quinone, to which the material on the funnel and the filter paper are added, is subjected directly to steam distillation, which drives the quinone into the receiver as magnificent golden-yellow crystals. Yield 14–16 g. These are dried, first for a short time between filter paper, and then in a non-evacuated desiccator over calcium chloride. Melting point 116°. On account of its great volatility it may not be kept for any length of time exposed to the air. (Test a sample.) Alcohol or petrol ether may be used for recrystallisation. The pure dry compound can be preserved for a long time.

[1] *Annalen*, 1838, 27, 268; 1842, 45, 354; 1882, 215, 125. *Ber.*, 1886, 19, 1467; 1887, 20, 2283.

[2] The distilled ether, which is yellow because quinone volatilises with it, can be made fit for other use by shaking with dilute alkali hydroxide solution which removes the quinone in the form of the dark brown "humic acid" salt.

p-Benzoquinone, commonly called simply quinone, is one of the most interesting of organic compounds, and is remarkable because of its colour, odour, and volatility. The steam distillation always involves not inconsiderable decomposition.

Quinone is produced in small yield by direct oxidation of benzene itself with silver peroxide, but better by the action of oxidising agents on a large number of its *p*-disubstitution products. Thus, in addition to quinol, *p*-aminophenol (experiment, p. 167), *p*-anisidine, *p*-toluidine, and sulphanilic acid as well as *p*-phenylenediamine and many of its derivatives yield quinone in this way.

More easily than benzene, *naphthalene* can be oxidised directly to (*a*)-quinone; this is the preparative method with *anthracene* and *phenanthrene*.

Although quinone is a derivative of dihydrobenzene and therefore not genuinely " aromatic ", its formation from a suitable precursor is favoured whenever an oxidising agent gives out energy.

The reactions of quinone consist essentially of various types of addition to the double bonds present in the molecule. This addition takes place:

(1) *at the C=C-double bond.*—Production of di- and tetra-bromides of quinone;
(2) *in the* 1:6-*positions at the two oxygen atoms.*—Addition of hydrogen to form quinol. The affinity in this reaction is so great that even aqueous sulphurous acid suffices for hydrogenation. (Experiment.) The dehydrogenation of quinol to quinone constitutes the converse process, an elimination from the 1:6-positions;
(3) *in the* 1:4-*positions at oxygen and carbon simultaneously*; derivatives of quinol are formed.

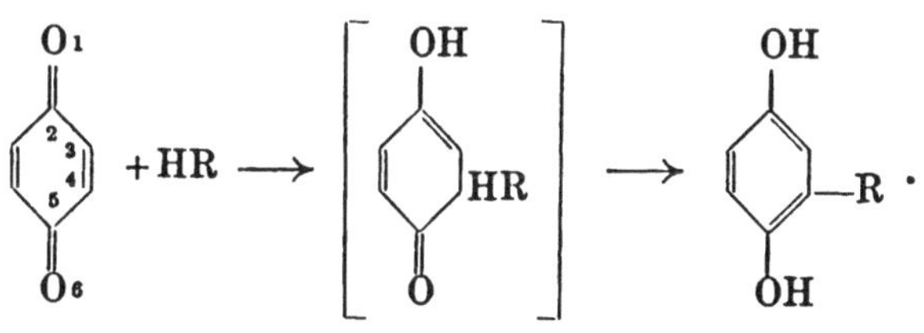

To this scheme most of the reactions of quinone conform, and all those which are of most importance, *e.g.* the addition of HCl, HCN, amines, thiophenol, thiosulphuric acid, acid chlorides, and acid anhydrides.

As an example the reaction with aniline may be selected. In accordance with the scheme formulated, the first reaction product obtained is *anilinoquinol* ($R = NH.C_6H_5$). The reaction does not stop at this stage, however. Between this first reaction product and unchanged quinone still present there promptly occurs a reciprocal hydrogenation and dehydrogenation characteristic of very many of the reactions of quinone.

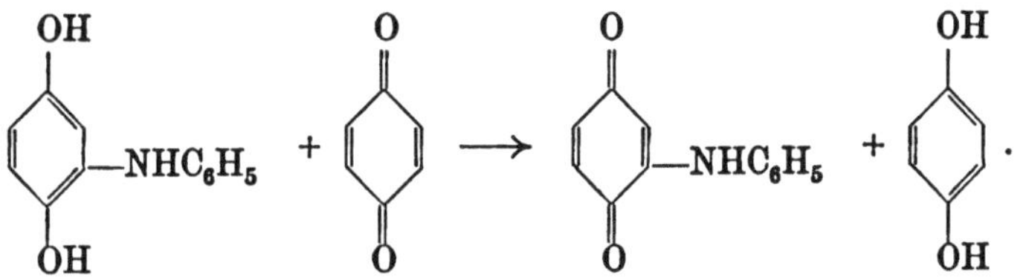

The great affinity of quinone for hydrogen leads to the formation of anilinoquinone which can now, in the same way, add a further equivalent of aniline to the half of the molecule not yet involved. The dianilinoquinol is converted into dianilinoquinone in the same way as the first reaction product was changed. (Write the equation.)

Experiment.—Quinol from Quinone. Suspend about 2 g. of quinone in 50 c.c. of water and while shaking frequently saturate the suspension with sulphur dioxide. Keep for some time and then extract the now colourless liquid twice with ether, dry the ethereal extract with calcium chloride, and evaporate the ether. The residue of quinol crystallises. Recrystallise it from a *little* water. Melting point 169°. Warm a sample with dilute sulphuric acid and a few drops of dichromate solution ; the odour of quinone is emitted.

Experiment.—Anilinoquinone.[1] Dissolve 4 g. of quinone in 400 c.c. of water. Cool the solution and add 1·72 g. of aniline dissolved in 10 c.c. of 20 per cent acetic acid. Leave the mixture in the cold for three hours with frequent shaking, then collect the reddish-brown crystalline precipitate at the pump, dry it *in vacuo*, and free it from the *monoanilino-compound* by repeated careful boiling with petrol ether (boiling point 80°–90°). From the petrol ether this compound separates on cooling in the form of small golden-brown needles. Melting point 119°. The insoluble portion consists of *dianilinoquinone.*

On the Formation of Quinone from Aniline.—The greenish-black intermediate product which is observed during the preparation of quinone is *aniline black.* This fact already shows that the conversion of aniline into quinone does not consist simply in the introduction of oxygen and elimination of nitrogen as the simplest equation would indicate. Quinone is rather the end-product of a whole series of complicated processes and is formed, as we shall show, by alternate dehydrogenation and hydrolysis (Willstätter [2]). Consequently, the NH_2-group of the aniline appears in the end-product as *ammonia.* That this is so can easily be shown by expelling dissolved ether from a little of the extracted

[1] H. and W. Suida, *Annalen*, 1918, **416**, 118.
[2] *Ber.*, 1907, **40**, 2665 ; 1909, **42**, 2147.

solution on the water bath, and then testing in the usual way for ammonia. It is probable that each time the aniline is dehydrogenated the first compound formed is the very unstable radicle-like *nitrogen-phenyl* (Bamberger, S. Goldschmidt).

$$C_6H_5.NH_2 \xrightarrow{-2H} C_6H_5N<.$$

In an alkaline or neutral medium this polymerises to azobenzene

$$2C_6H_5N< \longrightarrow C_6H_5.N{=}N.C_6H_5.$$

in an acid medium to a quinonoid substance *phenylquinonediimine.*

Phenylquinonediimine is at once further polymerised to the so-called emeraldine (in conformity with stage II on p. 313).

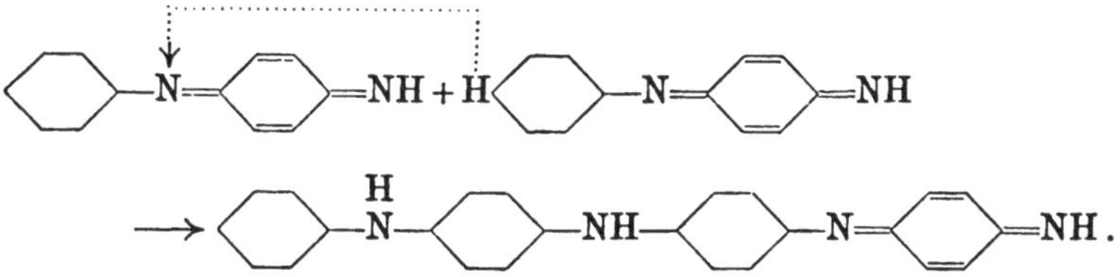

Emeraldine

According to a general rule, not only dihydric phenols, but also those diamines of the *p*-series which still contain one hydrogen atom attached to each nitrogen, are dehydrogenated to quinone or quinonediimine with great ease. Hence in the oxidation solution emeraldine is also immediately converted into the doubly quinonoid chain

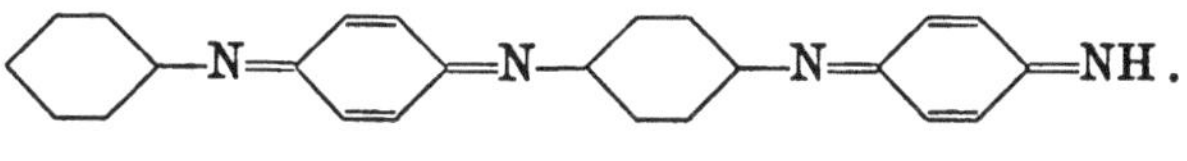

The polymerisation process repeats itself, and the resulting chain with eight benzene rings, three of which are quinonoid, is dehydrogenated exactly as before to a chain of four quinonoid rings. (Write out the formulae.)

Aniline black which is thus obtained is hydrolysed by acids like all derivatives of quinonediimine and like the *indamines*, to which aniline black belongs.

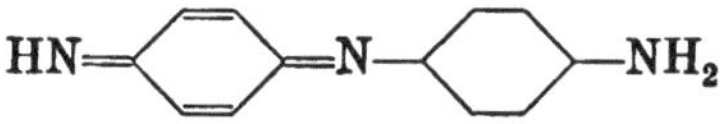

simplest indamine

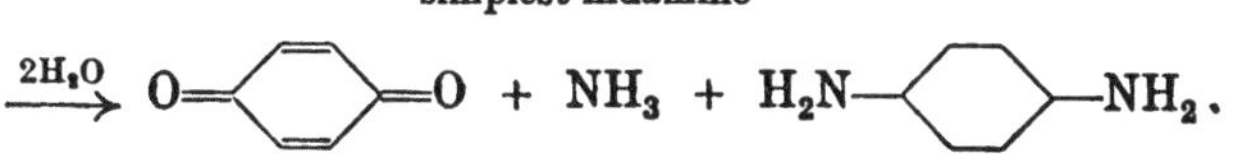

This phase of the whole process takes place slowly, and that is why the oxidation solution must be kept for a long time.

If this cleavage principle is applied to the molecule of aniline black it will be found that from it four molecules of quinone, three molecules of *p*-phenylenediamine, and one molecule each of aniline and ammonia arise. Since an excess of chromic acid is present, the *p*-phenylenediamine is readily dehydrogenated to quinonediimine, which is hydrolytically decomposed into quinone and ammonia. The single molecule of aniline begins the cycle anew.

Aniline black is also an important fast dye for cotton and is produced on the fibre at a high temperature (by " steaming ") from an aniline salt (hydrochloride) and an oxidising agent. It is not probable that the two kinds of this dye which are known have the same constitution, for they have different degrees of stability towards acid; technical aniline black probably contains phenazine ring systems (Bucherer, Green).

The oxidation of aniline by Caro's acid to nitrosobenzene, which proceeds in a quite different way, may be recalled here (p. 170).

Other Quinones.—Look up *o*-quinones and the three naphthoquinones; the best method of preparing α- and β-naphthoquinone is from 1:4- and 1:2-aminonaphthol respectively, which are obtained from azo-dyes of the two naphthols by reduction (see pp. 291, 292).

Experiment. Quantitative Determination of Quinone.[1]—Add 10 c.c. of a 10 per cent potassium iodide solution and 10 c.c. of 2*N*-sulphuric acid to 20 c.c. of the aqueous mother liquors from which the quinone has been filtered and immediately titrate the iodine which is liberated with 0·1*N*-thiosulphate solution. One c.c. of thiosulphate solution ≡ 0·0054 g. of quinone.

A more accurate method is given by Willstätter and Majima, *Ber.*, 1910, **43**, 1171.

Experiment.—Pour 1 c.c. of dilute sodium hydroxide solution into a test tube containing a small amount of quinone. The substance dissolves with production of a deep brown colour. Acidify the alkaline solution. Amorphous flakes are precipitated.

Quinone is very sensitive to alkalis, even to carbonate and ammonia, and is converted by them into a brown acid possibly identical with the natural humic acid of brown coal (lignite) (Eller). The mechanism of this reaction remains obscure.

Experiment.—On a small hand balance, having two sheets of paper counterpoised in the pans, weigh out equal quantities (about

[1] A. Valeur, *Compt. rend.*, 1899, **129**, 552.

0·5 g.) of quinone and quinol; dissolve the two substances separately in warm water and mix the solutions. Magnificent green needles of quinhydrone crystallise almost at once. After some time collect them at the pump, wash with water, and dry between filter papers over calcium chloride in a non-evacuated desiccator. Boil a few of the crystals with water and smell the vapours.

The combination of quinones and quinols in solution or when molten to form the deeply coloured *quinhydrones* is quite general. Other components of different origin, such as *p*-phenylenediamine and even quinol ethers, also take part in this remarkable union. The molecular proportions may also be other than 1:1. In general the quinhydrones only exist as such in the crystalline state: in solution they are to a very great extent dissociated into their constituents. Consequently they have no special reactions of their own, but react like a mixture of quinone and hydro-compound. Thus the simple quinhydrone is converted by oxidising agents into quinone, by reducing agents into quinol.

This labile behaviour of the quinhydrones has led to the view that in them the molecules are held together, not by normal, but by secondary valencies. The most interesting point about the quinhydrones is the *deepening of colour* which accompanies their formation. An explanation of this deepening may be found in the simultaneous presence of a quinonoid and a benzenoid ring in the same molecule. (An analogous bathochromic action caused by different stages of oxidation is encountered in the case of Prussian blue.)

2. *p*-NITROSODIMETHYLANILINE[1]

Dimethylaniline (40 g.; 0·33 mole) is dissolved in 250 c.c. of approximately 5 *N*-hydrochloric acid (one part of concentrated acid and one part of water) in a filter jar (capacity 1 l.). The jar is immersed in ice; 200 g. of ice are dropped in, and then, with good stirring—preferably with a mechanical stirrer—a cold solution of 25 g. of sodium nitrite in 100 c.c. of water is run in gradually from a dropping funnel (Fig. 47, p. 136). The temperature should not rise above 5° and no nitrous gases should be evolved. After the mixture has stood for one hour the orange-yellow hydrochloride is filtered thoroughly at the pump and washed several times with dilute hydrochloric acid (say 2 *N*). The salt is sufficiently pure for subsequent reduction and for the preparation of *dimethylamine*. A few grammes may be recrystallised from dilute hydrochloric acid, without heating to the boiling point, however.

[1] Baeyer and Caro, *Ber.*, 1874, 7, 810, 963.

Since it is best to use the moist preparation for the further experiments, the whole of the material is weighed and an aliquot portion is dried, first on porous plate and then in a vacuum desiccator or else on the water bath, so that the yield can be determined. It amounts to over 90 per cent of the theoretical.

The free base is obtained by rubbing the moist salt (5–10 g.) under sodium carbonate solution in a separating funnel and shaking with ether until the whole is dissolved. The beautiful emerald green solution is first concentrated on the water bath and then the solvent is allowed to evaporate from a flat dish exposed to the air. Large green plates of nitrosodimethylaniline remain. They melt at 80° and can be recrystallised from petrol ether (boiling point 60°–80°).

The tertiary aromatic bases share with the *phenols* the property of taking up a nitroso-group in the *p*-position when treated with nitrous acid in acid solution. As was mentioned elsewhere (p. 298), this reaction may be compared to the coupling of these compounds with diazobenzene.

Secondary amines such as methylaniline and diphenylamine also form nitrosamines very rapidly. By means of gaseous hydrochloric acid in anhydrous solvents they undergo rearrangement to the isomeric *p*-nitroso-bases.

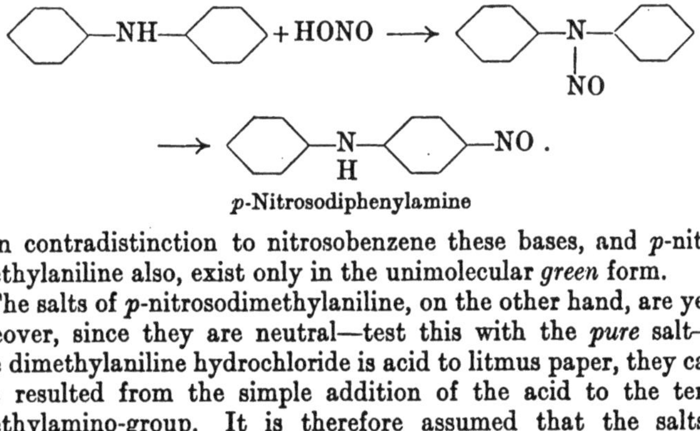

p-Nitrosodiphenylamine

In contradistinction to nitrosobenzene these bases, and *p*-nitrosodimethylaniline also, exist only in the unimolecular *green* form.

The salts of *p*-nitrosodimethylaniline, on the other hand, are yellow. Moreover, since they are neutral—test this with the *pure* salt—and since dimethylaniline hydrochloride is acid to litmus paper, they cannot have resulted from the simple addition of the acid to the tertiary dimethylamino-group. It is therefore assumed that the salts are formed by the addition of H and acid ion in the 1:7-positions, so that a quinonoid system results through a rearrangement:

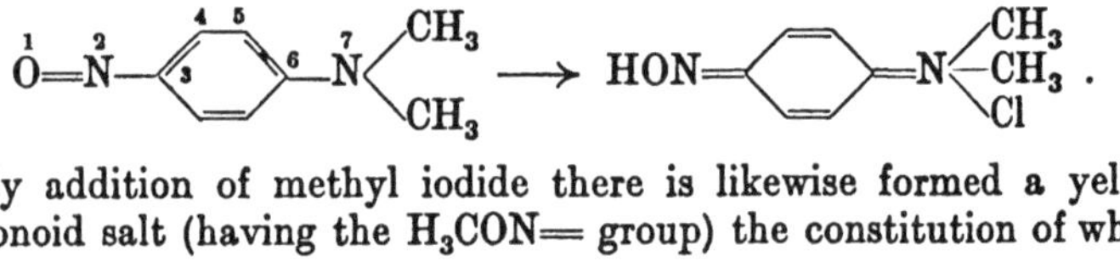

By addition of methyl iodide there is likewise formed a yellow quinonoid salt (having the $H_3CON=$ group) the constitution of which can be deduced with some degree of probability from the fact that the basic group is eliminated by alkalis not as trimethylamine but as *dimethylamine.*

DIMETHYLAMINE AND *p*-NITROSOPHENOL FROM *p*-NITROSODIMETHYLANILINE

A solution of 25 g. of sodium hydroxide in 500 c.c. of water is heated to boiling (porous pot!) in a distilling flask (capacity 1 l.) connected to a downward condenser which is provided with a receiver containing 60 c.c. of 2*N*-hydrochloric acid. Through the corked neck of the flask 18·6 g. of nitrosodimethylaniline hydrochloride—preferably the moist reaction product—are added in portions. Each addition is only made after most of the oily drops of the base from the preceding portion have dissolved, and finally the liquid is boiled until it has become reddish-brown. The *dimethylamine* produced is trapped in the hydrochloric acid in the receiver: at the conclusion of the distillation the contents of the receiver must still be acid. They are evaporated to dryness on the water bath in a small porcelain or glass basin, and finally the completely anhydrous salt can be recrystallised from a very small amount of absolute alcohol. Yield 5–6 g.

The *nitrosophenol* is precipitated from the cooled aqueous solution by acidification with dilute sulphuric acid and is extracted with ether in a separating funnel. After brief drying over calcium chloride the brownish-green solution is concentrated on the water bath. The sparingly soluble compound crystallises from the ether on cooling. Melting point 120°–130° (decomp.). Complete purification of nitrosophenol is difficult.

The nitroso-group in the *p*-position has a remarkable influence in making possible the hydrolytic removal of the dimethylamino group from the benzene ring. The reaction is used technically for the preparation of secondary amines. (*Trimethylamine* is obtained by heating ammonium chloride with formaldehyde.)

The tautomeric quinonoid formula O= 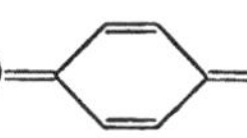 =NOH of *quinone monoxime* is also considered for *p*-nitrosophenol, although its whole chemical behaviour is in complete agreement with the phenolic structure. Like nitrosobenzene, nitrosophenol, when quite pure, is (very nearly) colourless and the solutions are olive-green. This would not be expected from the quinonoid formula.

The Liebermann Reaction.—Dissolve a small amount of nitrosophenol in a little molten phenol and add concentrated sulphuric acid. A magnificent cherry-red colour is obtained: when the melt is diluted with water and alkali is added the colour changes to blue.

Since phenol is converted to nitrosophenol by nitrous acid, even when the latter is combined in the form of the NO-group, labile nitroso-groups may be detected by the Liebermann reaction.

3. *p*-AMINODIMETHYLANILINE

In a short-necked, round-bottomed flask (capacity 500 c.c.) 100 g. of stannous chloride are dissolved in 120 c.c. of concentrated hydrochloric acid, and 38 g. (0·2 mole) of the crude *moist* nitrosodimethylaniline hydrochloride are gradually added in small portions with vigorous stirring or shaking. If the reaction does not set in at once the flask is warmed on the water bath; after a short time the salt which has been added should go completely into solution. The reaction must be so regulated that it goes on continuously without becoming altogether too violent.

At the end, the solution, which has become pale yellow, is cooled externally and internally (drop in some ice) and made alkaline with a solution of 150 g. of commercial sodium hydroxide in 300 c.c. of water;[1] most of the stannic acid which is at first precipitated goes into solution. The oily base set free is taken up in ether (irrespective of small amounts of undissolved stannic acid), the liquid is extracted once or twice more with ether, the ether is evaporated after drying for a short time over fused potassium carbonate, and the free base is

[1] The *electrolytic separation* of the tin is much more elegant. In cases where the reduction product cannot be extracted as base from the alkaline solution (amino-alcohols, amino-acids, and the like) this method is much to be preferred to the precipitation with hydrogen sulphide, and in the present case also it is to be highly recommended, if only for what it teaches.

The electrolysis is carried out in a medium-sized filter jar; two moderately thick carbon rods serve as electrodes. The cathode dips into the solution, the anode into 2*N*-sulphuric acid contained in a small porous cell, which also dips into the liquid. The electrodes are fixed a short distance apart and the current is obtained from two units, arranged in series, of an accumulator battery of the usual capacity. At a potential difference of 4 volts 1·5–2 ampères pass through the solution.

Application of Faraday's Law of Electrochemical Equivalents.—1 equivalent of $Sn^{++++} = \frac{118}{4}$ or 29·5 g. requires $\frac{96,500}{3600} = 26{\cdot}8$ ampère hours; with a current of 2 ampères, therefore, 13·4 hours.

Since, when the concentration of tin ions decreases, hydrogen is also evolved, the electrolysis lasts somewhat longer than the calculated time, but can perfectly well be allowed to proceed over night.

at once distilled in a vacuum. It passes over as an almost colourless liquid at 138°–140°/12 mm. Yield 18–20 g. (about 75 per cent). It solidifies on cooling. Melting point 41°. The free amine is unusually sensitive towards air. Already after a few hours the initially colourless preparation becomes brown. It can be preserved for some weeks if sealed in an atmosphere of nitrogen, but in air it can hardly be kept for a day. Its salts, on the other hand, are stable.

Hydrochloride.—A solution of the base in a small excess of hydrochloric acid (equal volumes of the concentrated 7 *N*-acid and water) may be evaporated in a porcelain basin on the water bath, and the residue may be dried completely in a vacuum desiccator over sulphuric acid and solid potassium hydroxide.

An almost general and very elegant way of obtaining the hydrochlorides of organic amines consists in neutralising them (until acid to Congo paper) with alcoholic hydrogen chloride and then gradually adding absolute ether with rubbing until precipitation of the salt takes place. Care must be taken not to precipitate the salt in an amorphous condition by too rapid addition of the ether. It is best to wait for the beginning of crystallisation, which usually shows itself by the appearance of a powdery coating on the places rubbed by the glass rod.

The base is acetylated by adding to it an equal weight of acetic anhydride, heating for a short time on the water bath, and then diluting with water. In order to isolate the still basic acetyl-compound the free acetic acid is just neutralised with sodium hydroxide. Melting point of the substance after recrystallisation from water 130°.

The diamine of which the preparation has just been described is of great importance from several points of view. The reactions concerned depend on the change which its salts undergo on oxidation, and therefore this change will be considered first.

Experiment.—Dissolve a few granules of the freshly prepared base (or of a salt) in a few drops of dilute acetic acid in a test tube and add, first, about 5 c.c. of water and a few small pieces of ice, and then a few drops of very dilute bromine water or a dichromate solution. A magnificent red colour appears. If somewhat more concentrated solutions are used and the oxidising solution is heated to boiling, the odour of *quinone* is perceived.

The typical transformation of all *p*-phenylenediamine derivatives by oxidising agents in acid solution consists in a change into a salt of the quinonediimine series. The dye just observed, so called "*Wurster's red*", was long regarded as a simple quinonimonium salt:

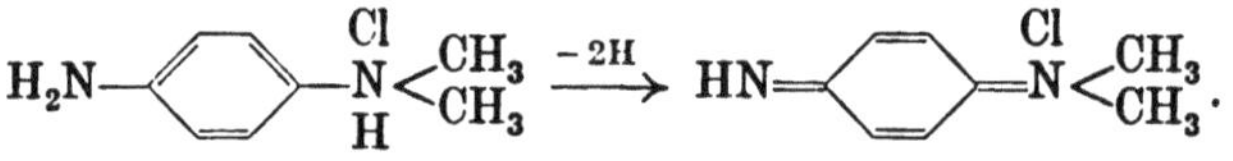

This, however, already appeared improbable when the (colourless) chloride of the simple quinonediimine became known (Willstätter).

Quinonediimine and its derivatives are reconverted by reducing agents into the corresponding phenylenediamines, It has been found that for the formation of Wurster's red the amount of oxidising agent required is equivalent to one, not to two H-atoms. Accordingly the reduction equivalent, which can be determined by titration with standard solution of stannous chloride, is also only half as great as was previously supposed. If a weighed amount of *p*-aminodimethylaniline salt is oxidised with dilute bromine solution of known titre, the point of maximum production of dye is reached when one equivalent of bromine has reacted with one mole of salt. If a second equivalent of bromine is added, the colour tone diminishes to *yellow*.

At this point the oxidation stage of quinonediimine has been fully reached; its (very unstable) salts have scarcely any colour. The production of colour only takes place when *quinonoid and benzenoid* systems are present together. The molecular union of the two substances at different stages of oxidation produces the intense absorption which is a prerequisite for the formation of a dye (Willstätter and Piccard). This union need not take place in the proportion 1:1, which obtains in the present case. The relations between quinhydrone and quinone-quinol are quite similar (p. 305).

In both cases the linkage between the molecules is labile and does not involve normal valencies. In general, "molecular compounds" are considered to be systems which are held together by the excess residual affinities of the components, by the mutual attraction of the molecular fields of force.

Willstätter's theory covers not only the class of *intermolecular* partially quinonoid (meriquinonoid) salts; it also provides a satisfactory conception of the true quinonoid dyes. The same principle, expressed *intramolecularly*, is encountered throughout, *i.e.* the two systems which are at different stages of oxidation (benzenoid and quinonoid) are here present in the same molecule. These relations are very clearly illustrated by *parafuchsine*.

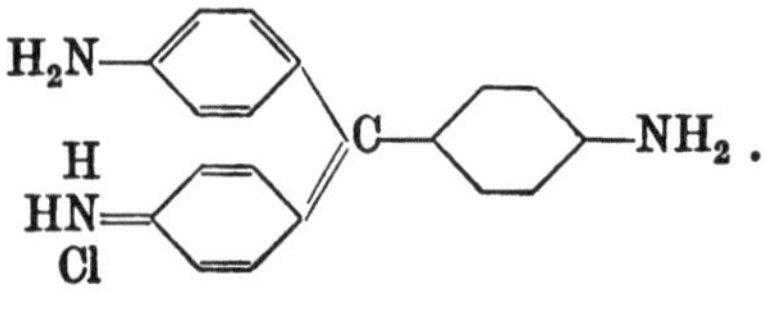

If we remove the NH_2-group from one benzene ring the resulting substance, "*Doebner's violet*", still has the character of a dye, since the above-mentioned conditions are still satisfied. But if the NH_2 is also lacking in the second benzene ring there results, as it were, a completely quinonoid (holoquinonoid) salt which, as we saw in the example from which we started, is no longer a dye. The tinctorial properties of the other quinonoid dyes, such as the indamines, safranines, methylene blue, etc., are to be explained in the same way. It is absolutely necessary to become familiar with the basis of this important theory through the following experiment.

Experiment.—Dissolve 1·3 g. of freshly prepared diamine base in 2 c.c. of glacial acetic acid which has been diluted with 10 c.c. of water and make the volume of the solution up to 95 c.c. in a measuring cylinder. Take 5 c.c. of this 0·1 *N*-solution in a conical flask (capacity 500 c.c.) and dilute further with 45 c.c. of ice-water. Before preparing the solution of the base, fill one burette with bromine solution (16 c.c. of saturated bromine water diluted with 280 c.c. of ice water) and another with approximately 0·02 *N*-stannous chloride solution, freshly prepared by dissolving 0·8 g. of tinfoil[1] or of thinly granulated tin in 4 c.c. of hydrochloric acid (1 : 1), and diluting to 500 c.c. with water which has previously been boiled.

Now keep the solution of the base cool in ice and rapidly run in bromine solution with continuous shaking. Observe that the fine red colour attains its maximum when about 25 c.c. of the bromine solution have been added, but that after the addition of 25 c.c. more the colour becomes much lighter. Because of secondary reactions a pure yellow colour is never produced ; as a rule, yet a little more bromine must be added rapidly. When the loss of colour has ceased run in stannous chloride solution immediately. After 25 c.c. have been added the beautiful colour of Wurster's red returns, only to disappear again on further reduction. The dye is also reformed by adding 5 c.c. of the original solution of the diamine salt.

Because of the great instability of the quinonoid salts it is necessary to work rapidly, using greatly diluted and cool solutions.

The *p*-quinonediimines are the parent substances of the indamine dyes and of the tricyclic quinonoid salts of the phenazine, phenthiazine, and phenoxazine groups which are related to them. (Detailed information on this subject is to be found in special works, *e.g.* of Nietzki-Mayer and of Bucherer.)

Let us consider the process of indamine formation, taking our quinone-

[1] Bear in mind that present-day " tinfoil " is almost always aluminium.

diimine base as the starting point. In accordance with a general reaction which is characteristic of all quinonediimines and also plays a part in the formation of aniline black from aniline (cf. p. 303), the dimethylquinonediimonium salt is capable of combining with great ease with one molecule of aniline or dimethylaniline:

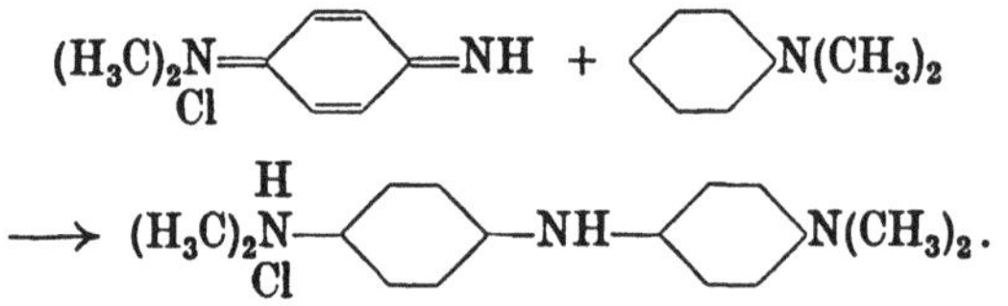

The new *p*-phenylenediamine derivative so formed is further dehydrogenated to a quinonoid indamine dye:

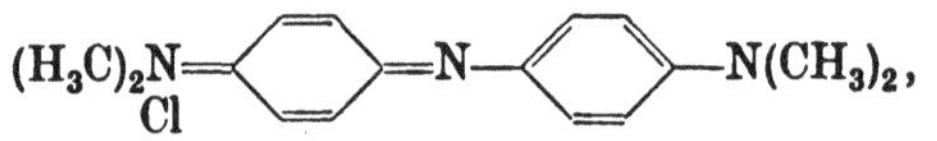

the constitution of which is in complete accord with Willstätter's theory. This dye, known as "*Bindschedler's green*", is of no practical importance since, like all indamines, it is easily hydrolysed, especially by acids:

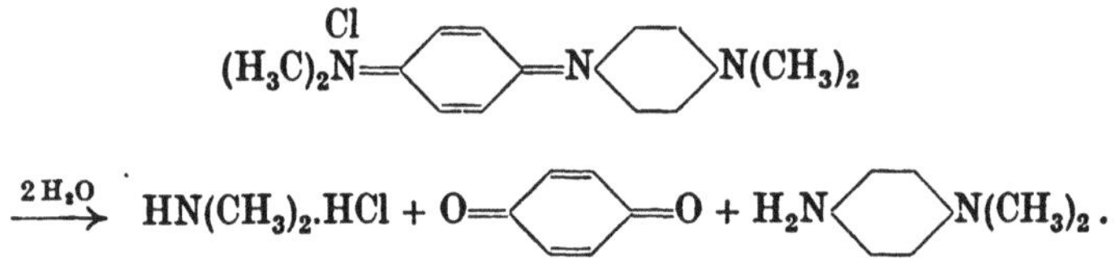

Preparation of Bindschedler's Green.[1]—Dimethyl-*p*-phenylenediamine (7 g.) and dimethylaniline (6 g.) are dissolved in a mixture of 40 c.c. of concentrated hydrochloric acid and 40 c.c. of water. Into the solution, which is kept cool in ice and is stirred mechanically or with a glass rod, a solution of 10 g. of sodium dichromate in 20 c.c. of water is slowly run from a dropping funnel. Zinc chloride solution (10 g. of salt in 20 c.c. of water) is now added and, especially on scratching, the beautiful zinc double salt of the dye crystallises. After half an hour it is collected at the pump and washed successively with cold water, alcohol, and ether. Yield 10–12 g. If well dried the dye can be preserved for a long time.

2–3 g. are placed in a distilling flask, along with 20 c.c. of 2*N*-hydrochloric acid, and distilled with steam through a water-cooled condenser. Characteristic yellow needles of *quinone* soon pass over.

[1] *Ber.*, 1883, 16, 464, 868; 1915, 48, 1087.

METHYLENE BLUE

If the oxidation which leads to the formation of "Bindschedler's green" takes place in the presence of *hydrogen sulphide* the process is varied by the entrance of sulphur. In principle, however, the mechanism is the same as before. The following formulae show that an addition of H—SH and a subsequent intramolecular addition of a mercaptan group are stages in the reaction :

I 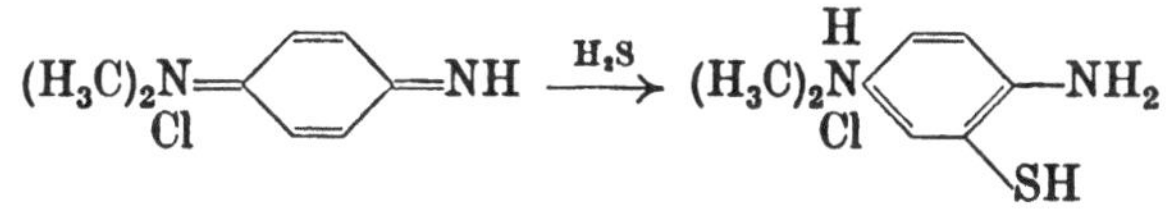

II 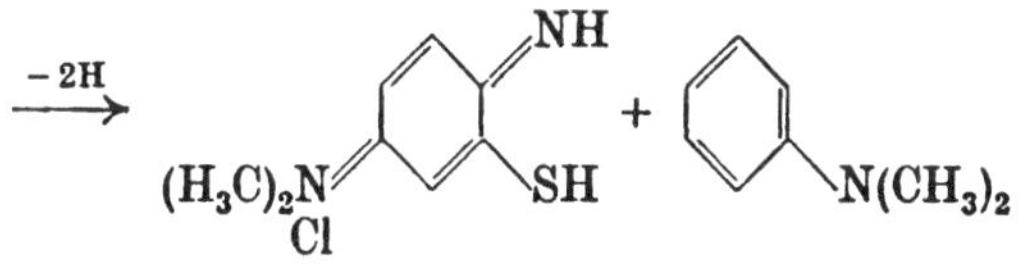

III 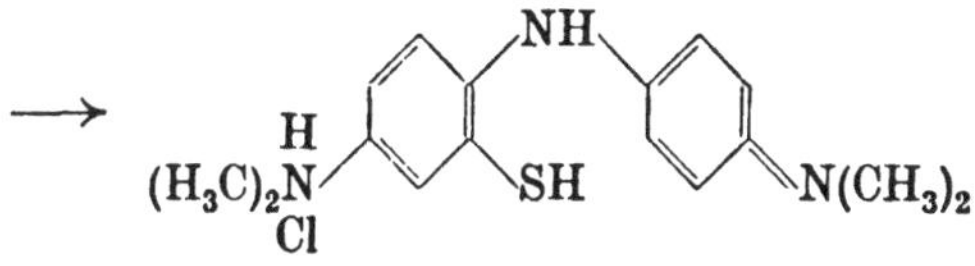

IV 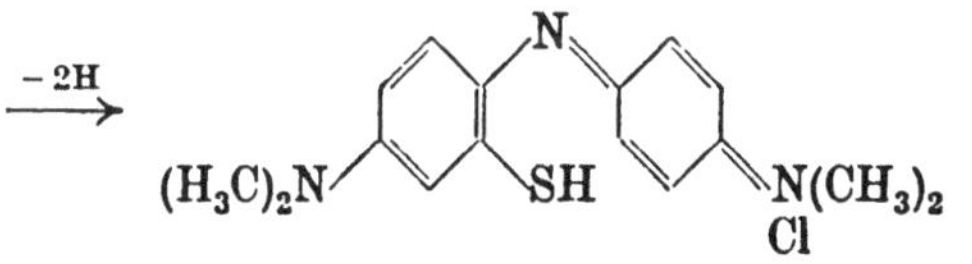

V 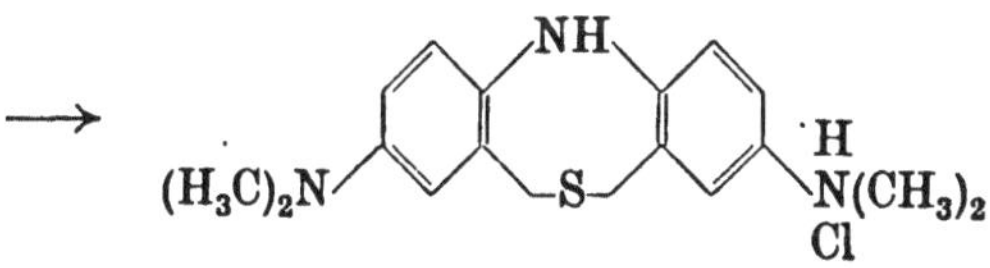

VI 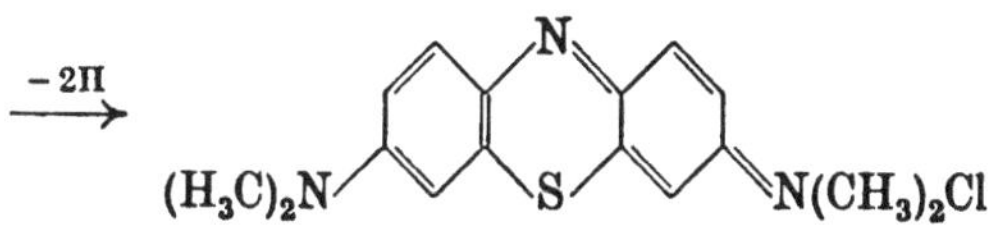

Methylene blue is also produced by the oxidation of dimethyl-*p*-phenylenediamine without the addition of dimethylaniline, and was discovered in this way by Caro in 1876. In explanation of this modification it is considered that the intermediate product (not isolated) at stage II. is probably converted into the substance IV., in which the other ring is quinonoid, by replacement of the NH-group by 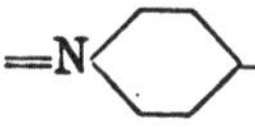 —$N(CH_3)_2$. The NH-group is eliminated as NH_3. All experience indicates that quinonoid linkages easily shift from one ring to another.

Bernthsen,[1] whose brilliant work elucidated the constitution of the dye, suggested a technical process in which, at stage I, *thiosulphuric acid* is added; its sulpho-group is eliminated as sulphuric acid in the course of the process.

Dimethyl-*p*-phenylenediamine (7·6 g.) is dissolved in 70 c.c. of *N*-hydrochloric acid, and 35 g. of zinc chloride in 50 c.c. of water are added. With good stirring 12 g. of aluminium sulphate in 20 c.c. of water and then 15 g. of sodium thiosulphate in 20 c.c. of water are poured in. To the solution thus obtained one-third of a solution of 16 g. of sodium dichromate in 30 c.c. of water is at once added and the temperature of the solution is raised as rapidly as possible to 40°. The addition of 6 g. of dimethylaniline dissolved in 8 c.c. of concentrated hydrochloric acid follows, and finally the remainder of the oxidising agent is poured in. Of course, all these solutions are prepared before the experiment is begun.

The temperature of the liquid is now raised, first quickly to 70° and then slowly to 85°; this causes precipitation of the dye. After a quarter of an hour the mixture is cooled to 50°; precipitated inorganic material is dissolved by adding 15 c.c. of concentrated sulphuric acid and, when the mixture has been cooled completely, the crude dye is filtered as dry as possible at the pump. The product is at once dissolved in 200–300 c.c. of boiling water; the solution is filtered and left to crystallise over night after 20 g. of concentrated zinc chloride solution (1 : 1) and 40 g. of finely powdered common salt have been added. The beautiful crystals, which have a red lustre, are filtered as dry as possible at the pump and washed twice with a little ice-cold water. Yield 10–12 g.

The aqueous solution of methylene blue is not decolorised by sodium hydroxide; the blue water-soluble ammonium base is formed. The dye is converted by reducing agents into the easily oxidisable leuco-base. The following experiment, which demonstrates the formation of methylene blue by the introduction of sulphur between the rings of Bindschedler's green, is very instructive.

Experiment.—Pass hydrogen sulphide slowly into the most concentrated aqueous solution of Bindschedler's green obtainable, until after some time the colour has changed to yellowish-red. Now add dilute hydrochloric acid and the solution of 0·3 g. of sodium

[1] A. Bernthsen, *Annalen*, 1885, **230**, 73; 1889, **251**, 1; see especially pp. 49, 69, 79; Fierz-David, *Farbenchemie*, 2nd Ed., 1922, p. 186.

dichromate. Precipitate the methylene blue formed by adding zinc chloride solution.

4. BASIC TRIPHENYLMETHANE DYES

(a) Malachite Green from Benzaldehyde and Dimethylaniline [1]

Preparation of the Leuco-Base.—Zinc chloride (10 g.) is fused in a porcelain basin, cooled, and powdered. It is then added to a mixture of 25 g. of dimethylaniline and 10 g. of benzaldehyde (both freshly distilled) and the whole is heated in a porcelain basin on the water bath with frequent stirring for four hours. By the addition of hot water the viscous mass is liquefied on the water bath and the hot liquid is poured into a half-litre flask into which steam is passed until drops of oil cease to distil. After the liquid has cooled the water is poured off and the residue is washed several times with water. When as much as possible of the water has been removed the material in the flask is dissolved by adding alcohol and warming on the water bath, and the solution is filtered. On leaving the filtrate over night in a cool place the base separates in colourless crystals which are collected at the pump, washed with alcohol, and dried in air on several folds of filter paper. A second crop of crystals can be obtained by concentrating the mother liquor. Should the base not crystallise, but separate as an oil—as often happens after the filtered solution has stood for a short time—it follows that too little alcohol has been used. In such cases somewhat more alcohol is added and the mixture is heated until the oil dissolves. Yield 20–24 g.

Oxidation of the Leuco-Base.—In a round-bottomed flask (capacity 1·5 l.) 11 g. of the dried preparation are dissolved in 80 c.c. of 2*N*-hydrochloric acid and the practically colourless solution is diluted with 800 c.c. of water. To the diluted solution, which is kept cool in ice and continually shaken, a suspension of 9 g. of lead dioxide in 30 c.c. of water is added slowly in small portions. Since commercial lead dioxide often reacts badly the oxidising agent should be prepared in the laboratory by the method given below.

To the solution of the dye so obtained (and filtered if necessary) a clear concentrated solution of zinc chloride (12 g.) is added, and then the zinc chloride double salt of the dye is precipitated by addition of *saturated* brine. (Test with a spot on filter paper to see whether

[1] Otto Fischer, *Annalen*, 1881, 206, 83; 1883, 217, 250.

the solution is nearly colourless.) To purify the dye, which has been collected at the pump and washed with brine, it is *redissolved* in hot water and, after cooling, salted out as above. Yield 9–10 g.

Lead Dioxide.—Good bleaching powder (50 g.) is vigorously shaken with water (750 c.c.); the mixture is filtered through a pleated filter and the filtrate is added gradually in small portions to a solution of lead acetate (25 g. in 100 c.c. of water) which is heated in a porcelain basin on the boiling water bath. When the originally pale precipitate has become dark brown a sample of the filtered solution is heated with more bleaching powder solution. If precipitation again occurs, more oxidising agent is added. Finally, the solution is decanted and the lead dioxide is repeatedly digested with hot water, collected at the pump, and thoroughly washed. The moist paste, dried as far as possible at the pump, is preserved in a closed glass bottle.

If no fresh bleaching powder is available, a hypochlorite solution is prepared by passing chlorine into an ice-cooled solution of 15 g. of sodium hydroxide in 250 c.c. of water, until the increase in weight is 14 g.

Assay of the Lead Dioxide Paste.—Although by drying and weighing a known amount of the moist lead dioxide its content of the pure substance can be approximately determined, the amount of active oxygen should nevertheless be ascertained with a view to greater accuracy. For this purpose 0·3–0·5 g. of the moist material (taken from the centre of the batch) is weighed out and transferred to a small conical flask (150 c.c.) into which a solution of 2 g. of potassium iodide in 5 c.c. of water is poured with vigorous shaking. The mixture is now cooled in ice and 15 c.c. of concentrated hydrochloric acid are added. After diluting the solution thus obtained to about 50 c.c. the liberated iodine is titrated with 0·1*N*-thiosulphate solution, prepared with sufficient accuracy by dissolving 6·2 g. of *pure* sodium thiosulphate ($Na_2S_2O_3 + 5H_2O$) in water and diluting in a measuring flask to 250 c.c. 1 c.c. of 0·1*N*-thiosulphate = 0·012 g. PbO_2.

(*b*) Crystal Violet from Michler's Ketone and Dimethylaniline

A mixture of 24 g. of dimethylaniline, 10 g. of Michler's ketone, and 10 g. of phosphorus oxychloride is heated in a dry open flask on the vigorously boiling water bath for five hours. The blue melt is then poured into water, made alkaline with sodium hydroxide, and submitted to steam distillation until oily drops of unchanged dimethylaniline cease to pass over. After the residue in the distilling flask has cooled and solidified, the alkaline liquid is removed by filtration and the colour base is washed with water and thoroughly boiled with a mixture of 500 c.c. of water and 4 c.c. of concentrated hydrochloric acid. The boiling blue solution is decanted through a folded filter and the undissolved colour base is repeatedly boiled with smaller amounts of acid of the *same concentration* as already used, until dissolution is practically complete. After cooling the solution of the dye, finely powdered common salt is added with stirring until the substance is precipitated. The precipitate is collected at the pump and crystallised from a little water. On cooling, the crystal violet separates from the solution in the form of stout crystals having a bronze lustre. The dye is collected by filtration and dried on filter paper in the air. Yield 14–15 g.

When boiling out the dye as little hydrochloric acid as possible should be used, since the much more soluble acid salt of the dye is formed if an excess of acid is employed.

5. FLUORESCEIN AND EOSIN[1]

Phthalic anhydride (15 g.) is intimately ground in a mortar with 22 g. of resorcinol, and the mixture is heated in an oil bath to 180°. For this purpose an internally glazed meat-extract jar is convenient; it is suspended in the bath by means of a wire triangle fastened round the neck of the jar. 7 g. of zinc chloride, previously dehydrated by fusing and then powdered, are stirred into the molten mass during the course of ten minutes. The temperature is then raised to 210° and the heating is continued until the mass, which becomes progressively more viscous, has completely solidified; for this one to two hours are required. By means of a sharp instrument, preferably a chisel, the brittle melt obtained on cooling is chipped out of the jar, powdered finely, and boiled for ten minutes in a porcelain basin with 200 c.c. of water and 10 c.c. of concentrated hydrochloric acid. In this way the unchanged starting materials and the basic zinc salt are dissolved. The fluorescein is then separated from the aqueous liquid by filtration, washed with water until the filtrate is no longer acid, and dried on the water bath. Yield, almost quantitative. Dissolve a particle of the preparation in a little ammonia and dilute in a beaker with one litre of water.

[1] Baeyer, Caro, *Annalen*, **1876**, **183**, **1**.

Eosin.—Into a flask containing 16·5 g. (0·05 mole) of fluorescein under 80 c.c. of alcohol, 36 g. (12 c.c.) of bromine are dropped with shaking from a tap funnel during the course of twenty minutes. When half the bromine has been added, and the fluorescein has been converted into dibromofluorescein, all the solid material disappears temporarily, since dibromofluorescein is soluble in alcohol, but later the sparingly soluble eosin begins to crystallise.

After standing for two hours the mixture is filtered and the precipitate, after being washed several times with alcohol, is dried on the water bath. The material so obtained still contains alcohol of crystallisation, which is removed by drying at 110°, when the colour becomes lighter. Observe the magnificent fluorescence obtained on greatly diluting an alkaline solution of a trace of the substance.

Ammonium Salt of Eosin.—Place a very stout filter paper on a flat-bottomed crystallising basin which is one-third full of concentrated aqueous ammonia solution, spread eosin to a depth of about 0·5 cm. on the paper, and cover the whole with a funnel. Very soon the light red crystals acquire a darker colour, and after about half an hour they are completely converted into the ammonium salt, which forms dark red crystals having a green iridescence. When a sample of the material dissolves wholly in water the reaction is known to be complete.

Sodium Salt of Eosin.—Grind 6 g. of eosin with 1 g. of anhydrous sodium carbonate, transfer the mixture to a moderate-sized, wide-necked, conical flask, moisten with a little alcohol, add 5 c.c. of water, and warm on the water bath until evolution of carbon dioxide ceases. To the aqueous solution of the sodium salt thus obtained now add 20 g. of alcohol, heat to boiling, and filter the hot solution. From the cooled filtrate there separate beautiful brownish-red crystals with metallic lustre, often only after long standing. Collect them at the pump and wash with alcohol.

Triphenylmethane Dyes : Theoretical Considerations

A. **The Basic Series.**—In general, aromatic aldehydes condense with aromatic amines in the presence of zinc chloride to form triphenylmethane derivatives (O. Fischer) ; phenols and phenyl ethers behave similarly in the presence of concentrated sulphuric acid (Baeyer). The products formed are the leuco-compounds of well-known dyes.

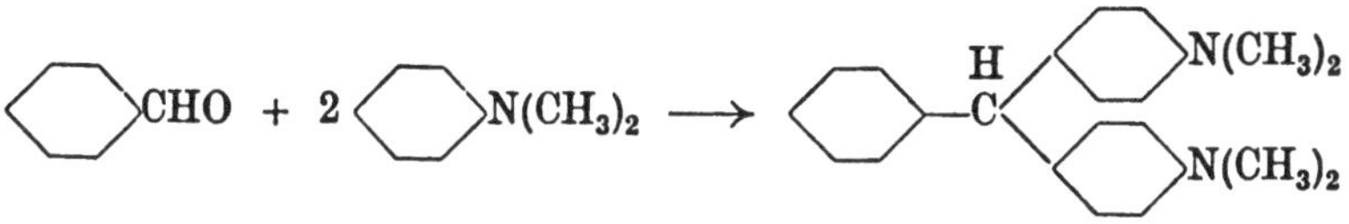

Leuco-malachite green

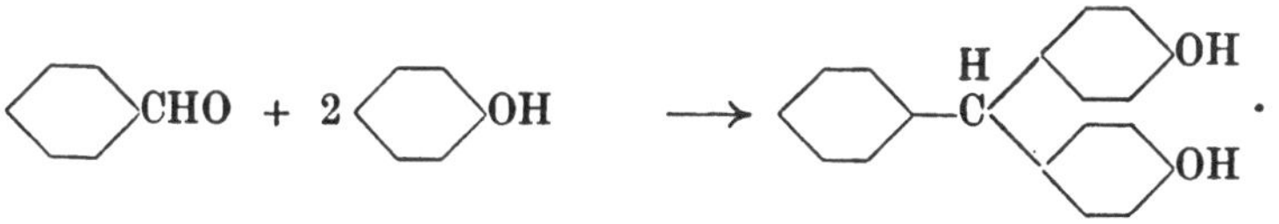

Leuco-benzaurin

When pure, these leuco-compounds dissolve in acids or alkalis to give colourless solutions. *Leuco-malachite green* is a weak base of the aniline type. The dye is formed by oxidation of the base in acid solution ; as in the case of dimethyl-*p*-phenylenediamine (p. 310), two hydrogen atoms are removed from *p*-positions

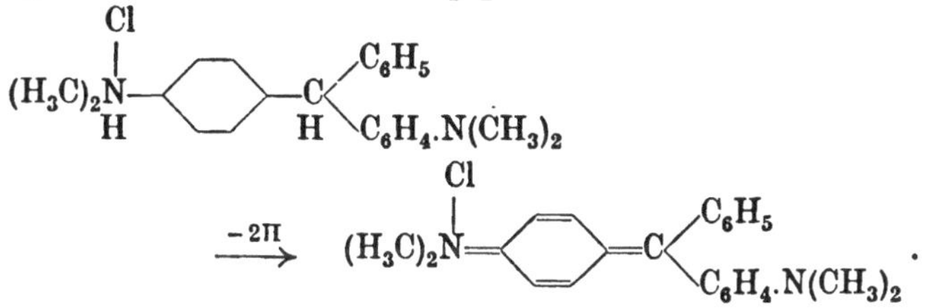

The formation of the acid dye *benzaurin* takes place in a quite analogous manner. *p*-Tetramethyldiaminobenzophenone (Michler's ketone), which is prepared technically from dimethylaniline and phosgene, can, like benzaldehyde, be condensed with dimethylaniline in the presence of phosphorus oxychloride. In this case, however, it is not a methane derivative, a leuco-compound, which is obtained, but, as the equation shows, the carbinol stage is at once reached. As will soon be shown, however, this stage in acid solution is equivalent to the formation of the dye.

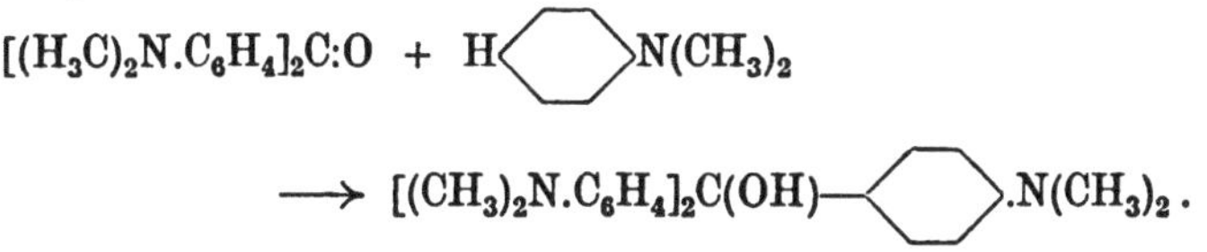

Michler's ketone is converted into the *carbinol base* of *malachite green* by the action of phenyl magnesium bromide (the Grignard reaction)

$$(H_3C)_2N.C_6H_4.CO.C_6H_4.N(CH_3)_2 + C_6H_5MgBr \longrightarrow (H_3C)_2N.C_6H_4.\underset{\displaystyle C_6H_5}{\underset{|}{C}}(OH).C_6H_4.N(CH_3)_2 .$$

The *technical preparation of crystal violet* and of its methyl-free parent substance, *parafuchsine*, the oldest triphenylmethane dye, and indeed the first of all the aniline dyes, is not so easily explained. As is well known, in this process aniline and *p*-toluidine are "united by oxidation" in an acid melt. (In the preparation of fuchsine itself, which contains a methyl group attached to one of the benzene rings, *o*-toluidine is an additional ingredient.) Although all the phases of this important synthesis have not yet been experimentally established, we may nevertheless explain it on the basis of a dehydrogenation similar to that involved in the formation of malachite green. Moreover, the union of several molecules of base proceeds exactly according to the principle on which indamines are formed (p. 312) (Bucherer).

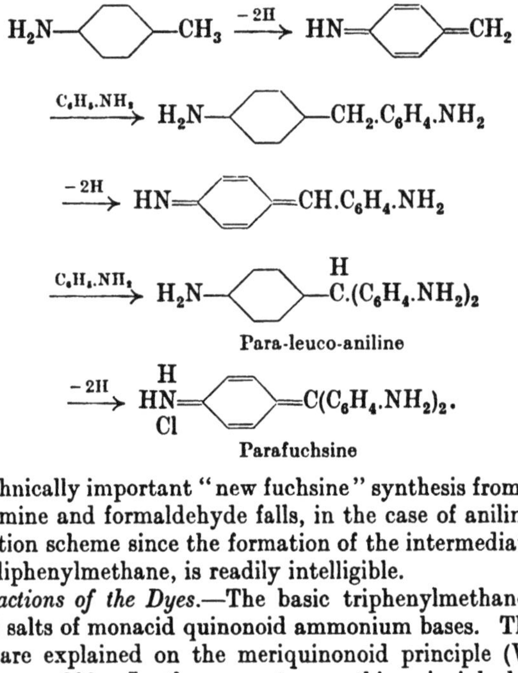

Para-leuco-aniline

Parafuchsine

The technically important "new fuchsine" synthesis from a primary aromatic amine and formaldehyde falls, in the case of aniline, into the above reaction scheme since the formation of the intermediate product, *p*-diaminodiphenylmethane, is readily intelligible.

The Reactions of the Dyes.—The basic triphenylmethane dyes are the neutral salts of monacid quinonoid ammonium bases. Their dyeing properties are explained on the meriquinonoid principle (Willstätter) discussed on p. 310. In the present case this principle holds intramolecularly. From *fuchsonimine*

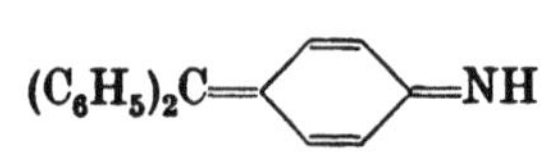

(fuchsone is the corresponding O-quinone) no dye salts are derived, but from the *p*-amino-derivative, on the other hand, which contains the benzenoid component $H_2N.C_6H_4$—, the dye known as *Doebner's violet* results.

The members of this class dye wool and silk directly, but cotton only when the latter is mordanted by tannin. They are fast neither to acids nor to alkalis for reasons which depend on important alterations in the compounds. If a little dilute hydrochloric acid is added to an aqueous solution of crystal violet the colour changes to green. One $N(CH_3)_2$-group takes part in the change and there is formed the salt with two equivalents of acid:

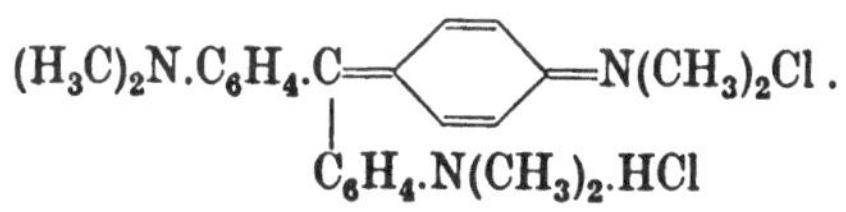

In this salt the influence on the colour of one dimethylaminophenyl radicle is suppressed in consequence of the saturation of the tervalent nitrogen; the substance produced is of the malachite green type.

When a little concentrated hydrochloric acid is added the green solution becomes yellow, because now the trichloride is formed and the influence of the second benzene ring is also suppressed, so that the (yellow) fuchsonimine type is formed. All triphenylmethane dyes dissolve in concentrated sulphuric acid with an orange-yellow colour exactly like triphenylcarbinol itself (*carbonium salts*, Kehrmann). By diluting the solution with water a colourless solution of the tri-acid benzenoid carbinol salt can be obtained.

The colour change produced by acids can be reversed by careful addition of alkali.

Carry out these and the other experiments described below.

Behaviour towards Alkalis.—Add a few drops of sodium hydroxide solution to the aqueous solution of a triphenylmethane dye. For a short time the colour remains unchanged, but soon fades and at the same time a faintly coloured flocculent precipitate separates. In all cases this is the carbinol to which repeated reference has already been made; when pure it is colourless. In the case of crystal violet the carbinol is $[(CH_3)_2N.C_6H_4]_3COH$.

As long as the solution which has been made alkaline retains its colour the *ammonium colour base*, which preserves the quinonoid structure of the dye, is still present (Hantzsch). This base is stable for a short time only and can in no case be isolated. The disappearance of the quinonoid system, which rapidly follows, occurs either through a rearrangement of the genuine base to the so-called *pseudo-* or *carbinol-base*, accompanied by the wandering of the OH-group:

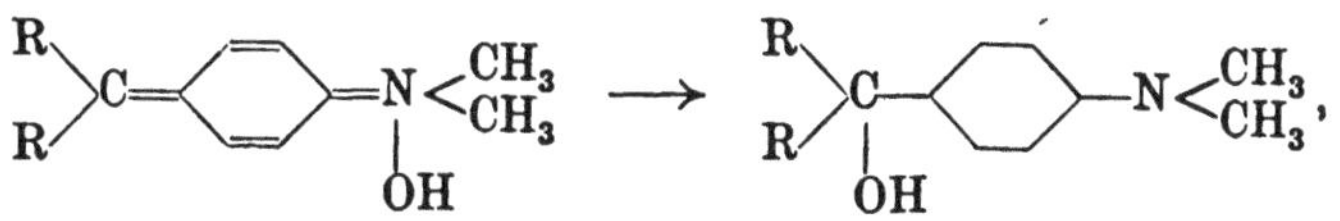

or, more probably, through the addition of water to the quinonoid system to form

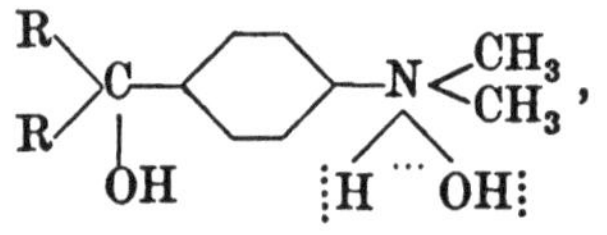

from which the water attached to the nitrogen is subsequently again eliminated.

Those dyes which, in contrast to malachite green and crystal violet, do not have the nitrogen in tertiary combination, for example, fuchsine, can lose water from their colour bases by direct elimination of its elements from their attachment to nitrogen and change into derivatives of *fuchsonimine*. Thus there is formed from parafuchsine the orange-coloured base called after Homolka :

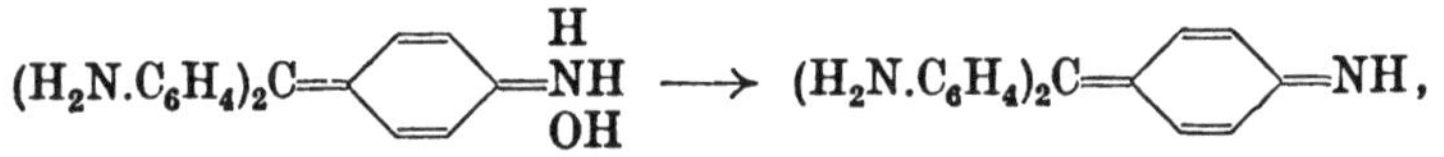

which changes into the carbinol by the addition of water and is reconverted into the dye by treatment with acid.

When the colourless carbinol base is treated with acid the quinonoid structure is re-formed and the *dye* produced. But on dissolving cautiously, with cooling, it can be seen that the colour attains its maximum depth only gradually. Hence the exceedingly unstable colourless carbinol salt is first formed ; it changes into the dye with spontaneous elimination of water :

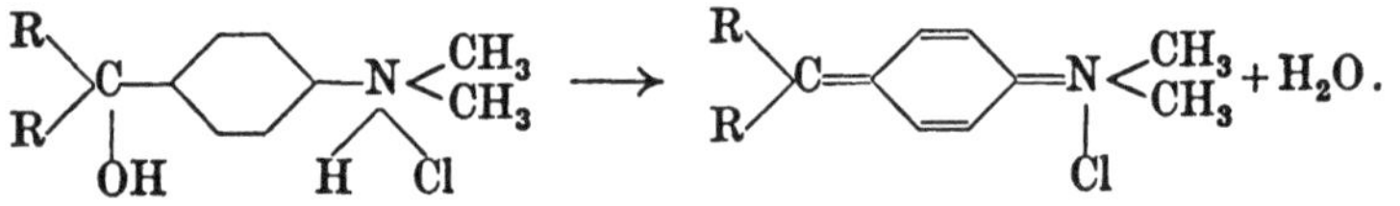

B. **The Phthaleins.**—If *two* molecules of a phenol and *one* molecule of phthalic anhydride are caused to interact under the conditions which prevail during the Friedel-Crafts synthesis (Chap. XI. p. 336), the tendency of the ketone first formed to combine with a second molecule of the phenol outweighs condensation to the anthraquinone derivative ; in this way are formed the phthaleins, discovered by Baeyer in 1871. This process may be discussed by taking phenolphthalein as an example.

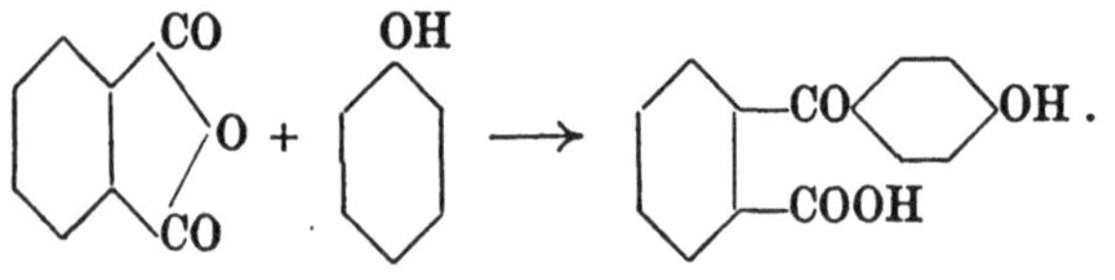

The intermediate product undergoes an aldol condensation with a second molecule of phenol, of which the *p*-position is involved along with the C=O-group in exactly the same way as Michler's ketone condenses with dimethylaniline, as described above.

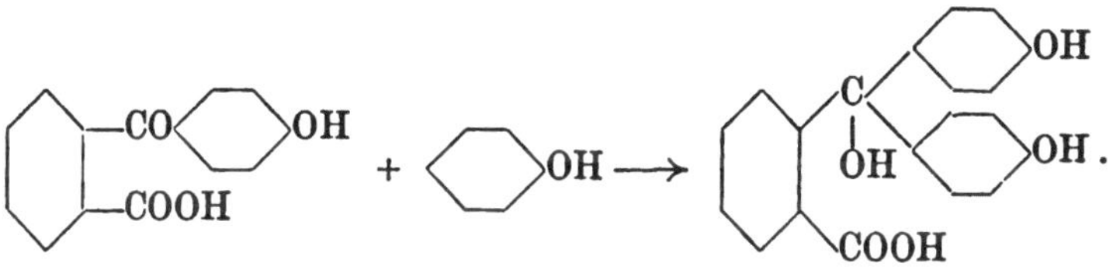

The *p*-dihydroxytriphenylcarbinol *o*-carboxylic acid, which cannot be isolated, loses water between the COOH- and OH-groups, which are favourably situated with respect to each other; the acid is thus converted into the lactone, *phenolphthalein*:

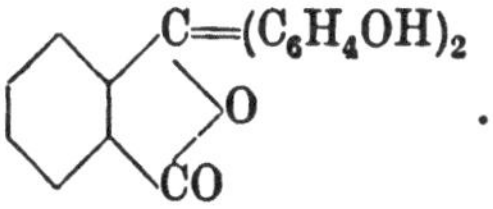

The colourless lactone is hydrolysed by alkalis and the intensely red alkali salts, well known in volumetric analysis, are formed. In them one benzene ring has assumed the quinonoid form with elimination of water in the manner expressed in the following equation:

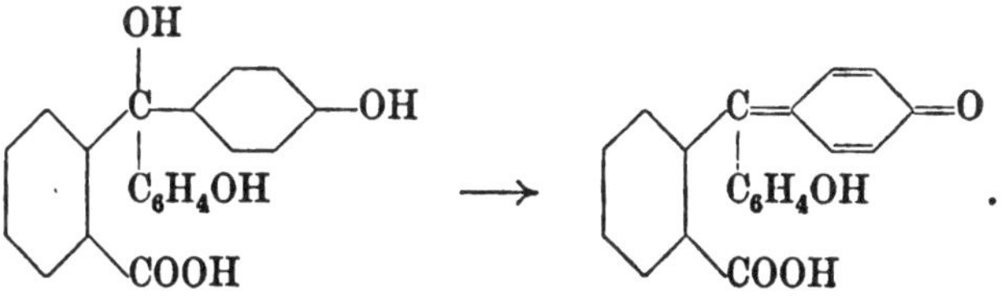

The red salts are the di-alkali salts of the quinonoid phenolcarboxylic acid formulated. This acid is not stable in the free state, but undergoes immediate isomerisation to the colourless lactone.

Phenolphthalein is a triphenylmethane derivative and can easily be connected with *fuchsone*, the parent substance of the dyes which belong to this series. Fuchsone is diphenylquinomethane and is obtained from *p*-hydroxytriphenylcarbinol by elimination of water (Bistrzycki):

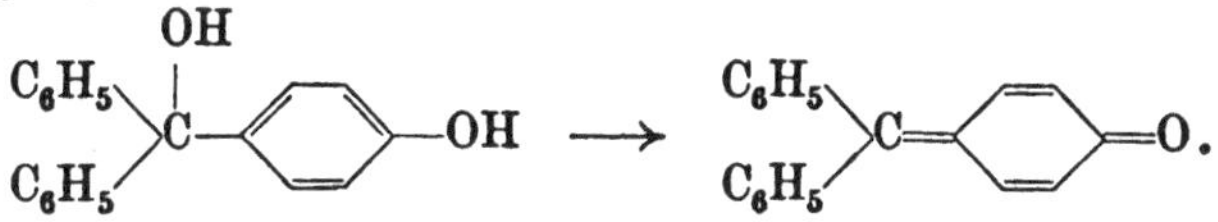

In complete accordance with Willstätter's theory, fuchsone is only yellowish-orange in colour. But if, in addition, one of the two free benzene rings has an OH-group in the *p*-position, we have the dye *benzaurin* already mentioned on p. 319. The *o*-carboxylic acid of this dye is the quinonoid form of phenolphthalein. In fact, the colour tones of these two substances are very similar. Phenolphthalein is decolorised by *concentrated* alkali; NaOH (KOH) is added on and the trisodium (or potassium) salt of the benzenoid carbinol form is produced. (Try these reactions with phenolphthalein.)

Fluorescein.—In this case the reaction undergoes an extension. The two OH-groups of the resorcinol molecule which are *ortho* to the position of condensation together form an oxygen bridge and hence a new ring (the xanthane ring) results by elimination of water:

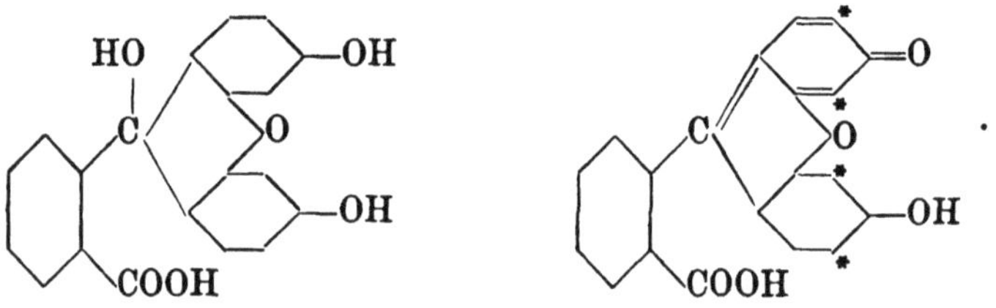

Since fluorescein is coloured it seems that the quinonoid formula on the right is the probable one even for the free compound, and that the lactone formula is doubtful.

In *eosin* the four bromine atoms are grouped in pairs in the *o*-positions indicated by asterisks in the formula. Eosin, also, must be regarded as quinonoid, especially because its reduction product, leuco-eosin, is colourless.

Experiment.—Dissolve some eosin in sodium hydroxide solution and boil with zinc dust until the mixture is colourless. Decant the solution. Acidify one portion of the decanted liquid and leave the other portion to stand in an open dish.

Like quinone itself, its derivatives are also converted by reducing agents into colourless hydrogenated products (in the case of the dyes they are called " leuco-compounds "). The following scheme expresses this process for eosin as well as for the other dye:

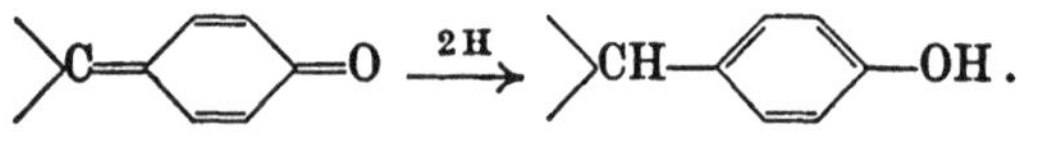

Many leuco-compounds are reconverted into the dye even by the oxygen of the atmosphere; leuco-indigo (p. 364) and leuco-eosin are examples of this.

The most magnificent dyes, which are chiefly used in the dyeing of silk, are relatives of eosin obtained from di- and tetra-chlorophthalic anhydride (phloxine, rose bengale). The (basic) rhodamines also belong to this class. They are produced by condensation of phthalic anhydride with *m*-aminophenols (in place of resorcinol); the dye containing the diethylated NH_2-group, in particular, is of great technical importance.

Finally *gallein* may be mentioned; in it pyrogallol is the phenolic component.

The conversion of the phthaleins into anthracene derivatives, the so-called *phthalideins*, need not be detailed here.

6. ALIZARIN

A mixture of 2 g. of potassium chlorate, 30 g. of commercial sodium hydroxide, 10 g. of finely powdered sodium β-anthraquinone-sulphonate ("silver salt"), and 40 c.c. of water is heated for twenty hours at 170° (oil bath) in an autoclave or in an iron tube with screwed-on cap. The cooled melt is repeatedly extracted with hot water and the extracts, after being combined and filtered, are acidified while hot with excess of hydrochloric acid, which precipitates the alizarin. When the mixture has cooled the precipitate is collected at the pump, washed successively with dilute hydrochloric acid and water, and dried.

For purification the crude product is boiled with glacial acetic acid (preferably in the extraction apparatus shown in Fig. 28). Fine red needles; melting point 289°. Sublimation in a vacuum from a sausage flask is also to be recommended; the "sausage" should be fixed low down and the bulb completely immersed in a nitrate bath (equal parts of potassium and sodium nitrates). Much poorer yields of alizarin are obtained by using an open round-bottomed flask at 180°–190°.

Alizarin or 1:2-dihydroxyanthraquinone is one of the most important dyes. Like indigo, the dye occurs in the plant (the madder root) as the glucoside of the leuco-compound. The cultivation of the madder plant, which, chiefly in southern France, extended over large areas, was brought to an end by the synthesis of the dye from the anthracene of coal-tar (Graebe and Liebermann, 1869). By distillation with zinc dust according to the method of Baeyer, these two chemists had previously obtained *anthracene* from alizarin.

Anthracene can be oxidised directly to its *meso*-quinone, *anthraquinone*, by means of chromic acid. For almost all its reactions the middle ring of anthracene provides the point of attack.

Experiment.—Dissolve 1 g. of purest anthracene in just sufficient good glacial acetic acid at the boiling point; without further heating add 3 c.c. of concentrated sulphuric acid and, drop by drop, 4 g. of sodium dichromate dissolved in a quite small amount of water.

(Neglect any turbidity or precipitate which appears after the addition of the sulphuric acid.) A very vigorous reaction occurs and the chromic acid is used up almost immediately; after all the dichromate has been added boil for five minutes longer. Dilute the solution. The *anthraquinone* separates in a flocculent condition. After collecting the substance at the pump, wash with water and dry. Recrystallise from glacial acetic acid. Fine pale yellow needles; melting point 285°.

When completely pure the compound is *colourless*. Compare with benzoquinone and naphthoquinone.

Anthraquinone is reduced by heating with sodium hydroxide and zinc dust: the deep red disodium salt of anthraquinol goes into solution.

Carry out the reduction and filter the red solution. Anthraquinone soon separates again from the filtrate on exposure to the air.

Details of the interesting desmotropic phenomena exhibited by the hydroxyanthracenes are given by K. H. Meyer (*Annalen*, 1911, **379**, 37). *Meso*-hydroxy- and dihydroxyanthracene exist in two forms, a genuine, coloured acidic *enol* which fluoresces in solution and a colourless, neutral *keto*-form.

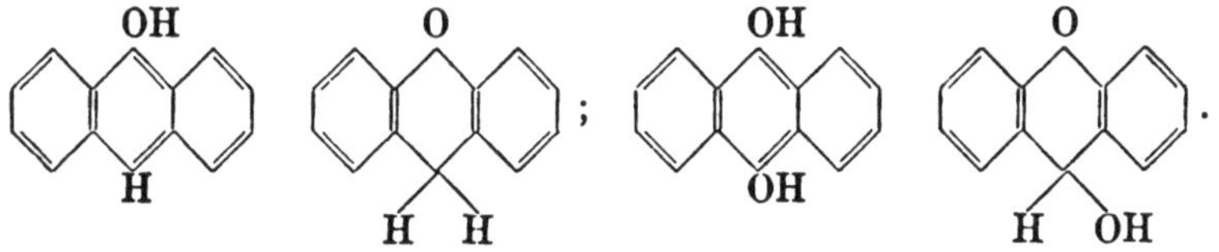

Anthranol (labile) ⟶ Anthrone (stable) Anthraquinol (stable) ⟵ Oxanthrone (labile)

For syntheses of anthraquinone from phthalic anhydride compare Chap. IX. 6, p. 340. As a dye alizarin is very important. This is due to its power of forming very stable beautifully coloured internally *complex* diphenoxides with the hydroxides of plurivalent metals (Cu, Sn, Cr, Fe, Al). The best known is the bright red aluminium lake "Turkey red".

Alizarin and the dyes related to it are called "mordant dyes" because they are applied to fibres impregnated with metallic hydroxide, *i.e.* mordanted.

Quinizarin.—1:4-dihydroxyanthraquinone is of no use as a dye; it has been found, as a general rule, that only those polyhydroxyquinones of the anthracene and naphthalene series (*naphthazarin*) which have their adjacent OH-groups in positions adjoining the carbonyl group are capable of forming colour lakes.

The fact that dihydroxyanthraquinones can be directly oxidised to higher phenols with fuming sulphuric acid is of technical importance.

Alizarin and quinizarin yield in this way the same 1:2:5:8-tetrahydroxyanthraquinone (*alizarin bordeaux*), which can be further oxidised to the important compound *anthracene blue* (1:2:4:5:6:8-hexahydroxyanthraquinone). This dye is obtained technically from 1:5- or 1:8-dinitroanthraquinone by means of a very interesting reaction, namely, by reductive-oxidative fusion with fuming sulphuric acid and sulphur (S_2O_3) in the presence of boric acid (R. Bohn, R. E. Schmidt). The production of *naphthazarin* from 1:5- or 1:8-*dinitronaphthalene* under similar conditions constitutes a more readily intelligible prototype of this complicated reaction.

It consists essentially in a rearrangement to the quinoneoxime, which is then hydrolysed and partially reduced.

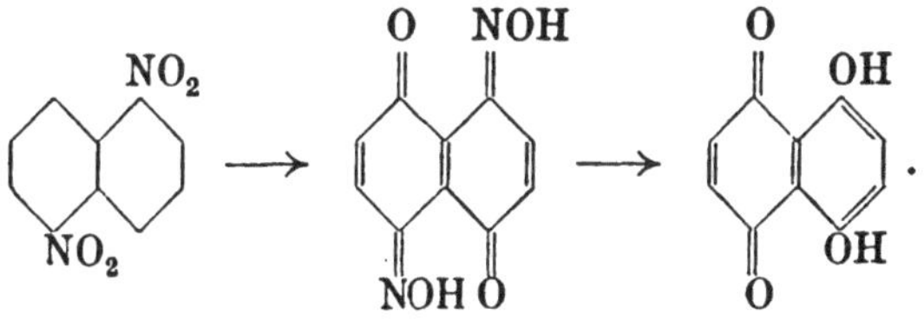

Naphthazarin

CHAPTER IX

THE GRIGNARD AND FRIEDEL-CRAFTS SYNTHESES. ORGANIC RADICLES

THE GRIGNARD REACTION

1. PREPARATION OF ALCOHOLS

(*a*) Benzohydrol from Benzaldehyde and Phenyl Magnesium Bromide

A small, dry, round-bottomed flask is fitted with a branched neck attachment and the oblique branch is connected with a reflux condenser closed at the upper end with a calcium chloride tube. Through the stopper of the upright branch a dropping funnel with a long outlet tube is inserted. The flask contains 3·2 g. of magnesium shavings, on to which a mixture of 20 g. of pure, constant-boiling bromobenzene and 50 c.c. of absolute ether is dropped *gradually* from the funnel. After about a quarter of the mixture has been added the operation is interrupted until evolution of heat and ebullition of the ether indicate the beginning of the reaction. Occasionally there is an obstinate delay; by holding a basin with warm water under the flask, or, better, by dropping in a particle of iodine, the reaction can be started with rapidity and certainty. When preparing the phenyl magnesium bromide solution it is important to keep the process in check by cooling from time to time and by regulating the inflow of bromobenzene so that the reaction just continues of its own accord. The bromobenzene which adheres to the dropping funnel is washed into the flask with a little absolute ether. When most of the metal has dissolved and the reaction becomes noticeably slacker, the solution is boiled for some time by dipping the flask into a basin of warm water until only a few particles of magnesium remain in suspension.

The solution is now cooled in ice-water and 10·6 g. of freshly distilled benzaldehyde mixed with 10 c.c. of ether are rapidly added drop by drop, at first with continued cooling.

To complete the reaction the solution is boiled for fifteen minutes under reflux and again cooled. External cooling is continued, 20–30 g. of ice are then added in one portion, followed by sufficient hydrochloric acid to dissolve the magnesium hydroxide which is precipitated (about 10 c.c. of concentrated acid plus 10 c.c. of water). The ether layer is now separated in a funnel and the aqueous layer is extracted with a little fresh ether. If a glass rod after dipping into the ethereal solution still smells of benzaldehyde, half of the ether is evaporated and the remaining solution is vigorously shaken for five minutes with a few cubic centimetres of 40 per cent bisulphite solution, then with a little sodium carbonate solution to remove dissolved sulphur dioxide; the ethereal solution is next dried for a short time with calcium chloride. On evaporating the ether the benzohydrol remains as an oil which soon solidifies. Yield, after pressing on porous plate, 12–14 g. The alcohol can be recrystallised from ligroin or from a little spirit. Beautiful colourless prisms; melting point 68°.

If the formation of the Grignard compound proceeds too violently the reaction product usually contains considerable amounts of *diphenyl* because of the reaction:

$$C_6H_5MgBr + BrC_6H_5 \longrightarrow C_6H_5.C_6H_5 + MgBr_2 .$$

(*b*) Triphenylcarbinol from Ethyl Benzoate and Phenyl Magnesium Bromide

Ethyl benzoate (15 g.) mixed with 15 c.c. of absolute ether is dropped into a Grignard solution prepared as just described from 6·4 g. of magnesium and 40 g. of bromobenzene. The conditions are the same as those observed in the preceding preparation; at the end the solution is boiled for half an hour and worked up as before. Colourless prisms of triphenylcarbinol, melting point 162°, are obtained by recrystallising the solid residue from hot alcohol. Yield over 20 g. For further information about this important alcohol see pp. 346-347.

2. SYNTHESIS OF A KETONE FROM A NITRILE. ACETOPHENONE[1]

An ethereal solution of phenyl magnesium bromide is prepared from 6·4 g. of magnesium and 40 g. of bromobenzene according to the

[1] Blaise, *Compt. rend.*, 1901, 133, 1217.

procedure given under 1 (*a*), and to this solution 8 g. of acetonitrile diluted with an equal volume of ether are added drop by drop. The mixture is then kept boiling on the water bath for one hour more, poured on to ice in a round-bottomed flask (capacity 1 l.), mixed with 100 c.c. of approximately 8 *N*-sulphuric acid, and submitted to distillation with steam. The ether and the acetophenone produced pass over with the steam. The distillate is extracted with ether, the extract is dried with calcium chloride, the ether is evaporated, and the ketone which remains is distilled. Boiling point 202°. Yield 10–12 g. = 45–50 per cent of the theoretical.

Here also a purer product is obtained by distillation *in vacuo*. Boiling point 88°/12 mm. In any case the acetophenone must be water-white and must crystallise when cooled with ice. Melting point 22°.

As a change *phenylacetone* may be prepared from benzyl magnesium chloride and acetonitrile. This ketone is purified by way of the bisulphite compound and distilled in a vacuum. The yield, with respect to acetonitrile, does not exceed 25 per cent.

Explanations relating to 1 and 2

Grignard Reagents.—Magnesium metal dissolves in alkyl halides in the presence of absolute ether to form organo-metallic compounds of the type R—Mg—Hal. Arylhalides act in the same way. In both series the iodides react most rapidly, then come the bromides, and lastly the chlorides. Often the reaction exhibits a somewhat stubborn unwillingness to begin: by addition of a little iodine or else of ethyl iodide this difficulty can be overcome. Occasionally it is necessary to activate the magnesium by heating with iodine (von Baeyer). The ether required for the reaction is bound, to the extent of two molecules, as a complex addition compound (Meisenheimer); it can be replaced by tertiary amines.

Grignard reagents are, quite generally, decomposed according to the following equation by substances which contain reactive hydrogen:

$$R.Mg.Hal + HR_1 \longrightarrow RH + R_1.Mg.Hal.$$

In all cases, therefore, the hydrocarbon RH, corresponding to the halide used, is formed.

The simplest example of this type is decomposition by water:

$$H_3C.Mg.I + HOH \longrightarrow CH_4 + HO.Mg.I.$$

Hence: *in all experiments with Grignard reagents moisture must be completely excluded.* Alcohols, phenols, carboxylic acids, primary and secondary amines, oximes, acetylene, etc., react in the same way as water.

Since one reactive hydrogen atom always liberates one molecule of hydrocarbon, the use of methyl magnesium iodide provides a serviceable method for determining quantitatively the proportion of active hydrogen by measuring the volume of methane liberated by a weighed amount of a substance under investigation (Zerevitinoff). The process is of considerable value for the determination of constitution. On the method of carrying it out see p. 72.

In virtue of their great capacity for forming addition compounds the Grignard reagents are primarily used for synthetic purposes. What was formerly attained with the rather unmanageable *zinc alkyls* is nowadays accomplished in a much wider domain by means of the easily made Grignard reagents (K. Ziegler has recently also used organic lithium compounds).

Quite generally addition to unsaturated systems takes place, *e.g.* to $>C{=}O$, $>C{=}N{-}$, $-C{\equiv}N$, $-N{=}O$, but $>C{=}C<$ and $-C{\equiv}C-$ do not react. The mechanism of the addition is as follows: the Grignard reagent is added on in the form of its two components R and MgHal; in the case of the C=O-double bonds the Mg-containing constituent always unites with the oxygen, R always with the carbon.

If the action of methyl magnesium bromide on acetaldehyde be taken as example, the following equation holds:

$$CH_3.\underset{H}{CO} + CH_3.Mg.Br \longrightarrow CH_3.\underset{H}{C}\begin{array}{l}\diagup CH_3\\ \diagdown O{-}MgBr\end{array}.$$

The addition product is decomposed by water, thus:

$$CH_3.\underset{H}{C}\begin{array}{l}\diagup CH_3\\ \diagdown O{-}MgBr\end{array} + H_2O \longrightarrow CH_3.\underset{H}{C}\begin{array}{l}\diagup CH_3\\ \diagdown OH\end{array} + HO.Mg.Br.$$

As a result, therefore, acetaldehyde is converted into isopropyl alcohol. Quite generally it may be said that the Grignard reaction consists in the addition to the unsaturated linkage—as H and R—of the hydrocarbon corresponding to the halide used. In effect a "synthesising hydrogenation" occurs.

The course of the following Grignard syntheses can thus be understood without further comment:

Formaldehyde → primary alcohols
Other aldehydes → secondary alcohols
Ketones → tertiary alcohols
Carbon dioxide → carboxylic acids
Nitriles → ketones (*via* the ketimine $\begin{array}{l}R_1\diagdown\\ R\diagup\end{array}\!C{=}NH$).

The reactions with the esters, chlorides, and anhydrides are somewhat more complicated.

Here also, at first, the usual addition to the C=O-group takes place :

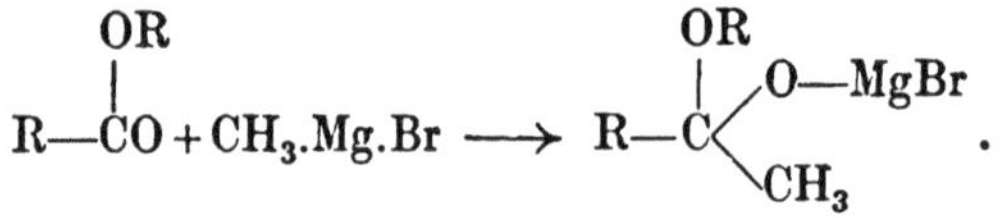

The product so formed reacts with a second molecule of the Grignard compound according to the following equation :

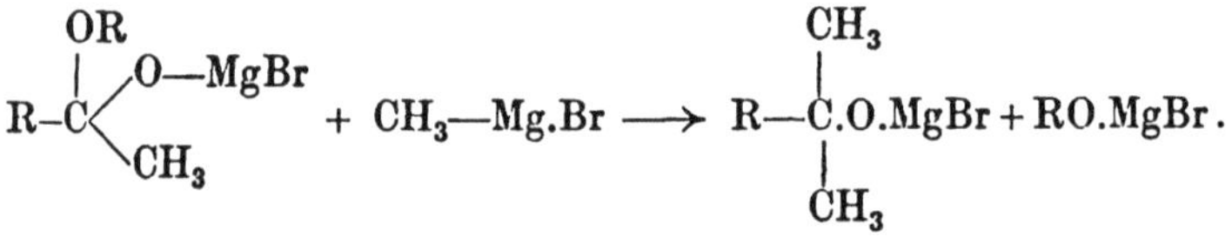

Finally, decomposition with water yields the tertiary alcohol. In the case of the esters of formic acid, used in excess, the reaction can be checked at the first stage, and by decomposing the product

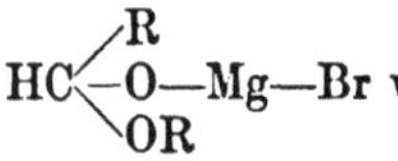 with water, *aldehydes* can be obtained.

The Grignard reagents also attack nitrogenous complexes in a similar way, as the example of the azides on p. 276 has shown.

Nitrosobenzene can be converted into diphenylhydroxylamine (p. 172) by the action of phenyl magnesium bromide.

The wide range over which the Grignard synthesis can be applied should be sufficiently evident from this short review.

There remains to be mentioned a secondary reaction which often takes place, undesired, in the preparation of Grignard reagents, but is also sometimes aimed at.

The Grignard compounds react with varying ease with organic halides in the manner of the Wurtz reaction according to the equation :

$$R.Mg.Hal + R_1Hal \longrightarrow R.R_1 + MgHal_2 .$$

Hence diphenyl, as was already mentioned, is always obtained as a by-product in the preparation of phenyl magnesium bromide.

The great variety of synthetic processes included in the Grignard reaction may be illustrated by the production of diphenylmethyl carbinol and consequently of asymmetrical diphenylethylene :

1. From benzophenone and methyl magnesium bromide :

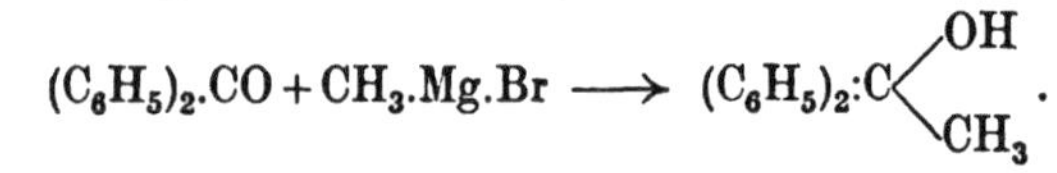

2. From ethyl acetate and phenyl magnesium bromide :

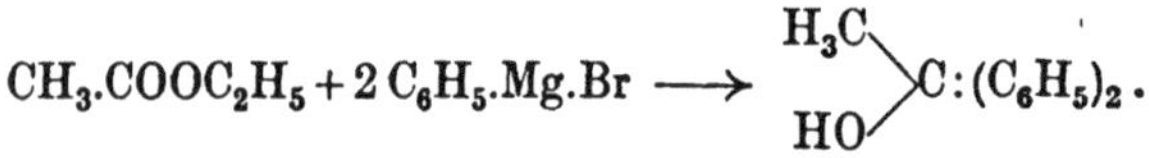

THE FRIEDEL-CRAFTS SYNTHESIS

Aluminium Chloride.—For the success of a Friedel-Crafts synthesis it is essential that the quality of the aluminium chloride used as catalyst should be beyond reproach. Commercial preparations are often partially decomposed and unfit for use in consequence of the entrance of moisture into insufficiently closed containers. In order to make sure, a small sample should be heated over a flame in a test tube held obliquely; the whole, or at least the bulk, of the chloride should sublime. Preparations which are not too seriously decomposed can be made fit for use by resublimation.

If it is necessary to prepare the aluminium chloride in the laboratory the following method should be used.[1]

Connect a wide (3·5–4 cm.) hard glass tube, *e.g.* a tube used for combustions by Dennstedt's method, by means of a cork with a wide-necked bottle in such a way that the end of the tube is practically flush with the cork. Through a second hole in the cork push a narrower, yet not too narrow, bent glass tube of which the shorter arm extends rather more than half-way into the bottle. The end of the longer arm extends upwards at a right angle. Along the combustion tube, to a depth equal to one-third of its diameter, arrange a layer of aluminium filings. The length of this layer will depend on the amount of aluminium chloride required and on the length of the combustion furnace used, but in any case the end of the heated layer must not be more than 8 cm. from the receiving bottle. Protect the cork from the heat by means of a sheet of asbestos cut to admit the tube and fixed slightly in front of the cork.

At the other end connect the tube by means of the shortest possible rubber tubes with two wash-bottles containing concentrated sulphuric acid, which in turn are connected to an efficient supply of hydrogen chloride. The whole apparatus must, of course, be perfectly dry.

Lay the tube in a combustion furnace in a fume chamber and pass a current of hydrogen chloride through until the air is expelled, then slowly heat the whole length of tube containing the aluminium. When, as the temperature rises and the passage of fumes into the receiver shows that aluminium chloride is forming, increase the

[1] It is recommended that aluminium chloride obtained during preparative work in the inorganic department of the laboratory should be employed for syntheses of the type here described.

stream of hydrogen chloride; at the same time also heat more strongly, and take care that at this stage an extremely vigorous current of hydrogen chloride ensures that there is no time for the aluminium chloride to condense inside the cork and so to choke the apparatus; choking must be carefully avoided. No serious reduction in the yield is caused by the escape of clouds of aluminium chloride from the exit tube of the bottle. Continue the experiment until all but a small residue of the metal has volatilised.

Keep the chloride obtained in this way in a bottle provided with a *tightly* fitting ground glass stopper.

3. SYNTHESIS OF A KETONE

(a) Benzophenone from Benzoyl Chloride and Benzene

Freshly prepared, finely powdered aluminium chloride (35 g.), weighed in a dry corked test tube, is poured with frequent shaking during the course of ten minutes into a dry flask containing 50 c.c. of benzene, 35 g. (0·25 mole) of benzoyl chloride, and 100 c.c. of pure carbon bisulphide (or 70 c.c. more benzene). The flask is then attached to a long reflux condenser and warmed on the water bath at 50° until only small amounts of hydrogen chloride are being evolved (2–3 hours). The solution acquires a deep brown colour. The carbon bisulphide (or benzene) is now removed by distillation through a downward condenser and the residue is cautiously poured, while still warm, into a capacious flask containing 300 c.c. of water and small pieces of ice. After rinsing out the reaction flask with a little water and adding 10 c.c. of concentrated hydrochloric acid to the reaction mixture, steam is passed through for about twenty minutes. The material which remains in the flask is then cooled and taken up in ether. After the ethereal solution has been dried with calcium chloride and the ether has been evaporated, the residue is distilled from a flask with low side tube. Boiling point 297°. Melting point 48°. Yield about 35 g. A purer product is obtained by vacuum distillation in a sausage flask.

Benzophenone Oxime

To a solution of 4 g. of benzophenone in 25 c.c. of alcohol, cooled solutions of 3 g. of hydroxylamine hydrochloride in 6 c.c. of water and 5 g. of potassium hydroxide in 5 c.c. of water are added and the whole is heated under reflux on the water bath for two hours. The product is then poured into 50 c.c. of water, any unchanged ketone is removed by filtration after shaking to cause aggregation, the filtrate is acidified with dilute sulphuric acid, and the free oxime recrystallised from alcohol. Melting point 140°.

Beckmann Rearrangement to Benzanilide

A weighed quantity of the oxime is dissolved in some cold ether, free from water and alcohol, and to the solution 1·5 parts of finely powdered phosphorus pentachloride are gradually added. The ether is then removed by distillation and water is added to the residue with cooling: the ensuing precipitate is recrystallised from alcohol. Melting point 163°.

The interesting intramolecular rearrangement[1] which is brought about in this way involves an exchange of the positions of C_6H_5 and OH according to the scheme:

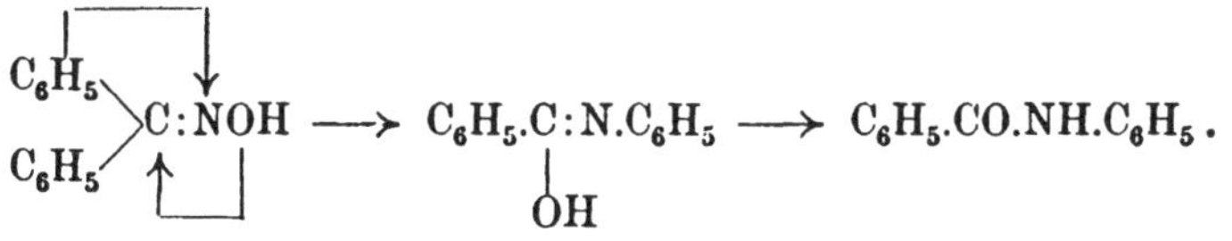

Under the influence of the catalyst (PCl_5, concentrated H_2SO_4) a compound rich in energy is converted into its stable isomeride; the change which occurs is similar to that which was discussed on pp. 177-178 in connection with the relationship between hydrazobenzene and benzidine. A comparison with the benzilic acid rearrangement also suggests itself.

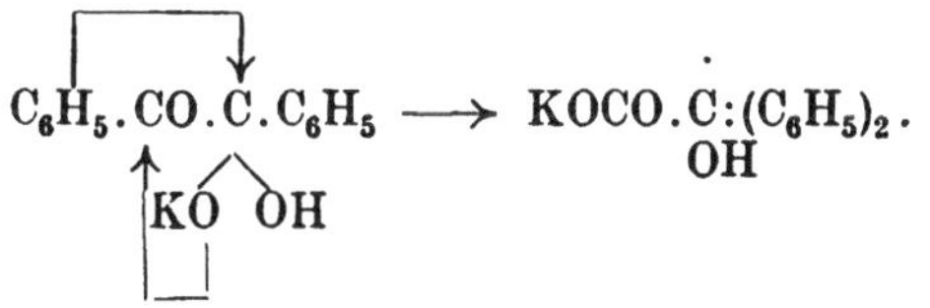

There is also some relationship to the degradation reactions of Hofmann and Curtius.

In this connection the *spatial isomerism of the oximes* must also be mentioned. A theory for this type of isomerism had already been suggested by Werner and Hantzsch and is similar to that already discussed in the case of the diazotates. According to this theory, those oximes in which the C-atom bearing the isonitroso-group is also united to two different radicles exist in a *syn*- and an *anti*-form:

[1] E. Beckmann, *Ber.*, 1886, 19, 988; 1887, 20, 1507, 2580; *Annalen*, 1889, 252, 1, 44.

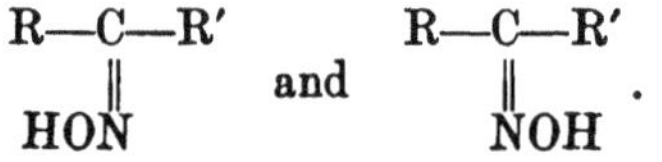

In a model the isomerism is seen to be of the same kind as that of maleic and fumaric acids.

The *syn*-form of aldoximes is readily dehydrated to the *nitrile*, the *anti*-form is not.

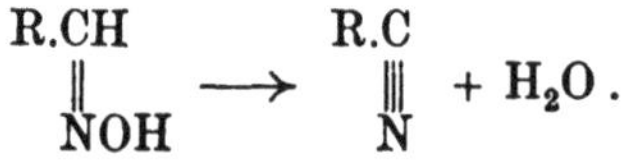

For a long time it was supposed that the results of the Beckmann rearrangement of ketoximes afforded proof of their configuration; the OH-group was thought to change places with the adjacent substituent, for the rearrangement of the two stereoisomeric ketoximes leads to isomeric amides. Recently it has been shown, however, that just the opposite is the case, as the following formulae indicate (Meisenheimer, *Ber.*, 1921, **54**, 3206):

In excellent agreement with the theory, *benzil* yields two stereoisomeric monoximes and three dioximes:

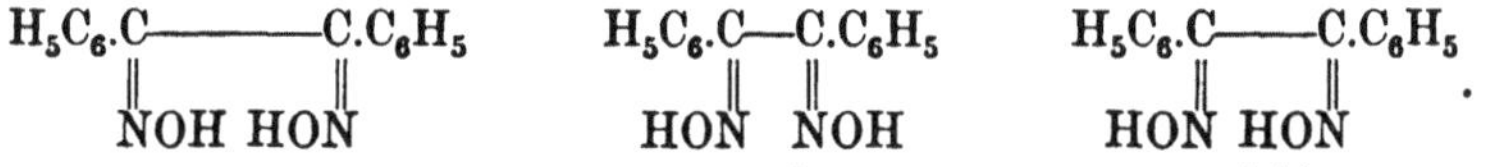

syn- *anti*- *amphi*-form

(*b*) ACETOPHENONE FROM BENZENE AND ACETIC ANHYDRIDE [1]

A *stirrer* with mercury seal (Fig. 32, p. 38) is fitted into the middle neck of a three-necked round-bottomed flask (so-called Tscherniak flask; capacity 500 c.c.) and the other two necks are provided respectively with a *reflux condenser* and a *dropping funnel*. The flask contains 100 c.c. of sodium-dried benzene and 80 g. of *freshly sublimed* aluminium chloride. Pure *acetic anhydride*, 25 g., is then run in with vigorous stirring during the course of half an hour. The mixture becomes warm and a very brisk evolution of hydrogen chloride takes place. After heating to boiling on the water bath with continuous stirring for half an hour more, the cooled solution is poured on to ice in a separating funnel and the aluminium hydroxide precipitated is dissolved by concentrated hydrochloric acid. Some ether is added, the benzene layer is separated, the aqueous layer is

[1] R. Adams, *J. Amer. Chem. Soc.*, 1924, 46, 1889.

extracted with ether, and the combined benzene and ether solutions are shaken with sodium hydroxide solution. When the solution has been dried with calcium chloride and the solvents have evaporated the acetophenone is distilled, preferably in a vacuum. See p. 330. Yield 24–25 g.; calculated on the acetic anhydride, 80–85 per cent of the theoretical.

When *acetyl chloride* is used instead of acetic anhydride hardly half this yield is obtained. Comparison of the two reactions is instructive.

4. TRIPHENYLCHLOROMETHANE FROM BENZENE AND CARBON TETRACHLORIDE[1]

In the apparatus used for the preparation of benzophenone 60 g. of fresh active aluminium chloride are gradually added to a mixture of 80 g. of pure dry carbon tetrachloride and 200 g. of benzene. At first the flask is kept cool with water and the reaction is not allowed to become too violent. The copious fumes of hydrogen chloride are absorbed as described in similar cases, *e.g.* in the preparation of bromobenzene (p. 91). When all the aluminium chloride has been added and the main reaction is over, the mixture is heated under reflux on the boiling water bath for half an hour longer and the brownish-yellow reaction mixture, after cooling, is poured, with constant shaking, into a sufficiently large separating funnel containing a mixture of 100–200 g. of ice with 200 c.c. of concentrated hydrochloric acid. If the ice melts before the whole of the mixture has decomposed, more ice and an equal amount of concentrated hydrochloric acid are added. The purpose of the hydrochloric acid is to prevent the hydrolysis of the triphenylmethyl chloride. When the two layers of liquid in the funnel have separated—perhaps after the addition of fresh benzene—the aqueous portion is removed, and extracted again, if necessary, with benzene. The combined benzene solutions are dried with calcium chloride and then *as much benzene as possible* is evaporated on the water bath. The residue is digested with an equal volume of ether and cooled for some hours in ice. Then the crystalline sludge is filtered at the pump on a wide filter plate and the solid, after being pressed down well, is washed several times with a little ice-cold ether. A second, less pure, crop of crystals is obtained by concentrating the mother liquor (ultimately

[1] M. Gomberg, *Ber.*, 1900, 33, 3144.

in vacuo), digesting the residue with a little cold ether, and filtering at the pump. Yield 110–120 g.

For purification the still yellow crude product is dissolved in *very little* warm benzene, four volumes of light petrol are added, and crystallisation is induced by cooling in ice and stirring with a glass rod.

Distillation in a high vacuum also yields a very pure preparation (Lecher).

5. GATTERMANN AND KOCH'S ALDEHYDE SYNTHESIS:[1] *p*-TOLYLALDEHYDE

KETONE SYNTHESIS BY THE METHOD OF HOESCH

Finely powdered, freshly prepared aluminium chloride (45 g.) and pure cuprous chloride[2] (5 g.) are added, not too rapidly, with water-cooling, to 30 g. of freshly distilled toluene (boiling point 110°) in a wide-necked vessel (round-bottomed flask or meat-extract jar). The vessel is then closed with a three-holed cork. Into the middle hole is pushed a glass tube which serves as guide for a stirrer (having a glass paddle-wheel). An inlet and an outlet tube are fixed in the other two holes (Fig. 52). After the apparatus has been clamped it is placed in a basin of water at 20° and the mixture is mechanically stirred, while a moderately strong current of carbon monoxide and hydrogen chloride is passed in through a forked tube. The carbon monoxide is taken from a gas-holder (capacity 10 l.) and passed successively through wash-bottles containing potassium hydroxide solution (1:1) and concentrated sulphuric acid; it is regulated at a rate which is twice that of the

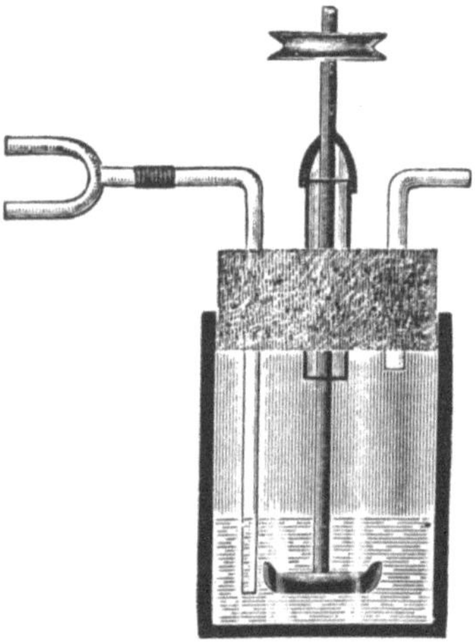

FIG. 52.

[1] *Ber.*, 1897, **30**, 1622; 1898, **31**, 1149; *Annalen*, 1906, **347**, 347.

[2] Heat an aqueous solution of 50 g. of copper sulphate and 24 g. of common salt to 60°–70° and pass in sulphur dioxide until there is no further increase in the precipitate formed. Filter at the pump, wash the cuprous chloride with aqueous sulphurous acid and with dilute hydrochloric acid, and dry it in a desiccator over potassium hydroxide and sulphuric acid.

hydrogen chloride. The gas which escapes is led directly into the flue of a fume chamber. After about one to two litres of carbon monoxide have been passed in (one hour) the temperature is raised to 25°–30° and then the rest of the gas is passed in during the course of four to five hours more. If, before this time has elapsed, the reaction mixture should become so viscous that the stirrer no longer runs, the operation may be stopped at this point. The syrupy material is then poured on to crushed ice in a large flask and the aldehyde produced, along with some unchanged toluene, is distilled over with steam. By extracting the distillate with ether and *prolonged* shaking of the extract with bisulphite solution (as in former cases) the solid bisulphite compound of the aldehyde is obtained. It is collected at the pump, washed with ether, and decomposed with saturated sodium carbonate solution. The free aldehyde is again distilled with steam and extracted from the distillate with ether. When the ether is evaporated 20 to 22 g. of pure *p*-tolylaldehyde, boiling point 204°, are obtained.

Apply to the reaction product the tests given in the section on aldehydes.

Carbon Monoxide.—If a steel cylinder of carbon monoxide is not available in the laboratory, the gas is prepared from the most concentrated technical formic acid, which is dropped into an excess of warm concentrated sulphuric acid. The temperature and the rate of dropping of the acid are regulated according to the rate of evolution of the gas, which is washed with potassium hydroxide solution, preferably in a Péligot tube in order to prevent sucking back, and collected in a gas-holder, from which it is taken for the reaction. Remember the *toxicity* of carbon monoxide (fume chamber !).

2:4-Dihydroxyacetophenone from Resorcinol and Acetonitrile[1]

Finely powdered anhydrous zinc chloride (2 g.) is added to a solution of 5·5 g. of resorcinol and 3 g. of acetonitrile in 25 c.c. of absolute ether, the mixture is kept cool in ice and saturated with hydrogen chloride. After standing for some hours in a closed vessel the product acquires the consistency of porridge. It is kept cool externally while 25 c.c. of ice-water are added, some ether is poured in, and the ether layer is separated. The ketimine which is present in the aqueous solution as hydrochloride is hydrolysed by heating

[1] K. Hoesch, *Ber.*, 1915, **48**, 1122; 1917, **50**, 462.

for half an hour on the water bath. The *resacetophenone* crystallises from the solution on cooling. Yield 4–5 g. It can be recrystallised from water. Melting point 145°.

6. QUINIZARIN FROM PHTHALIC ANHYDRIDE AND QUINOL [1]

A mixture of 5 g. of pure quinol, 20 g. of phthalic anhydride, 50 c.c. of pure concentrated sulphuric acid, and 5 g. of boric acid is heated on an oil bath for three hours at 150°–160°, and then for one hour at 190°–200°. While still hot the solution is stirred into a porcelain basin containing 400 c.c. of water, then heated to boiling and filtered hot at the pump. This operation is repeated, and then the precipitate is boiled with 250 c.c. of glacial acetic acid and filtered while hot at the pump. The filtrate is poured into a beaker and diluted with an equal volume of hot water. The crude quinizarin which separates on cooling is collected at the pump, washed repeatedly with water, and dried, first on the water bath and then in an oven at 120°. It is recrystallised from 150 c.c. of boiling glacial acetic acid. Melting point 194°. The substance forms large orange-yellow leaflets which, after collection at the pump, are washed with a little glacial acetic acid and then with ether. Especially fine crystals are obtained from toluene or xylene. Like alizarin, quinizarin dissolves in alkalis with a deep violet colour. Quinizarin can be sublimed without decomposition. Yield 2–2·5 g.

Sections 3, 4, 5, 6 : Theoretical Considerations

Both acid chlorides and alkyl chlorides react with aromatic compounds in the presence of aluminium chloride, zinc chloride, or ferric chloride in such a way that HCl is eliminated and the acyl or alkyl group enters the ring :

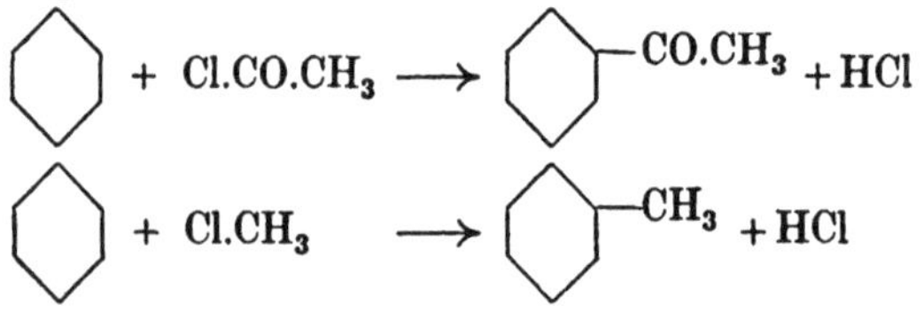

Whereas the first reaction, which results in the synthesis of ketones, is much used because of the ease with which it usually proceeds, the introduction of alkyl groups proves much less satisfactory since, in the first place, the substitution goes further and, in the second, the

[1] Grimm, *Ber.*, 1873, 6, 506 ; Baeyer, *Ber.*, 1875, 8, 152. Liebermann, *Annalen*, 1882, 212, 10 ; G.P. 255,031 (Friedländer, XI, 588).

alkyl groups may at the same time be partially split off again. The method of Fittig is usually to be preferred in this case.

Under the conditions of the Friedel-Crafts reaction, substances containing an olefinic double bond react in such a way that the acid chloride is added on at this bond and a saturated β-chlorinated ketone is formed. When heated this substance loses HCl and yields the unsaturated ketone:

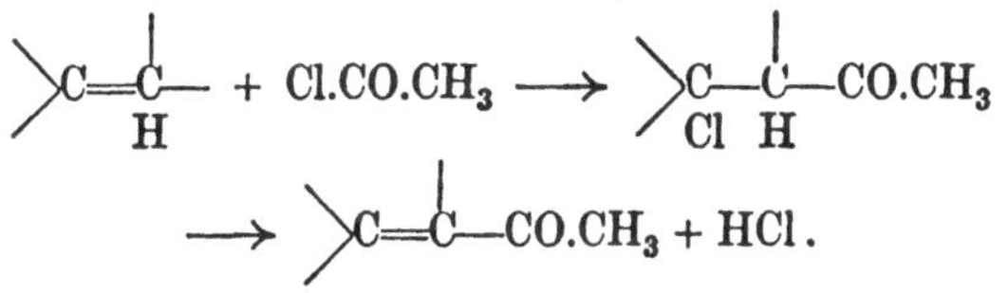

Hence we are justified in assuming an analogous course for the reaction in the aromatic series also (cf. *Ber.*, 1922, **55**, 2246).

The function of the aluminium chloride is catalytic, and its amount is therefore independent of stoicheiometrical proportions. But since, in the ketone synthesis, a fairly definite complex addition compound with one molecule $AlCl_3$ is formed, at least one mole must be used.

On account of the great reactivity of the substances taking part in the reaction, the choice of a solvent in the Friedel-Crafts synthesis is limited. The most important are carbon bisulphide, well purified petrol ether, chlorobenzene, and nitrobenzene.

The method of action of the aluminium chloride is not yet clearly understood. Since with acyl and alkyl chlorides it forms complex addition products which can be isolated, it is possible that in these products the bond between the chlorine and the rest of the organic molecule is loosened, and that so the additive power is increased.

It is also possible, however, that the aluminium chloride increases the reactivity of the hydrocarbon by combining with it.

In addition to the hydrocarbons, the phenolic ethers undergo this synthesis with special ease.

The following individual applications may be quoted:

The reaction of phthalyl chloride with benzene, leading to *phthalophenone*, the parent substance of the phthaleins:

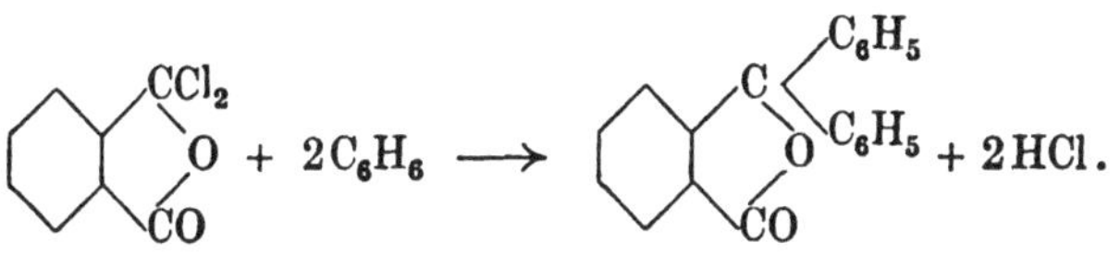

The intramolecular ketone synthesis from the chloride of hydrocinnamic acid:

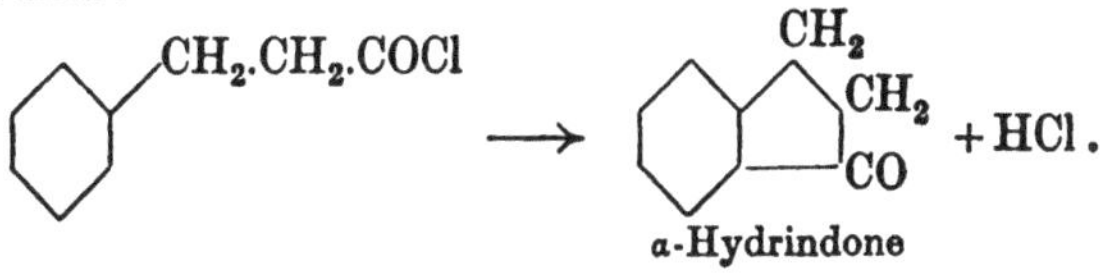

α-Hydrindone

The direct synthesis of ketones from hydrocarbons and phosgene, *e.g.*

$$2C_6H_6 + COCl_2 \longrightarrow C_6H_5.CO.C_6H_5 + 2HCl.$$

The introduction of simple or substituted formamide radicles by the use of *urea chlorides* (cyanic acid and cyanic esters + HCl).

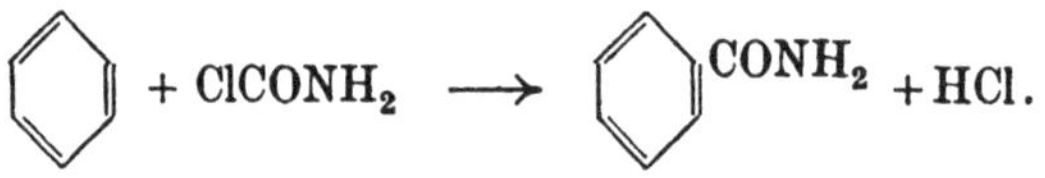

In this way aromatic carboxylic acids can be prepared by the Friedel-Crafts method.

The range of the reaction was extended by the elegant aldehyde synthesis of Gattermann and Koch. If a mixture of carbon monoxide and hydrogen chloride is allowed to act in the presence of aluminium chloride (and cuprous chloride) on toluene (benzene is less suitable), the reaction occurs which might be expected with formyl chloride if this substance were capable of existence.

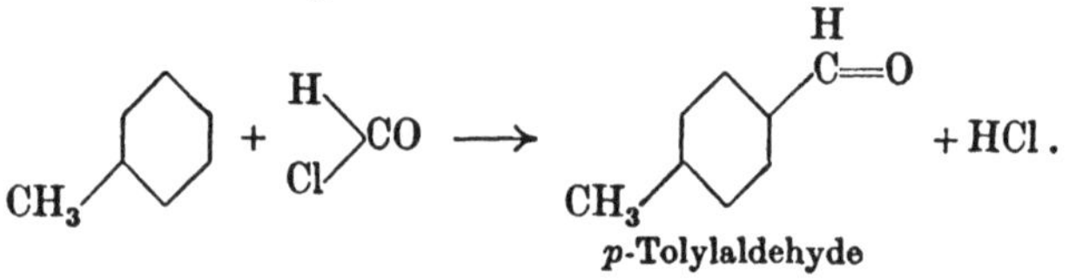

p-Tolylaldehyde

It seems as if the chloride of formic acid is temporarily produced in the form of a complex compound with cuprous chloride.

By substituting *hydrogen cyanide* for carbon monoxide it is further possible to prepare a very great variety of *phenolic aldehydes* and of *their ethers*. Usually the aldehyde group takes up the position *para* to the substituent already present. In this case the cuprous chloride is unnecessary. Since hydrogen cyanide and HCl combine to form formiminochloride $\begin{matrix}H\\Cl\end{matrix}\!>\!C{=}NH$, it is evident that here an aldimine is first formed and then, when the reaction mixture is worked up, is converted into the aldehyde; NH_3 is eliminated by the action of water. Enols of the aliphatic series (ethyl acetoacetate, acetyl acetone) react fundamentally in the same way. The use of mercury fulminate, from which by the action of hydrogen chloride the beautifully crystalline chloride

$$(C{=}NO)_2Hg + 4HCl \longrightarrow HgCl_2 + 2\begin{matrix}H\\Cl\end{matrix}\!>\!C{=}NOH$$

of formhydroxamic acid can be isolated, leads, in the aromatic series, to the formation of aldoximes (Scholl).

The ketone synthesis of Houben and Hoesch, which is based on the principle of the Gattermann reaction, proceeds very smoothly and gives very favourable results, especially in the case of polyhydric phenols. In this synthesis *nitriles* are used. Here it is the iminochlorides $R{-}\underset{Cl}{C}{=}NH$ which are converted into ketimines and then into ketones, a process which is analogous to that which occurs when hydrogen cyanide is used. The formulation follows at once from what has been said.

All three chlorine atoms of *chloroform* take part in the Friedel-Crafts reaction; the product of the reaction with benzene is the important hydrocarbon *triphenylmethane*, the parent substance of the well-known class of dyes. Paraleucaniline, $[(p)\ NH_2.C_6H_4]_3CH$, has been converted into triphenylmethane by reductive hydrolysis of its tris-diazo-compound (E. and O. Fischer).

We might expect that the reaction would lead to the formation of tetraphenylmethane from benzene and *carbon tetrachloride* in the presence of aluminium chloride, but this is not so. In this case the fourth Cl-atom remains in the reaction product. *Triphenylchloromethane* $(C_6H_5)_3CCl$ has acquired extraordinary importance because, when applied in the Wurtz reaction, it made possible the discovery of the first free organic radicle (Gomberg, 1900). Compare p. 344.

Often in the Friedel-Crafts synthesis the acid chloride can be replaced by the *anhydride*. The preparation of acetophenone (p. 336) is an example of this modification.

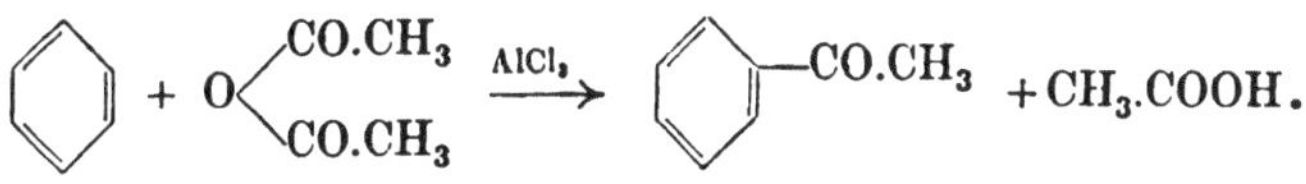

It has become especially important because of the behaviour of phthalic anhydride, which can be condensed with benzene in the presence of aluminium chloride to o-*benzoylbenzoic acid.*

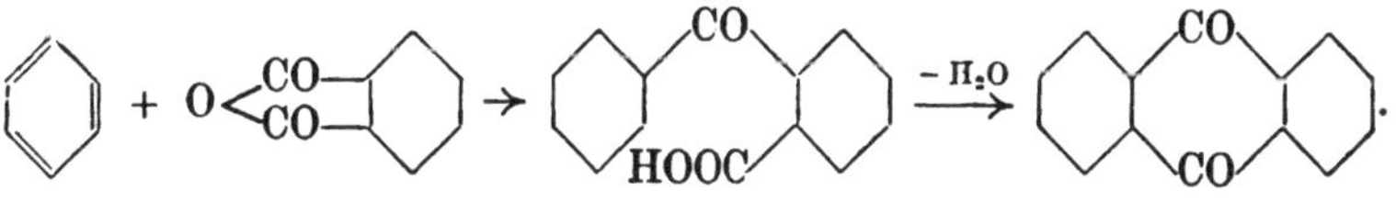

Since, as the equation shows, this substance is converted by concentrated sulphuric acid with loss of water into *anthraquinone*, a very important route to a much-studied group is opened up (Baeyer). Thus β-methylanthraquinone, which serves as an intermediate for valuable vat dyes, is prepared technically in this way from phthalic anhydride and toluene.

Concentrated sulphuric acid here performs the same function in the first phase as does aluminium chloride, and by the use of the acid the anthraquinone derivative is obtained in a *single* operation. The synthesis of quinizarin, described above, provides a preparative illustration of this elegant reaction:

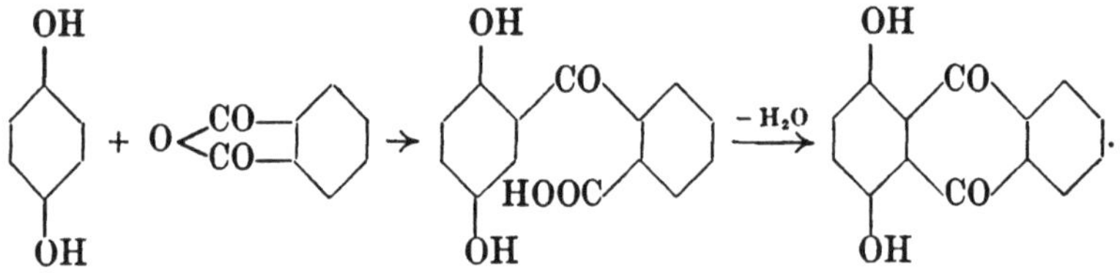

Catechol, on condensation with phthalic anhydride, yields the isomeric *alizarin*, although in much poorer yield. The combination of phthalic anhydride with pyrogallol leads to *anthragallol* (1:2:3-trihydroxyanthraquinone). This substance is made on a technical scale.

ORGANIC RADICLES

7. HEXAPHENYLETHANE

Preparation of a Solution of Triphenylmethyl.—Pure triphenylchloromethane (2 g.) is dissolved in a glass-stoppered bottle (cap. 25 c.c.) in 20 c.c. of benzene to a colourless solution. Zinc dust (5 g.) is added and the mixture is vigorously shaken for five minutes. The well-known dissociation experiment of Schmidlin is first performed with the golden-yellow or orange-yellow solution of the radicle. About 2 c.c. of the clear solution are poured into a large test tube, diluted with 2 c.c. of benzene, and shaken up. The colour disappears but soon returns again. By renewed shaking with air the radicle can again be converted into the colourless peroxide. The interesting experiment can be repeated several times. If immediate decolorisation does not occur at the first shaking, too much of the solution of triphenylmethyl has been used, and the experiment is repeated with half the quantity. The rest of the main solution is filtered through a folded paper and by shaking with air the unsaturated hydrocarbon is isolated as *peroxide* which separates in colourless crystals. After standing for some time they are collected at the pump and washed with ether. Melting point (with production of red colour and decomposition) 183°.

Schmidlin's experiment here described shows very clearly the equilibrium between hexaphenylethane and triphenylmethyl. The disappearance of the colour on shaking the substance with air indicates that the yellow radicle, present in equilibrium, is removed as (colourless) peroxide. The re-establishment of the equilibrium by renewed dissociation of (colourless) hexaphenylethane proceeds so slowly that the formation of the yellow radicle in the decolorised solution can be observed without difficulty.

As the odd number of hydrogen atoms already shows, triphenylmethyl, $C_{19}H_{15}$, known in solution only, contains a tervalent carbon atom. In contrast to the colourless hexaphenylethane, which can be isolated in the crystalline state, triphenylmethyl is intensely yellow. Its absorption spectrum exhibits characteristic bands (examine the spectrum in a spectroscope).

Triphenylmethyl is an extremely reactive substance. Its solutions are decolorised by air, which converts it into the colourless *triphenylmethyl peroxide*:

$$2\,(C_6H_5)_3C + O{=}O \longrightarrow (C_6H_5)_3C.O{-}O.\dot{C}(C_6H_5)_3\,.$$

Halogens react in the same way:

$$2\,(C_6H_5)_3C + Br{-}Br \longrightarrow 2\,(C_6H_5)_3C.Br\,.$$

Hydrogen chloride reacts in light to form *triphenylmethane* and *triphenylchloromethane.* The reaction is reversible (Schlenk). NO, NO_2, and many organic radicles add themselves to triphenylmethyl. The combination with quinone is analogous (Schmidlin).

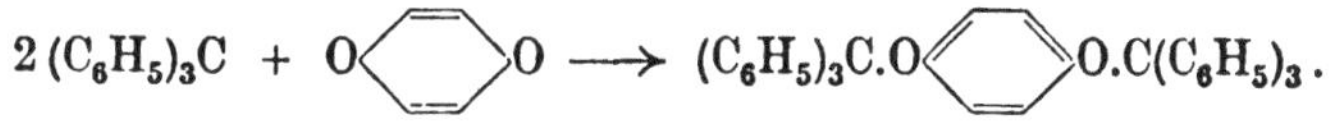

In addition, metallic *sodium* adds itself to the free valency, producing the very interesting orange-coloured *sodium triphenylmethyl* $(C_6H_5)_3C.Na$ (Schlenk).

When the molecular weight of hexaphenylethane is determined cryoscopically in benzene solution, the value corresponding to this hydrocarbon is obtained with close approximation. Indeed, only 2–3 per cent of the dissolved molecules are split into the triphenylmethyl halves.

The relationship

$$\underset{\text{colourless}}{O_2N{-}NO_2} \rightleftarrows \underset{\text{brownish-red}}{2\,NO_2}$$

is in many respects extraordinarily similar to the one considered here.

In both cases the degree of dissociation increases with rising temperature. Ebullioscopic determination of the molecular weight shows that, in boiling benzene, 30 per cent of the substance is present as triphenylmethyl.

The dissociation of hexaphenylethane can also be demonstrated *colorimetrically.* Whereas, in general, coloured solutions undergo no change in intensity of colour on dilution, since the number of coloured molecules observed in the colorimeter remains the same (Beer's law), the intensity must increase if the coloured molecules become more numerous as a result of progressive dissociation following dilution (Piccard).

Experiment.—Verify Beer's law by wrapping two test tubes in black paper, pouring the same volume (1–2 c.c.) of a dilute solution of a dye into each, and checking the equality of colour by looking down both tubes against a white surface. Then dilute the contents of one tube with 5–10 c.c. of water and again compare their colour.

The decomposition of hexaphenylethane is to be attributed to the inadequate binding force between the two ethane carbon atoms which are each over-much occupied by three phenyl groups. If these are progressively replaced by diphenyl groups the combining power of the fourth valency becomes smaller and smaller, and is finally reduced to zero in p-*tridiphenyl-methyl* (Schlenk).

This hydrocarbon 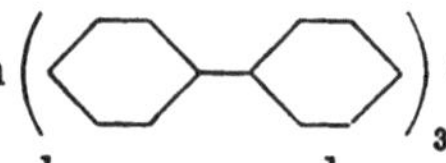 C, indeed, exists only as a free radicle, and has even been prepared as such in the solid state in the form of magnificent red-violet crystals.

Of the further results of the investigation of the carbon radicles, a field still actively cultivated, only the so-called *metal ketyls* will be mentioned. These addition products of the alkali metals to ketones are also intensely coloured (Schlenk), *e.g.*

$$(C_6H_5)_2{:}C{=}O + Na \longrightarrow (C_6H_5)_2{:}C{-}ONa\,.$$

They have already been discussed on p. 215.

The Triphenylmethyl Ion.—A solution of triphenylchloromethane in a dissociating solvent conducts the electric current (Walden). Since, on electrolysis, triphenylmethyl is liberated at the cathode, it follows that the solution contains the ions $(C_6H_5)_3\overset{+}{C}$ and $\overset{-}{Cl}$. The intensely yellow solution of hexaphenylethane in liquid sulphur dioxide also conducts electricity, and hence also contains ionised triphenylmethyl (possibly as a complex ion with SO_2). Such a solution does not exhibit the typical band spectrum and does not react with oxygen. Consequently the sharp distinction between radicle and ion exists here in the same way as is known, for instance among the metals, between atom and ion.

The triphenylmethyl ion is also, very probably, present in the orange-yellow products of the salt and complex salt type which are produced from triphenylcarbinol with concentrated sulphuric acid and from triphenylchloromethane with metallic chlorides ($ZnCl_2$, $AlCl_3$, $SnCl_4$, $SbCl_5$).

Experiment.—Dissolve a few granules of triphenylcarbinol or of triphenylchloromethane in 0·5 c.c. of concentrated sulphuric acid by rubbing with a glass rod. Add a little water: the deep orange-yellow solution is completely decolorised. Simultaneously the unchanged carbinol is precipitated.

In the same way the above-mentioned complex salts of triphenylchloromethane are readily decomposed by water. In both cases a hydrolysis occurs which causes the triphenylmethyl ion to lose its charge, and the carbinol to be re-formed.

The formation from neutral substances (triphenylcarbinol) of coloured, salt-like reaction products which are more or less easily decomposed by water is a phenomenon called " halochromism ". The halochromic salts of triphenylcarbinol are regarded as *carbonium* salts: this follows at once from the above discussion. A quinonoid formula, by which various authors explain the colour, seems less probable. Recently the attempt has been made to attribute complex formulae to the carbonium salts (Hantzsch), in accordance with Werner's scheme for ammonium salts. Such formulae express the fact that, in the ion, the charge is not localised at the methane carbon atom, but spread over the field of force of the whole radicle. The simplest carbonium salt of the group, the yellow perchlorate (K. A. Hofmann), would accordingly have the following structural formula :

$$\left[\begin{matrix} C_6H_5.\dot{C}.C_6H_5 \\ C_6H_5 \end{matrix}\right]\ldots OClO_3.$$

8. TETRAPHENYLHYDRAZINE

Diphenylamine (34 g., 0·2 mole) is dissolved in 200 c.c. of pure acetone in a bottle (capacity about 400 c.c.) having a well-fitting glass or rubber stopper. (Commercial acetone is usually stable towards permanganate. If not, add powdered potassium permanganate until the colour persists even on boiling the mixture for about half an hour under reflux, then distil. Acetone so treated is fit for use as a solvent for oxidations.[1]) The solution is kept cool in ice-water and vigorously shaken while *very finely* powdered permanganate is gradually added. Each portion is added only after the colour produced by the preceding has disappeared. After about 16 g. of permanganate have been used in the course of an hour and a half more of the oxidising agent is added without external cooling of the mixture until the colour persists for half an hour. Part of the diphenylamine is oxidised down to phenylisonitrile (odour, evolution of carbon dioxide). By means of a few drops of alcohol or of formaldehyde solution the mixture is now decolorised, and the manganese dioxide, after being collected at the pump, is pressed down well and washed twice with a little warm acetone. The acetone is then distilled under slightly reduced pressure from the water bath at 35° ; if the receiver is kept cool most of the solvent can be recovered.

[1] Cf. F. Sachs, *Ber.*, 1901, 34, 497.

The rest of the acetone is removed in a good vacuum at a bath temperature of 20°.

To the tetraphenylhydrazine which has crystallised 20–30 c.c. of ether are added with ice cooling. In this way oily material is dissolved. After some time the compound is filtered dry at the pump and washed clean with drops of ether. From 20 to 24 g. (60–70 per cent of the theoretical) of almost colourless crude product are thus obtained, which is pure enough for the following experiments. By recrystallisation from a little benzene (about two to three parts) absolutely pure tetraphenylhydrazine, melting point 144°, is obtained. The solution should be boiled for a short time only. If one-third of its volume of boiling alcohol is added with shaking to the hot solution, more crystallises than from benzene alone. The pure preparation is collected at the pump, washed, first with benzene-alcohol 1:1, then with alcohol alone, and immediately dried in a vacuum desiccator. The mother liquor may be evaporated *in vacuo* and the residue digested with cold ether as before. When pure and well dried the substance can be preserved unchanged for years if protected from light and acids.

Experiment.—Dissolve about 0·5 g. of tetraphenylhydrazine in 5 c.c. of xylene and warm the solution slowly over a small flame. The initially colourless solution becomes intensely olive green, even before the boiling point of the solvent is reached. The colour is that of the free radicle which, at this temperature, very rapidly undergoes further alteration, but, as the next experiments show, can be *fixed* as *diphenylnitrosamine* if nitric oxide is present.

Experiment.—Pour concentrated sulphuric acid on to a few centigrammes of tetraphenylhydrazine. A beautiful red colour at first appears; it changes in a short time to intense blue-violet.

The dye which is formed here is identical with that which is produced from *diphenylamine* in the well-known test for nitric acid (and other oxidising agents), namely, *diphenyldiphenoquinone-diimonium sulphate* (Kehrmann).

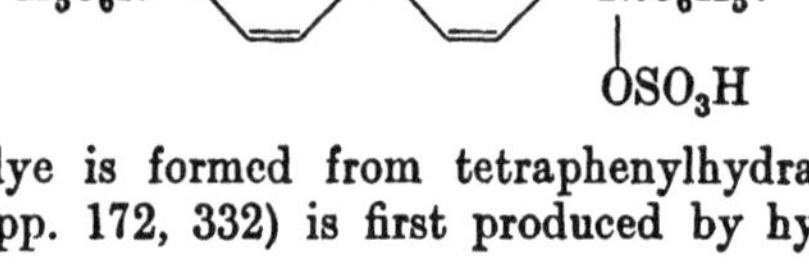

When the dye is formed from tetraphenylhydrazine, *diphenylhydroxylamine* (pp. 172, 332) is first produced by hydrolysis at the N—N-linkage.

THE FIXATION OF DIPHENYLNITROGEN BY NITRIC OXIDE[1]

An apparatus for producing pure nitric oxide is set up as follows. As in the apparatus for preparing hydrogen chloride, a filter flask (capacity 750 c.c.) is fitted with a dropping funnel, from which $4N$-sulphuric acid is dropped into concentrated sodium nitrite solution, containing 70 g. of $NaNO_2$ in 150 c.c. of water. For this amount of nitrite 250 c.c. of the $4N$-acid are required. The side tube of the flask leads first to a wash-bottle with concentrated sodium (or potassium) hydroxide, and then to one with concentrated sulphuric acid. By means of a short rubber tube a T-tube is attached to the second wash-bottle. One branch of the tube is connected to a carbon dioxide Kipp, the other to the reaction vessel. The rubber tube carries a screw clip so that, at the end of the experiment, the nitric oxide generator can be removed.

Tetraphenylhydrazine (5 g.) is dissolved in 40 c.c. of toluene in a small round-bottomed flask fitted with a two-holed cork. A delivery tube, which is connected to the apparatus just described and reaches to the bottom of the flask, passes through one hole, a short glass tube through the other. With the second wash-bottle disconnected from the rubber tube, and the latter closed by means of the clip, the dropping in of sulphuric acid is begun, and at the same time the air in the other part of the apparatus and the reaction vessel is driven out by carbon dioxide. The flask is clamped in such a position that, immediately afterwards, it can be placed on a vigorously boiling water bath kept ready for the purpose. When all the air in the first part of the apparatus has been expelled by nitric oxide, *i.e.* when the gas in the exit tube of the second wash-bottle has become quite colourless, the two parts of the apparatus are rapidly connected by means of the rubber tube, the clip is unscrewed, and the nitric oxide, accompanied by a gentle current of carbon dioxide, is passed into the flask. As soon as brown fumes (NO_2) begin to escape from its exit tube the flask is placed on the boiling water bath and a rather rapid stream of nitric oxide is passed into the toluene solution for half an hour. At the end of this time the solution is yellow. The flame under the bath is now lowered, the clip screwed down, the nitric oxide generator disconnected, and all the nitric oxide which remains in the rest of the apparatus is driven out by a brisk current of carbon dioxide (test with potassium iodide-starch paper). Then

[1] *Annalen*, 1911, 381, 211.

the toluene is *completely* removed by distillation in a vacuum, and the crystalline diphenylnitrosamine which remains is purified by recrystallisation from a little alcohol or from petrol ether. Melting point 66°.

Experiment.—In order to show that this formation of aromatic nitrosamines is reversible, boil a small amount of the pure compound just prepared in xylene and hold a piece of moist potassium iodide-starch paper over the mouth of the tube.

A close analogy to the carbon radicles is provided by certain nitrogen compounds, which are likewise free unsaturated complexes partaking of the character of atoms and having an abnormal valency. With nitrogen, as with carbon, the existence of radicles also depends on the presence of aromatic rings. *Tetraphenylhydrazine* corresponds to hexaphenylethane.

The linkage between the two N-atoms is here firmer than in the carbon prototype. Dissociation into the *diphenylnitrogen* radicles in solution is not perceptible to the eye below about 80° :

$$(H_5C_6)_2N—N(C_6H_5)_2 \longrightarrow 2\,(C_6H_5)_2N.$$

However, if positive substituents are introduced into the benzene rings, hydrazine derivatives result which considerably surpass hexaphenylethane in their degree of dissociation. Already the colourless p-*tetra-anisylhydrazine* is appreciably dissociated at room temperature into the green p-*dianisylnitrogen* radicles $(H_3CO.C_6H_4)_2N$ and the corresponding hydrazine containing four $—N(CH_3)_2$-groups instead of four methoxy groups is dissociated into the yellow *bis*-p-*dimethylamino-diphenylnitrogen* radicles, in cold benzene to the extent of 10 per cent, in nitrobenzene to the extent of 21 per cent.

In contrast to their inorganic prototype, nitric oxide, these radicles are not sensitive towards *oxygen*. On the other hand they take up *nitric oxide* with great ease, so that this property is of general use for their recognition.

$$(C_6H_5)_2N + NO \longrightarrow (C_6H_5)_2N.NO.$$

It will be noticed that by this reaction the *nitrosamines* of the corresponding diarylamines are formed. The nitrogen radicles also combine with triphenylmethyl and other radicles, with mutual saturation of the free valencies.

$$(C_6H_5)_2N + (C_6H_5)_3C \longrightarrow (C_6H_5)_2N - C(C_6H_5)_3.$$

As regards stability, the nitrogen radicles are inferior to those of carbon. They undergo a change which is observed throughout this division of chemistry. It consists in dismutation, *i.e.* the normal degree of saturation is restored by the removal, by one molecule, of an atom of *hydrogen* from the other. Along with the secondary amine a product poorer in hydrogen is formed, namely, a phenazine derivative.

$$2\,(H_3CO.C_6H_4)_2N \longrightarrow (H_3CO.C_6H_4)_2NH + (H_3CO.C_6H_4)_2N \text{ less H}$$

$$2[(H_3CO.C_6H_4)_2N \text{ less } 1\,H] \longrightarrow$$

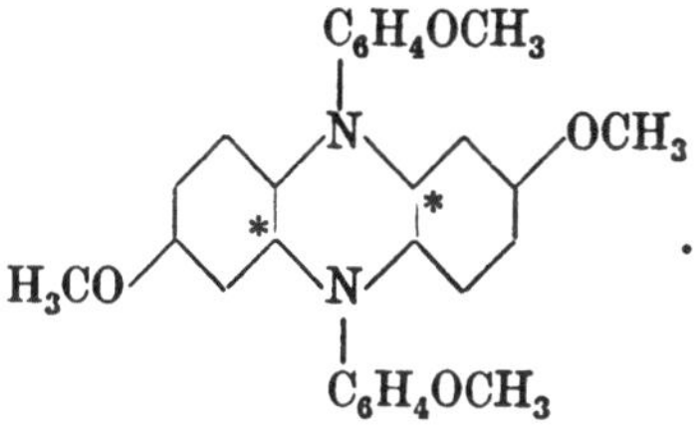

The hydrogen has therefore been taken from the positions marked with the asterisks. The simplest example of this dismutation of radicles occurs in the hydroxyl radicles, resulting from the discharge of the hydroxyl ion.

$$2\,OH \longrightarrow HOH + O\,;\ 2\,O \longrightarrow O_2.$$

Mention should also be made of radicles with bivalent nitrogen which are derived from hydrazines, the so-called *hydrazyls*. They are deeply coloured compounds which are obtained by dehydrogenation of tertiary hydrazines and form equilibrium mixtures with colourless *tetrazanes*, which dissociate into these free radicles (S. Goldschmidt), *e.g.*

$$\underset{\displaystyle C_6H_5\quad C_6H_5}{(C_6H_5)_2N.N\text{——}N.N(C_6H_5)_2} \rightleftarrows 2\,(C_6H_5)_2N\text{—}N.C_6H_5.$$

The quadrivalent radicle NO_2 also has an analogue in the aromatic series. When diphenylhydroxylamine is dehydrogenated with silver oxide the beautifully crystalline garnet-red diphenylnitrogen oxide is formed.

$$HO\text{—}N(C_6H_5)_2 \longrightarrow O{=}N(C_6H_5)_2.$$

This compound is surprisingly similar to NO_2, not only in colour, but also in many reactions. It, however, lacks all tendency to lose the radicle state and to dimerise like nitrogen peroxide. In this respect it resembles nitric oxide, whereas the organic relatives of nitric oxide are more like nitrogen peroxide.

The subject of free radicles is fully dealt with by P. Walden in *Chemie der freien Radikale*, Leipzig, 1924.

CHAPTER X

HETEROCYCLIC COMPOUNDS

1. PYRIDINE DERIVATIVES

(a) Hantzsch's Collidine Synthesis [1]

Ethyl Dihydrocollidinedicarboxylate.—A mixture of 33 g. of ethyl acetoacetate and 10 g. of aldehyde ammonia is heated in a small beaker on a wire gauze for three minutes at 100°–110°. During that time the mixture is stirred with a thermometer. Two volumes of 2*N*-hydrochloric acid are then added to the warm reaction product, which is vigorously stirred without further heating until the initially liquid mass has solidified. The solid is now finely ground in a mortar, collected at the pump, washed with water, and dried on porous plate. The crude material can be used directly for further treatment. A sample, recrystallised from a little alcohol, forms colourless plates having a bluish fluorescence. Melting point 131°.

Ethyl Collidinedicarboxylate.—The crude ester is mixed with an equal weight of alcohol, the mixture is kept cool in water, and gaseous nitrous acid [2] is passed in until the dihydro ester has dissolved and a sample of the liquid forms a clear solution in dilute hydrochloric acid.

The solution is now poured and rinsed into a moderate-sized (500–750 c.c.) separating funnel containing 100 g. of ice, and the acid is buffered by slowly adding finely powdered sodium carbonate. The ester separates as an oil and is taken up in ether, but the funnel must not be stoppered and shaken until evolution of carbon dioxide has ceased. Extraction with ether is repeated and the combined extracts, after being shaken again with water in order to remove

[1] *Annalen*, 1882, **215**, 1.

[2] *Cautiously* powder 50 g. of arsenious oxide and treat with a mixture of 75 c.c. of concentrated nitric acid (d = 1·4) with 30 c.c. of water in a round-bottomed flask. The acid is run in slowly through a dropping funnel passing through the two-holed cork; the delivery tube passes through the other hole and is connected to an empty wash-bottle. Heat the flask gently on a wire gauze.

most of the alcohol, are dried for a short time with potassium carbonate. The ether is then evaporated and the residue is distilled in a vacuum. Boiling point 175°–178°/12 mm. Yield, 15 g. of ester of collidine dicarboxylic acid from 20 g. of dihydro-ester.

Potassium Collidinedicarboxylate.—Purified potassium hydroxide (30 g.) is dissolved in 100 c.c. of absolute alcohol by boiling for a long time in a round-bottomed flask (capacity 250 c.c.) under reflux on a wire gauze, 15 g. of ethyl collidinedicarboxylate are added, and the solution is refluxed for three to four hours on a vigorously boiling water bath. The salt produced is sparingly soluble in alcohol and gradually separates in the form of a crystalline crust. When the hydrolysis is over the crystals are separated from the cooled solution at the pump and washed, first twice with alcohol and then with ether. Yield 12–14 g.

Collidine.—The carboxyl group is removed by heating the potassium salt with slaked lime. The salt is thoroughly mixed with twice its weight of calcium hydroxide by grinding in a mortar, and the mixture is poured into a short combustion tube (about 60 cm. long) closed 10 c.c. from the end with an asbestos plug. By means of another loosely fitting asbestos plug the mixture is kept in position and one end of the tube is tightly corked. The other end is connected with a receiver by means of an adapter and the tube is placed with the closed end uppermost in an inclined combustion furnace. *After a moderately deep channel has been produced over the mixture by tapping* the tube is warmed by small flames. Then, beginning at the elevated end of the tube, the flames are raised more and more until, with closed tiles, a bright red heat is finally reached. The collidine which distils is dissolved in ether, and the solution dried with a little potassium hydroxide; the ether is evaporated and the collidine distilled. Boiling point 172°. Yield 3–4 g.

If a nitrogen cylinder is available, the mixture of potassium salt and lime is heated in a slow stream of the gas.

The synthesis of the pyridine ring from ethyl acetoacetate, aldehydes, and ammonia proceeds extraordinarily readily. The mechanism is as follows. In the first phase the aldehydes react with the acetoacetic ester to form *alkylidene bis-acetoacetic esters*. The 1:5-diketone derivatives so formed undergo ring closure by introduction of a molecule of ammonia and elimination of two molecules of water:

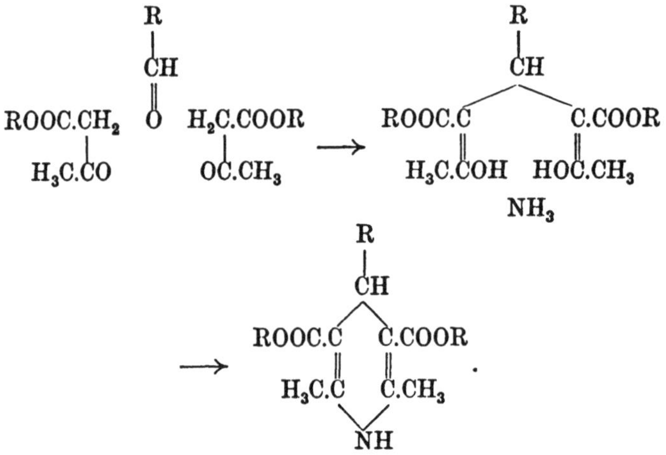

If the synthesis is carried out without ammonia the six-ring synthesis of Knoevenagel takes place. This synthesis, which is catalytically induced by bases such as diethylamine and piperidine, consists in ring closure of the intermediate product, of which the di-enol form is shown above.

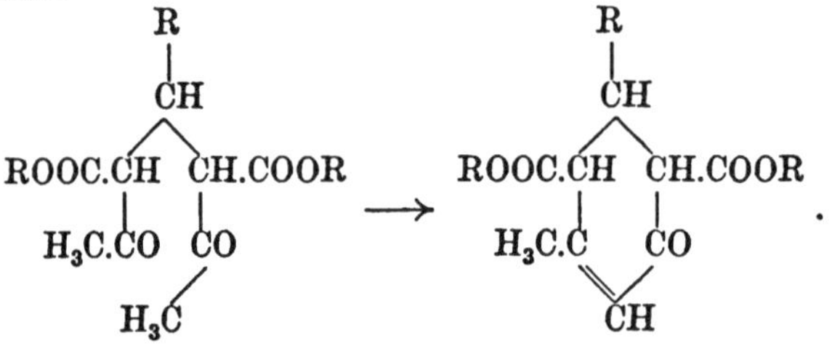

In Hantzsch's synthesis the condensation product is a derivative of *dihydro*pyridine and is only converted into a true pyridine derivative by dehydrogenation. It is only by the removal of the two hydrogen atoms in the 1:4-positions that the heterocyclic ring system analogous to benzene is produced. The corresponding conversion of ethyl Δ 2:5-dihydroterephthalate into ethyl terephthalate takes place much more readily.

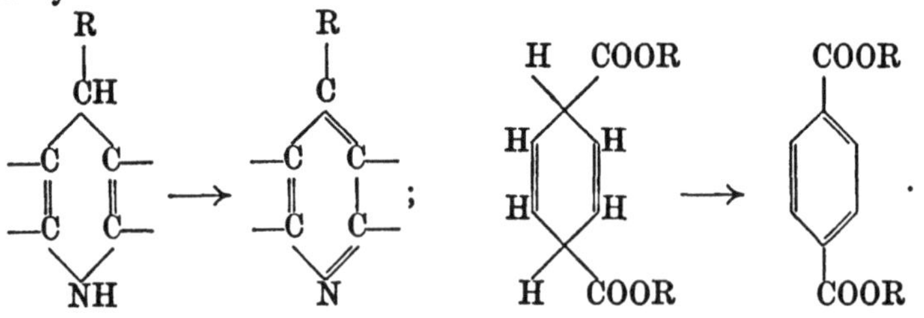

That the true pyridine derivative is more basic than the dihydro-compound is connected with the fact that in the latter the NH-group is united to two doubly bound C-atoms. Yet also pyridine and its derivatives are but weak bases.

Pyridine (and quinoline) which in so many respects are "aromatic" and comparable to benzene, lose this character completely on hydrogenation to *piperidine* (and hydroquinoline), which are entirely of the same nature as secondary aliphatic amines. The completely hydrogenated heterocyclic bases undergo degradation reactions which have become important particularly in the investigation of the constitution of alkaloids. A. W. Hofmann's method of opening rings by means of "exhaustive methylation" may be illustrated with piperidine. By thermal decomposition of the quaternary ammonium base a C—N-linkage is broken and at the same time water is eliminated.

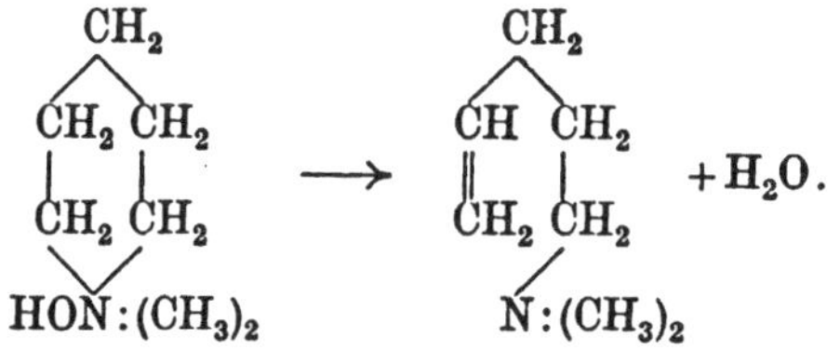

The open-chain unsaturated tertiary amine is again exhaustively methylated and its quaternary ammonium base, in its turn, decomposed as before.

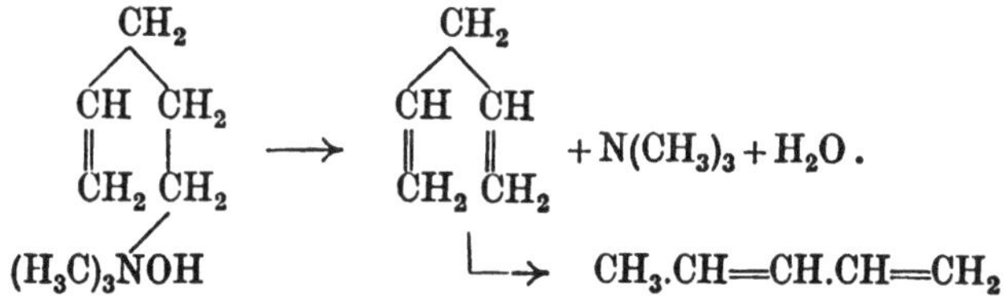

In this way piperidine is converted into the hydrocarbon "piperylene", α-methylbutadiene. The cause of the shift in the double bond is the same as in the rearrangement of *eugenol* to *isoeugenol* and of *β*- to *α-dihydromuconic acid.* Formulate this degradation reaction with *β-methylpyrrolidine.*

A second method is due to J. von Braun and consists in the addition of *cyanogen bromide* to tertiary cyclic bases.[1] In the unstable addition product a C—N-linkage is broken and at the same time the bromine wanders to a new position. A brominated derivative of cyanamide is produced and this, on hydrolysis, yields a secondary amine which can be broken down further, *e.g.*

[1] *Ber.*, 1907, 40, 3914; 1909, 42, 2219; 1911, 44, 1252.

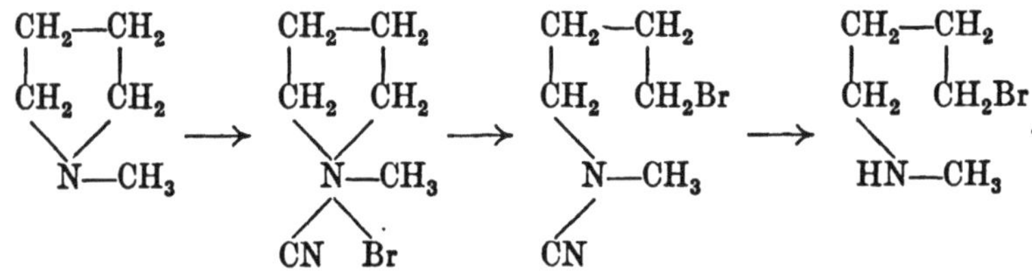

The demethylated secondary cyclic base is also produced as a by-product.

Another process for the degradation of *secondary* amines, due to the same author, may be illustrated with piperidine.

By treatment with phosphorus pentachloride, *N*-benzoylpiperidine is decomposed into benzonitrile and 1:5-dichloropentane.

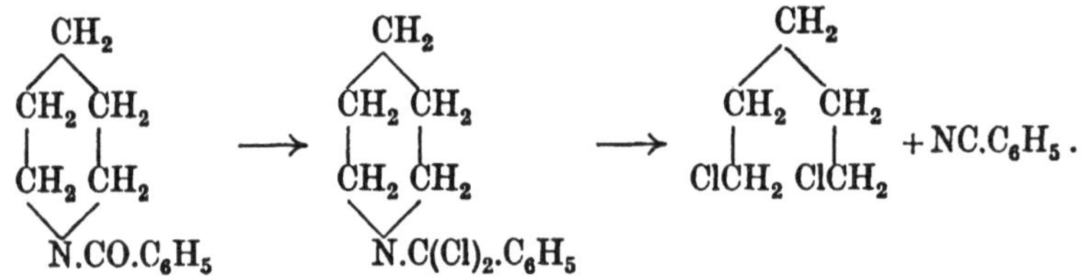

(*b*) *a*-Aminopyridine [1]

Sodamide (10 g.) is ground under xylene in a mortar and added to 16 g. (0·2 mole) of pyridine which has been dried over powdered potassium hydroxide or barium oxide, distilled, and mixed with 30 c.c. of sodium-dried xylene. The mixture is then heated under reflux in an oil bath at 140°–150° for seven hours. Moisture must be completely excluded. After cooling, 20 c.c. of cooled sodium carbonate solution are gradually and cautiously added, the mixture is shaken up, and the layers of liquid are then separated with a funnel. The aqueous layer is extracted several times with benzene; the extracts are combined and dried for a short time over solid potassium hydroxide; the solvents are removed by distillation. For purification the high-boiling aminopyridine is distilled in a vacuum (sausage flask); the first portion of the distillate consists chiefly of xylene. The base boils at 93°/11 mm. and 96°/13 mm. Yield 6–7 g. A further small amount can be obtained from the first and last portions of the distillate by fractionation. *a*-Aminopyridine crystallises easily and can be recrystallised from ligroin. Melting point 57°.

[1] Tschitschibabin, *Chem. Zentr.*, 1915, I, 1065; Wibaut, *Rec. trav. chim.*, 1923, 42, 240.

The very remarkable reaction which consists in the introduction of the NH_2-group into an aromatic ring by the action of sodamide is due to F. Sachs (*Ber.*, 1906, **39**, 3006), who studied various examples of the process in the naphthalene and anthraquinone series.

In the case of pyridine, Tschitschibabin's synthesis proceeds with special ease. An intermediate addition product of NH_2Na to the —N=C double bond with the grouping —NNa—C(NH_2)— is doubtless formed. The net result is as follows :

$$C_5H_5N + NaNH_2 \longrightarrow C_5H_4N.NH_2 + Na + H,$$

and the *a-aminopyridine* thus formed behaves, in its reactions, like a tautomeric compound. Many of its derivatives, especially the cyclic ones, are derived from a diimino-form which can arise as a result of the following rearrangement :

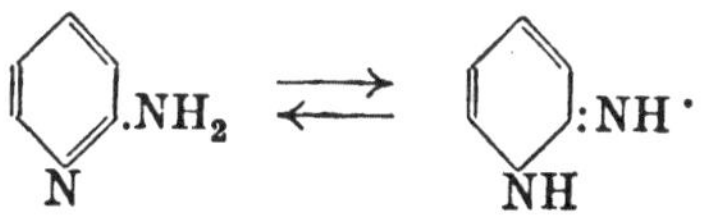

2. QUINOLINE

(a) Skraup's Quinoline Synthesis [1]

Concentrated sulphuric acid (45 c.c.) is poured with shaking into a flask (capacity 1·5 l.) containing a mixture of 20 g. of nitrobenzene, 31 g. of aniline, and 100 g. of *anhydrous* [2] glycerol. The flask is then fitted with a long wide reflux condenser and heated on a wire gauze. As soon as the sudden escape of bubbles of vapour from the liquid shows that the reaction has set in, the flame is immediately removed and the main reaction,[3] which is sometimes extremely violent, is allowed to proceed to completion without external heating. When the reaction has subsided the mixture is kept boiling for three hours more on a wire gauze or sand bath, then a little water is added and the unchanged nitrobenzene is completely removed from the acid liquid by a current of steam. While still warm the liquid in the flask is made alkaline with concentrated sodium hydroxide, and the liberated quinoline, along with unchanged aniline, is likewise

[1] *Monatsh.*, 1880, 1, 316; 1881, 2, 139. M. Wyler, *Ber.*, 1927, 60, 398. Darzens, *Bull. Soc. chim.*, 1930, 47, 227.

[2] Heat commercial glycerol in a porcelain basin in the fume chamber until a thermometer hung with the bulb in the liquid registers 180°.

[3] The main reaction can be moderated by adding only half of the sulphuric acid at the beginning, heating cautiously to gentle boiling with a small flame, and, after one hour, adding the remainder of the acid quite slowly drop by drop. The mixture is then kept boiling as above for three hours more.

distilled with steam. The distillate is extracted with ether, the ether is evaporated, and the crude bases are dissolved in a mixture of 50 c.c. of concentrated hydrochloric acid with 200 c.c. of water. After warming the clear solution so obtained, 30 g. of zinc chloride in 50 c.c. of 2 *N*-hydrochloric acid are added. A quinoline-zinc chloride double salt crystallises as the solution cools, and after the mixture has been left in ice for some time the crystals are collected at the pump, washed with cold 2 *N*-hydrochloric acid, and decomposed with concentrated sodium hydroxide solution. Once more the quinoline is distilled in steam, extracted from the distillate with ether, dried in the ethereal solution with solid potassium hydroxide, and ultimately distilled after the ether has been evaporated. Boiling point 237°. Yield 24–25 g. The preparation is water-clear.

(*b*) Quinaldine Synthesis of Doebner and Miller [1]

To a mixture of aniline (31 g.) and commercial concentrated hydrochloric acid (60 c.c.) in a one-litre flask, 45 c.c. of paraldehyde are added (or 60 c.c. of acetaldehyde carefully dropped in through a long reflux condenser, while the flask is cooled in ice). The mixture is left at room temperature; the condensation takes place gradually with slight evolution of heat. The liquid is then boiled under reflux for three hours, made strongly alkaline with sodium hydroxide, and submitted to steam distillation. From the distillate the crude base is extracted with ether and the extract is dried with solid potassium hydroxide. After the ether has been evaporated unchanged aniline is removed by boiling under reflux for a quarter of an hour with 10 c.c. of acetic anhydride, cooling, making distinctly alkaline with saturated sodium carbonate solution, and distilling again with steam. After working up in the usual way the *quinaldine* is purified by distillation in a vacuum. Boiling point 115°–120°/12 mm. The last portion of the distillate consists of a small amount of bases of higher boiling points. Yield 18–20 g.

The quinaldine can also be separated from the crude mixture of bases by conversion into the zinc chloride double salt in the manner described in the case of quinoline. This furnishes a somewhat smaller yield of a purer preparation.

The first quinoline derivative obtained by the process (*a*) was the dye *alizarin blue* (Prud'homme, 1877). *β-Nitroalizarin* was heated

[1] *Ber.*, **1881, 14,** 2816; **1883, 16,** 1664; **1884, 17,** 1712.

with glycerol and sulphuric acid. The constitution of the product was established by Graebe :

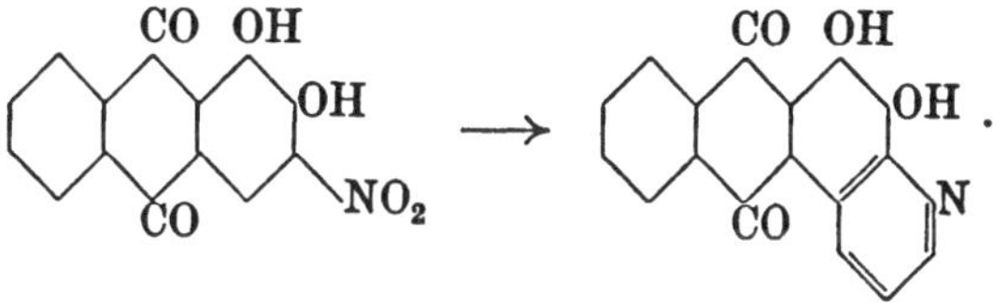

During the process the NO_2-group is reduced to NH_2.

Skraup's synthesis involves the elimination of water. Thus *acrolein* will be formed. It may react with aniline to form an azomethine (Schiff's base) (I), but more probably the base adds itself to the C=C double bond (II) :

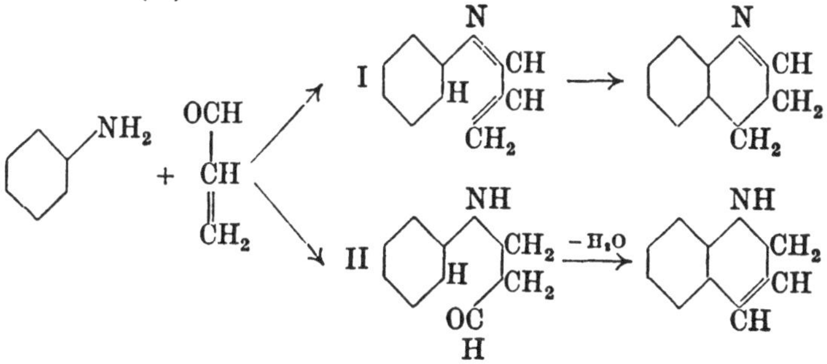

On either hypothesis a *dihydroquinoline* is formed. The extra hydrogen is removed by the nitrobenzene present.

A second similar synthesis, due to Doebner and Miller, leads to the formation of *substituted* quinolines. The simplest example is the production of quinaldine from aniline and paraldehyde by heating with concentrated hydrochloric acid. The course of the reaction is closely related to that of the Skraup synthesis by route II. There the aniline reacts with acrolein, here with *crotonaldehyde*, which is easily formed under the conditions which prevail :

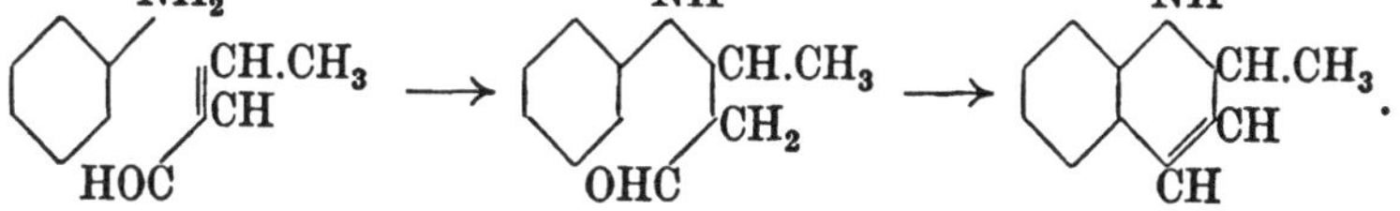

Here also there are two hydrogen atoms in excess, but they are removed by secondary reactions (leading to the formation of hydrogenated products). The well-known drug *atophan* (used in gout), α-phenylquinoline-γ-carboxylic acid, is the product of an analogous condensation of aniline with benzaldehyde and pyruvic acid :

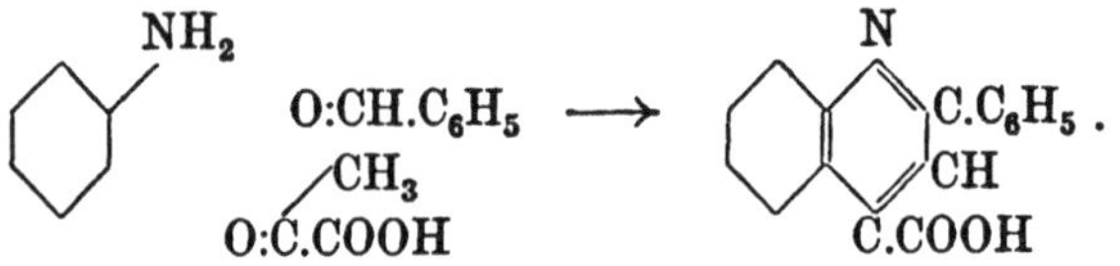

Atophan can also be obtained by alkaline condensation of isatin with acetophenone. (Formulate.)

The CH_3-group in quinaldine, like that of ketones, can condense with aldehydes and similar substances. With phthalic anhydride the yellow dye *quinophthalone* is formed.

3. INDIGO

Phenylglycine.[1]—Chloroacetic acid (19 g.) is exactly neutralised in the cold with 100 c.c. of 2*N*-sodium hydroxide solution, 18·6 g. of aniline are added, and the mixture is boiled for a short time under reflux until the aniline has reacted and dissolved. On cooling, the phenylglycine separates as an oil which soon crystallises when rubbed. The crystalline material is kept cool in ice for some time, collected at the pump, and washed with a little ice-cold water. Yield 22–24 g. of dry substance.

In order to prepare the *potassium salt*, 20 g. of phenylglycine are made exactly neutral to phenolphthalein with 2*N*-potassium hydroxide solution (about 70 c.c. are required), and the clear solution is then evaporated to dryness on the water bath. For the indoxyl fusion the residue of salt must be dried in an oven at 100° for several hours.

Indoxyl Fusion.[2]—A mixture of 15 g. of sodium hydroxide and 20 g. of potassium hydroxide is fused and carefully dehydrated by heating to about 500° in a nickel crucible. When the mass has barely solidified it is just remelted by gentle heating and poured into a Jena glass conical flask (capacity 100 c.c.) which is at a temperature of 220° in an oil bath. If this procedure is adopted there need be no fear that the glass will crack.

Sodamide (10 g.) is added to the melt in the flask and dissolves with slight evolution of ammonia. The pure potassium phenylglycine (20 g.) which has been completely dried at 100° in an oven is then introduced in spoonfuls during the course of five to ten minutes with cautious stirring by means of a glass rod. (Eyes and hands must be

[1] J. Houben, *Ber.*, 1913, 47, 3988.
[2] According to directions supplied by Dr. J. Pfleger, Frankfurt a. M.

protected.) The temperature of the bath is maintained at 200°–220°. Two minutes after the last portion of potassium salt has been added, the flask, which has been loosely stoppered with a cork, is removed from the bath and left to cool. When quite cold the flask is broken up and the melt, in small pieces, is dissolved in 500 c.c. of water in a beaker (capacity 1 l.). The liquid is poured rapidly through a large folded filter into a round-bottomed flask or filter flask[1] (capacity 1·5 l.), and air is drawn through the solution by means of a water pump until a drop of the aqueous suspension of indigo, placed on filter paper, produces a sharply defined ring of precipitated indigo, outside which the liquid no longer becomes blue on exposure to the air.

After oxidation has thus been completed the indigo is collected at the pump, washed with hot water, transferred to a beaker by means of a jet of water, boiled with 10 per cent hydrochloric acid, again collected at the pump, washed with hot water, and dried. The yield reaches 60 to 70 per cent of the theoretical.

A simple qualitative test of the purity of the indigo obtained can be carried out as follows: A little of the material is heated for some time (with shaking) in a test tube with pyridine and some drops of the liquid are then poured on to a filter paper. If the indigotin is pure the pyridine is not coloured, whereas impurities which may be formed when working on a small scale confer on it a more or less dirty brown colour, as shown by a "spot" test. If it is desired to purify the whole of the indigo with pyridine, the dye is collected at the pump after boiling with the liquid, washed with pure hot pyridine, boiled once more with hydrochloric acid, collected at the pump, washed with hot water, and dried.[2]

The synthesis of indigo given here is the one now usually employed on a technical scale; it shows how the dye is ultimately made from coke and lime (yielding acetic acid *via* acetylene), chlorine and aniline. The cultivation of indigo-yielding plants has thus become unnecessary, although the biological method of production has not yet suffered the same fate as befell the cultivation of madder through the synthesis of alizarin.

The synthesis of indigo by means of alkaline fusion of phenylglycine was discovered by Heumann as early as 1892, but a satisfactory yield was only obtained through the addition of sodamide (J. Pfleger).

The constitution of the dye was first elucidated by the classical researches of A. Baeyer. The numerous syntheses cannot here be

[1] Filtration is not absolutely essential, but yields a purer product.
[2] G. P. 134,139. Höchst Dye Works.

discussed in detail; only the most elegant, which was indeed employed for some time on a technical scale, can be described.[1]

In this process *o-nitrobenzaldehyde* is condensed in alkaline solution with acetone. The so-called *o-nitrophenyllactic acid ketone* thus formed loses acetic acid, and by further loss of a molecule of water is changed into *indolone,* which corresponds to half of the indigo molecule. Possibly *o*-nitrostyrene is an intermediate stage in this transformation. Indolone cannot exist in the free state, and hence polymerises at once to the dye:

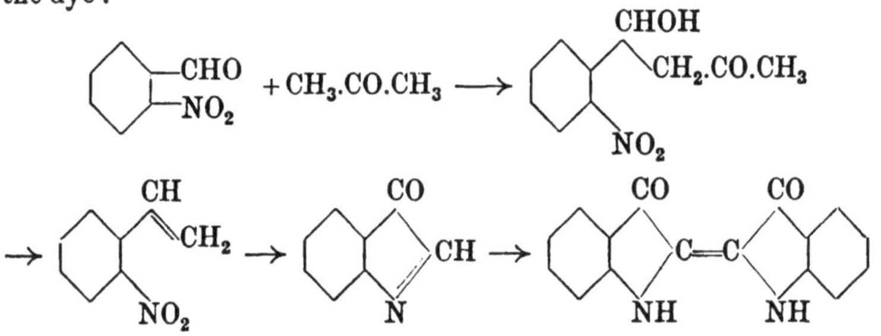

The transfer of the oxygen from a nitro-group to a carbon atom in the *ortho*-position may not seem very likely, but several similar reactions are known. Thus *o-nitrotoluene* is converted by alkali into *anthranilic acid* (Binz), *o-nitrobenzaldehyde* by sunlight into *o-nitrosobenzoic acid* (Ciamician):

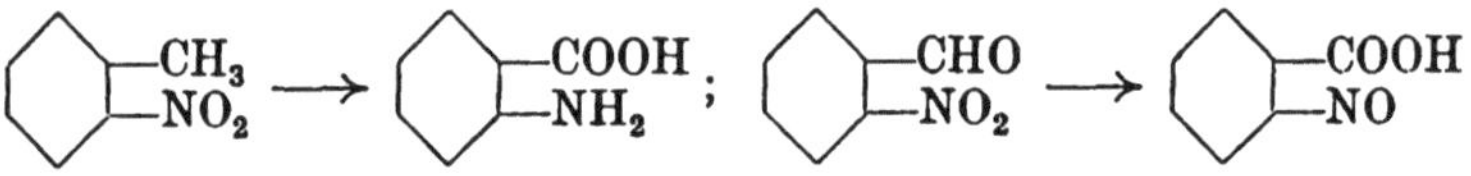

and *γ*-nitroanthracene into anthraquinonoxime (Meisenheimer):

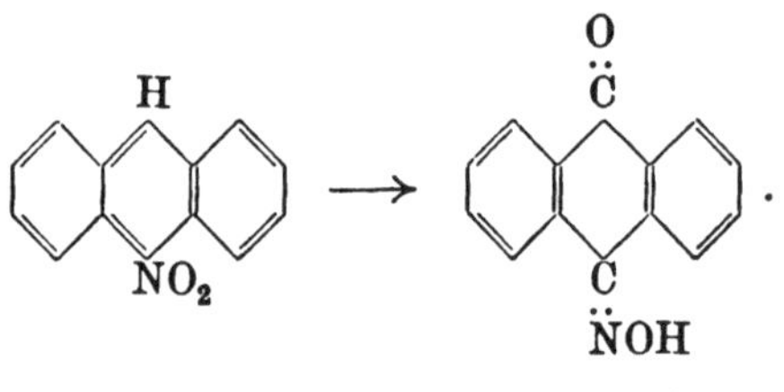

This simple indigo synthesis of Baeyer should not be omitted if *o*-nitrobenzaldehyde is available.

Experiment.—Dissolve 1 g. of *o*-nitrobenzaldehyde in 3 c.c. of pure acetone, add about an equal volume of water which leaves a clear solution, and then, drop by drop, *N*-sodium hydroxide solution. Heat is developed and the solution becomes dark brown.

[1] *Ber.*, 1882, 15, 2856.

After a short time the dye separates in crystalline flakes. Collect the precipitate at the pump after five minutes and wash, first with alcohol, then with ether. Indigo so prepared is specially pure and has a beautiful violet lustre.

The starting material for the first large-scale technical production of indigo was *naphthalene*, which was oxidised with fuming sulphuric acid (in the presence of mercuric sulphate) to *phthalic acid*. The *phthalimide* obtained from this acid was partially hydrolysed, to the *phthalamido acid*, and then underwent the Hofmann degradation, to *anthranilic acid*. Combined with chloroacetic acid, the latter yielded *phenylglycine-*o*-carboxylic acid*, which on fusion with alkali furnishes indoxyl. Formulate.

The synthesis of the red dye *thioindigo* (Friedländer), and its technically important derivatives, proceeds in a corresponding way. The starting material is *thiosalicylic acid* :

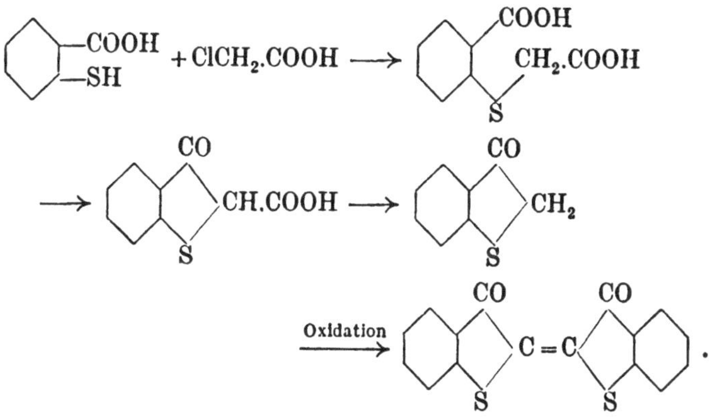

In the synthesis due to Heumann and Pfleger the fusion yields the potassium derivative of indoxyl, which is dehydrogenated to indigo even by atmospheric oxygen; at the same time hydrogen peroxide is produced (see p. 166).

Hardly any other organic compound has been so much examined from every point of view as indigo, and we must therefore confine ourselves here to a few of the most important reactions.

The Chemistry of Indigo Dyeing.—On account of its insolubility the dye itself cannot be applied directly to the fibre. Yet an indirect process of great antiquity is available, for Tyrian purple has been identified as 6:6′-dibromoindigo[1] by Friedländer. The indigo is made soluble by reduction in alkaline solution and so converted into the alkali salt of its dihydro compound; in technical language, indigo is a *vat* dye. Primitive peoples have, from remote times, brought about this

[1] The two bromine atoms are in the *m*-positions to the nitrogen.

conversion *biologically*, *i.e.* by means of reducing bacteria. In industry *ferrous hydroxide* or *zinc dust* was used, now replaced by *sodium hydrosulphite.*

Experiment.—Grind about 50 mg. of the indigotin prepared, in a small mortar, with a few drops of water to a fine sludge, wash this into a small conical flask with a jet of water from a wash-bottle, and after adding a small excess of alkaline sodium hydrosulphite solution warm to 30°–40°. Soon a greenish-yellow, then brownish solution, the *vat*, is produced. At the surface of this solution there forms, as a result of contact with the air, a fine blue film, the so-called "bloom". Dilute with water to 25–30 c.c., introduce a previously moistened strip of linen into the solution, and stir with a glass rod for about a minute. Then withdraw the cloth, press it, and hang it over two parallel stretched strings or over two thin glass rods. Already after five minutes the material is coloured deep blue.

Precipitate the dye again from the *vat* by drawing air through the solution. This process is also suitable for purifying indigo.

Chemically, conversion into a *vat* consists in a 1:6-addition of hydrogen and recalls exactly the conversion of quinone into quinol. Like quinol, "indigo white", also a dihydric "phenol", is a weak acid, the alkali salts of which are coloured intensely yellow.

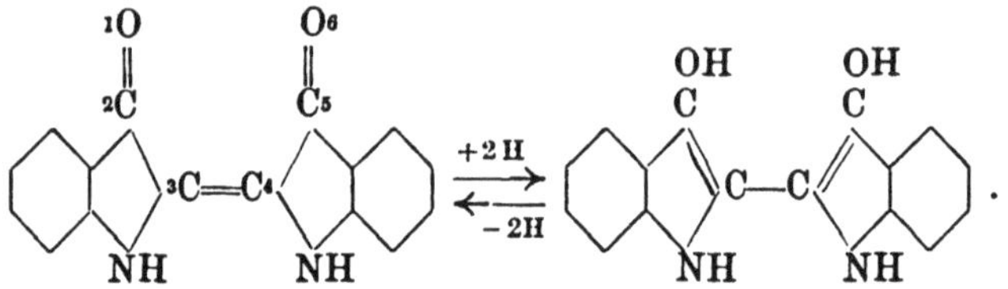

From its alkali salt, which is partially hydrolysed, the large "indigo white" molecule is adsorbed by the fibre and then, in this finely divided condition, dehydrogenated by the oxygen of the air, so that the dye now remains as a fast blue pigment. The oxidation is analogous to that of indoxyl.

The *vat dyes* are characterised by quite unusual fastness; apart from the true indigoids, the most important members of this group are found in the anthraquinone series. (Indigoids are dyes containing ring systems of the type present in indigo connected by a double bond.) Almost without exception these dyes contain condensed rings possessing great chemical stability. The blue dye *indanthrene*, which is obtained from the technically very important β-aminoanthraquinone by fusion with alkali (with accompanying elimination of hydrogen), may be taken as an example of an anthraquinone vat dye (R. Bohn):

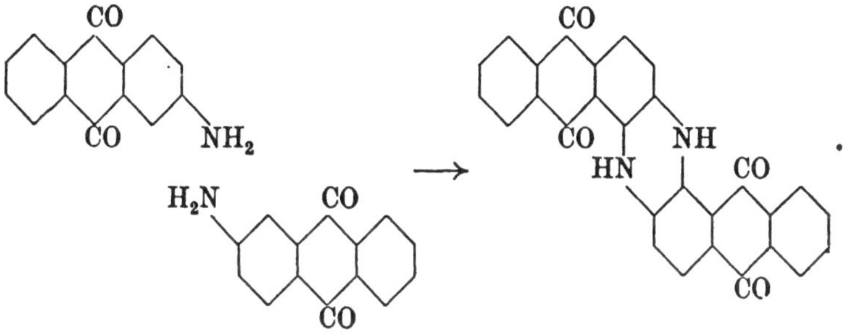

Although indigo on conversion into "indigo white" takes up hydrogen, and energetic reduction even breaks the double bond and yields indoxyl and indole, it can also be *dehydrogenated* by a not less remarkable reaction. The hydrogen is removed from the NH-groups of the indole ring most readily by lead peroxide (Kalb).

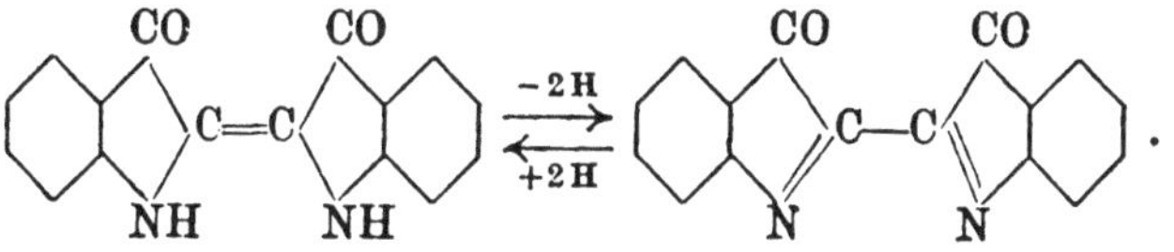

The *dehydroindigo* which is thus formed is much more soluble than indigo itself. It is a beautifully crystalline brownish-red substance which is very easily *hydrogenated*, even by quinol, for example, in the manner shown in the above equation (read from right to left). The quinol, of course, is converted into quinone.

Experiment.—Heat as much finely powdered indigo as amply covers the tip of a knife with about twice as much lead peroxide, a few granules of calcium chloride, and 5 c.c. of benzene in a test tube held in a boiling water bath for five minutes. Filter the reddish-brown solution obtained and divide the filtrate into two portions in test tubes. To one portion add *quinol* dissolved in very little alcohol, to the other dilute solution of stannous chloride in hydrochloric acid. The dehydroindigo is converted into the dye which separates in blue flocks.

Conversely a suspension of finely divided indigo in chloroform can be converted into a beautiful reddish-brown solution of dehydroindigo by adding first a little calcium hydroxide and then bromine drop by drop.

The beautiful preparation should be *isolated* by the former of Kalb's two procedures.[1]

[1] L. Kalb, *Ber.*, 1909, 42, 3649.

The best-known *oxidative* transformation of indigo is that into *isatin*. This is a normal oxidation at a double bond:

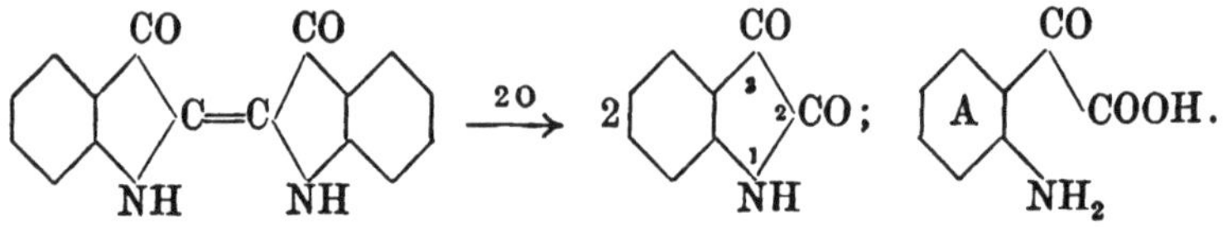

Isatin is the inner anhydride (lactam) of a γ-amino-α-ketocarboxylic acid, *isatinic acid* (A), and is converted into a salt of this acid by the action of alkali. The keto-group in position 3 can condense with many other substances, and for this reason isatin is manufactured on a technical scale and converted into valuable indigoid vat dyes. The magnificent thioindigo scarlet, which is obtained from isatin and a sulphur analogue of indoxyl (the so-called *hydroxythionaphthene*), may be taken as example.

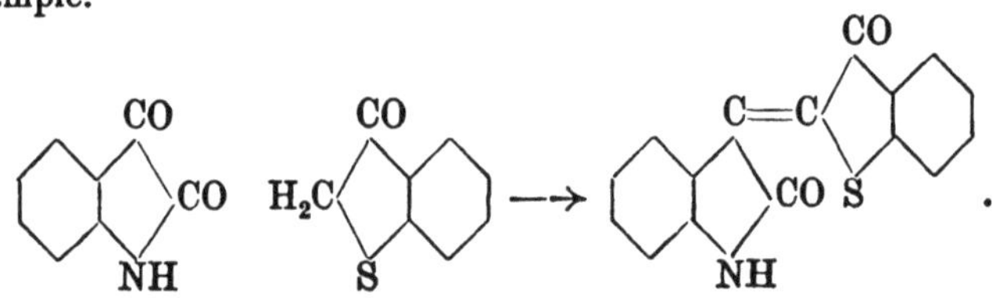

Sandmeyer's valuable isatin synthesis involves the removal of sulphur from diphenylthiourea (I) (p. 160) with basic lead carbonate. Hydrogen cyanide is combined with the reactive diphenylcarbodiimide (II) so obtained, and the nitrile (III) produced is converted by means of hydrogen sulphide into the thioamide (IV). Concentrated sulphuric acid brings about ring closure and the product is the α-anil of isatin (V). Then, by hydrolysis with dilute sulphuric acid, aniline is removed:

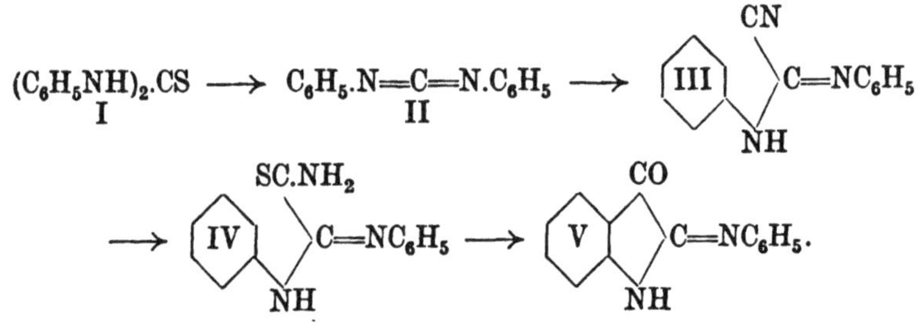

Sandmeyer's synthesis may be recommended also as a preparative method.

It may be recalled that the first synthesis of indigo, by Baeyer, started from isatin chloride. (Formulate it.)

CHAPTER XI

HYDROGENATION AND REDUCTION

1. CATALYTIC HYDROGENATION WITH PALLADIUM

(a) WITH A COLLOIDAL CATALYST BY SKITA'S METHOD [1]

FIG. 53 shows the manner in which the apparatus is arranged.[2] The confining liquid in the gas-holder is water. A Rabe turbine, a

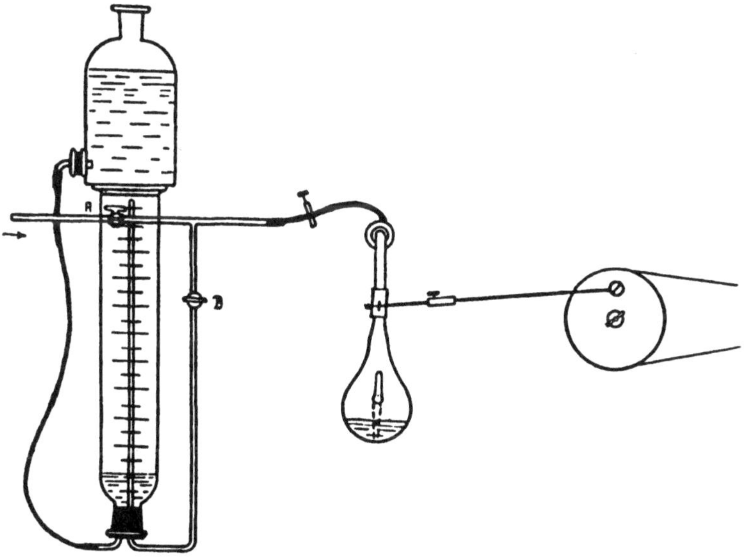

FIG. 53.

small electro-motor or a hot-air engine, which turns an eccentric attached by means of a rigid wire to the hydrogenation bulb, serves as a shaker. (If, however, the wire is in two parts, as the figure

[1] *Ber.*, 1909, 42, 1627; 1912, 45, 3312, 3519.
[2] The stands for the gas-holder, hydrogenation vessel, and eccentric are not shown in the figure.

shows, the apparatus can be dismantled more conveniently. The parts are connected by means of a brass sleeve provided with a screw.) All the stands are weighted with heavy pieces of iron. Into the side of the bulb itself (Fig. 54) an inlet tube [1] is sealed and to this tube a small piece of rubber tubing carrying a screw clip is tightly fixed. The horizontal tube of the bulb, which is at the same time the axis about which the bulb oscillates, turns in a brass sleeve (cork borer) held in position by short pieces of rubber tubing drawn over the glass. A strong clamp grips a cork through which the sleeve is pushed and thus maintains stability. By means of a thick-walled rubber tube carrying a spring clip the glass tube is directly connected to the gas-holder. The measuring cylinder of the latter is fixed in a large condenser clamp.

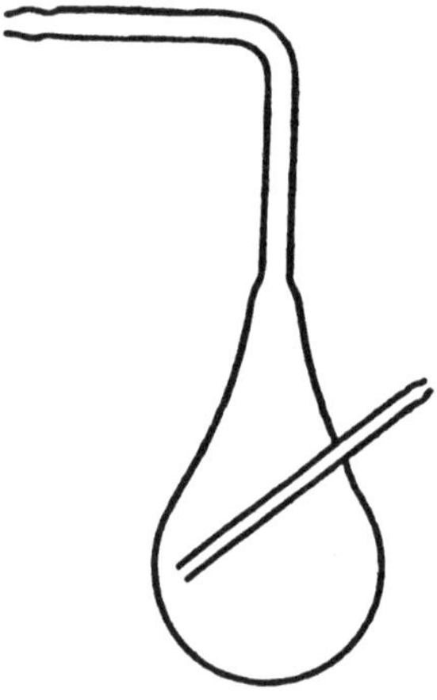

FIG. 54.

Before hydrogenation is begun the whole apparatus is tested for gas-tightness as follows: The screw-clip on the inlet tube is closed, and by opening the cocks A and B the gas-holder is provisionally filled with hydrogen. After closing the cocks the level of the water in the cylinder is marked while it is at the same height as that of the water in the container, and the empty bulb is shaken for a quarter of an hour. If the level is unchanged after this period (with unaltered room temperature) hydrogenation can proceed.

The solution of the catalyst, *i.e.* 50 mg. of gum arabic in 1 c.c. of water and 3 c.c. of a 1 per cent palladium chloride solution, is poured into the bulb and a solution of 5 g. of purest cinnamic acid in 30 c.c. of 80 per cent methyl alcohol is added. The cock B is closed, A is opened, and hydrogen (washed with potassium permanganate solution) is passed in from a cylinder through the inlet tube until all the air has been displaced from bulb and tubes. Previously the gas-holder (capacity 1 litre) and the tubes to the cock B have been filled with water, which is now driven out by hydrogen after A has been

[1] In place of an inlet tube the bulb may be provided with an opening into which a delivery tube can be fixed by means of a (clean) rubber stopper. In that case the apparatus shown here (Fig. 53) permits of the bulb being filled with hydrogen from the upper tube, so that the lower inlet tube becomes unnecessary.

closed, B opened, and the water container lowered. The clip on the inlet tube is now closed, the hydrogen cylinder (or Kipp) is disconnected, the level of the liquid read as when testing for gas-tightness, and the shaking apparatus is set in motion.[1] The container stands on the gas-holder so that a slight pressure is maintained (Fig. 53).

Although the amount of catalyst used here is very small (18 mg. Pd), sufficient hydrogen to saturate the ethylene double bond is nevertheless taken up in three hours (at 20° and a barometric pressure of 740 mm., 840 c.c.). The flocculated palladium is removed by filtration, the methyl alcohol is evaporated, and the hydrogenated acid is recrystallised as described on p. 225.

If undiluted methyl alcohol is used as solvent a secondary catalytic action of the palladium results in the production of ester.[2] In this case 5 g. of potassium hydroxide are added, the solution is concentrated by heating, and hydrocinnamic acid precipitated with dilute hydrochloric acid.

Calculation of the Amount of Hydrogen Required

One mole of substance requires for each double bond 22·4 litres of hydrogen under standard conditions. The volume of a mole of hydrogen under the experimental conditions which prevail can be calculated from the formula

$$V = V_0 \frac{760\ T}{273\ b},$$

where b is the observed barometric pressure less the vapour pressure of water at the temperature concerned, and T is the absolute temperature. Generally (for $b=760$ mm. and $t=15°$) it amounts to 24 litres. Five grammes of cinnamic acid (mol. wt. 148) = 5/148 mole; the volume of hydrogen required is therefore $24 \times 5/148$ litre = 811 c.c.

(*b*) HYDROGENATION WITH A CATALYST DEPOSITED ON A CARRIER

Preparation of the Catalyst.[3]—Barium sulphate (10 g.) precipitated while hot and well washed is made into a paste in a porcelain basin with 20 c.c. of hot water, and 0·8 g. of palladium chloride

[1] If it is necessary to refill the gas-holder during the experiment owing to consumption of gas, the cylinder is connected to A, the clip on the connecting rubber tubing of the horizontal tube of the bulb is screwed down, the cocks A and B are opened, and hydrogen is carefully passed in.

[2] Cf. E. Waser, *Helv. Chim. Acta*, 1925, **8**, 117.

[3] G.P. 252,136; Friedländer, **10**, 1205.

dissolved in 50 c.c. of water along with 1 c.c. of 40 per cent formaldehyde solution are added. The mixture is made alkaline to litmus with sodium hydroxide solution and heated at the boiling point until the supernatant liquid is clear and colourless. Then the solid is separated by filtration, washed with hot water till neutral, and dried in a vacuum over potassium hydroxide. It is reduced to fine dust by powdering and is kept in a glass bottle closed with a well-fitting stopper.[1]

Hydrogenation of Linolenic Acid.—Linolenic acid (1·5 g.) (for preparation see p. 140) is dissolved in 10 c.c. of glacial acetic acid which has previously been boiled with a little chromic anhydride and then redistilled, 0·2 g. of the catalyst is added and the mixture is shaken with hydrogen. When the calculated amount (how much?) of hydrogen has been taken up, the acetic acid is removed by distillation in a vacuum at 40°, and the residue is recrystallised from a little alcohol, in which the stearic acid is sparingly soluble at 0°. Yield quantitative. Melting point 69°.

Apply Baeyer's permanganate test to the starting material and to the hydrogenation product.

2. CATALYTIC HYDROGENATION WITH NICKEL. CYCLOHEXANOL[2]

Half a porous plate is broken into pieces the size of small peas and the fragments, after being mixed with a solution of 40 g. of chlorine-free nickel nitrate in 20 c.c. of water, are dried, with stirring, on the water bath. The impregnated material is then heated at dull redness in a nickel crucible until evolution of oxides of nitrogen ceases, and filled into a combustion tube (leaving at each end 10 cm. empty).

The tube is then heated in a slightly sloping bomb furnace, or, better, in an electric resistance furnace,[3] while a stream of hydrogen is passed through which is washed with saturated permanganate solution and dried in two wash-bottles containing concentrated sulphuric acid. An adapter is fitted to the other end of the tube and

[1] The catalyst prepared by the method given on p. 64 is similarly suitable for hydrogenating non-acidic substances.

[2] Sabatier, *Compt. rend.*, 1903, 173, 1205.

[3] Such furnaces can easily be constructed from simple materials See, for example, H. Rupe, *Helv. Chim. Acta*, 1918, 1, 454; *Chem. Fabr.*, 1929, 50, 519. Cf. also *Handbuch der Physik*, H. Geiger and K. Scheel, 11, 370 *et seq.*

afterwards, when the nickel oxide has been reduced, is kept with its point dipping into a vessel containing concentrated sulphuric acid (to prevent the entrance of air).

Before the tube is heated it must be completely filled with hydrogen; test a sample of the issuing gas by collection in a small test tube and ignition.

The tube is now heated while a current of hydrogen passes, and the temperature is maintained at 300°–310° until water ceases to appear in the adapter; one and half to two hours are required. Then the contents of the tube are allowed to cool in a slow stream of hydrogen.

Meanwhile 25 g. of freshly distilled phenol are poured into a small distilling flask, the side tube of which is attached low down on the neck. Through a *cork* in the neck a delivery tube reaches to the bottom of the bulb whilst the side tube is pushed through another *cork* into the combustion tube. The end of the side tube reaches the heated zone. In order to restrict the amount of air which enters the tube when the distilling flask is attached the previous inlet tube is quickly replaced by a cork (fitted in advance), hydrogen is then passed through the phenol, and then the side tube of the flask is quickly fixed in the combustion tube. After a test of the issuing gas has shown the absence of air, a rather slow stream of hydrogen is passed while the furnace is gradually heated to 185°–190° and the distilling flask is dipped as deeply as possible into an oil bath at 140°,[1] and after the two temperatures have been reached a *vigorous* stream of hydrogen is passed through. The hydrogenation product is collected in a small filter flask kept cool in ice. To this flask a second receiver containing a little ether is attached, which is likewise kept cool and serves at the same time as a bubble counter for gauging the consumption of hydrogen.

After three hours, on the average, the phenol has been completely evaporated and driven over the catalyst. When the receivers have been disconnected the hydrogen current is slowed down and the tube allowed to cool. The contents of the receivers are washed with a little ether into a small dropping funnel and shaken with 10 c.c. of 50 per cent sodium hydroxide solution in order to remove

[1] Vapour pressure of phenol:

t.	120°	131°	139°	145°
mm.	100	150	200	250

The stoicheiometrical ratio of 1 mole of phenol to 3 moles of hydrogen would normally require a phenol pressure of 760/4 = 190 mm., but in the present case complete saturation is not attained.

non-hydrogenated phenol. The solution is then decanted from the sodium phenoxide, which is washed with ether and dried with a little potassium carbonate after removal of small amounts of cyclohexanone by thorough shaking with 40 per cent bisulphite solution. When the ether has been evaporated in a bath at 50° the residue is fractionated. As the boiling point rises, a little cyclohexane at first passes over with the ether; heating is then continued over a naked flame and pure cyclohexanol distils as a colourless liquid at 160°–161°. The yield is 18–20 g. (75 per cent of the theoretical). If the experiment is performed with care and precision this amount can easily be obtained.

If the catalyst has shown itself to be active the experiment can be directly followed by the *hydrogenation of* (thiophene-free [1]) *benzene*; after the furnace has cooled the connections are changed as above described. In the same time and at the same temperature 40 g. of benzene can be hydrogenated. The hydrocarbon is evaporated from a water bath at 26°–28° in the same way as in the case of phenol. Unchanged benzene is removed from the reaction product by thorough shaking with 10 per cent fuming sulphuric acid (cf. Chap. IV. 1, p. 181). Cyclohexane boils at 81° and, like benzene, solidifies in ice.

During the last two decades catalytic hydrogenation has acquired extraordinarily great importance for all branches of organic chemical practice. Sabatier first showed (1901) that a great variety of unsaturated substances are hydrogenated by passing their vapours together with hydrogen over heated, finely divided *nickel*. The application of the method to the hydrogenation of liquids is due to Normann, who showed that fatty oils are changed into fats of higher melting point by addition of hydrogen with the aid of a nickel catalyst suspended in them. (Technical process for *hardening fats*.) According to the same principle the *hydrogenation products* of *naphthalene* (*tetralin* and *decalin*) are prepared on an industrial scale (Schröter). When the degree of subdivision of the nickel is increased by deposition on a carrier (kieselguhr, asbestos, barium sulphate) the activity is so increased that the hydrogen can be added to unsaturated substances even at room temperature and in solution (Kelber).[2] H. Rupe [3] has described the preparation of a particularly active nickel catalyst.

The Sabatier method was followed by the catalytic hydrogenation

[1] Pure commercial benzene is shaken in a machine for six to eight hours with one-tenth of its volume of concentrated sulphuric acid; the liquids are separated with a funnel and the benzene, after being shaken with sodium hydroxide solution, is distilled. Test with isatin and sulphuric acid.

[2] *Ber.*, 1916, 49, 55; 1924, 57, 136.

[3] *Helv. Chim. Acta*, 1918, 1, 453.

with *colloidal palladium* and albumin as protective substance (Paal); this procedure was brought to perfection by the introduction of gum arabic and gelatin under conditions which ensure the highest possible degree of dispersion of the metal sol, and also permit the hydrogenation to be carried out in acid solution and with high concentrations of the catalyst.

The hydrogenation of the carbon-carbon double bond in dissolved substances and at low temperatures claims by far the most interest. It was first carried out successfully over a wide range by Willstätter with *platinum black* after Fokin had obtained a few satisfactory results. Willstätter was also the first to hydrogenate aromatic rings (benzene, naphthalene) under such conditions. Adams's platinum oxide, which is promptly reduced by hydrogen to very finely divided metal, is widely used for the hydrogenation of resistant double bonds because of its high activity.

Before metallic catalysts were used as carriers there was no possibility of adding elementary hydrogen directly to the carbon-carbon double bond. In such catalysts we have agents by means of which it is possible to saturate with hydrogen practically all unsaturated systems, and it is indeed the olefinic double bond which is most readily attacked by catalytically activated hydrogen. Such hydrogen does not react so quickly with the carbonyl group of aldehydes and ketones, whilst it leaves carboxyl and ester groups intact.

The solvents which, in the scientific laboratory, have by far the greatest importance for catalytic hydrogenation in the cold are glacial acetic acid, ethyl acetate, the alcohols, ether, and water. Since the solubility of hydrogen in all solvents is low, the catalyst, in suspension or colloidal solution, must be continuously kept in contact with the gas phase by shaking, so that it can always take up further amounts of hydrogen and give them up again to the substance undergoing hydrogenation. Instead of the bulb-shaped hydrogenation vessel described above (Willstätter and Waser), a "shaking duck" may be used with equally good results. Frequently a hydrogenation, after proceeding well at the start, comes to a stop before the uptake of hydrogen is complete; in many cases the catalysts can then be reactivated by shaking with air (Willstätter). In this connection it must be borne in mind that a mixture of hydrogen and air is ignited by the finely divided metallic catalyst, and that, therefore, before reactivation is undertaken, the hydrogen present must be replaced by nitrogen or, more simply, removed by evacuation.

Only *completely pure* substances should be subjected to catalytic hydrogenation. This rule is based on the fact that the catalyst is inactivated especially by substances containing sulphur and often also by those containing halogen; not infrequently incalculable influences interfere with hydrogenation. The most certain means of avoiding such troubles is to use pure materials and also pure solvents.

The same catalysts which permit the addition of elementary hydrogen to a double bond are able to accelerate the opposite process—*dehydrogenation* or elimination of hydrogen—when the temperature is altered. Thus *cyclohexane* is decomposed into benzene and hydrogen when passed over nickel or palladium black at about 300° (Sabatier, Zelinsky). The equilibrium

$$C_6H_6 + 3\,H_2 \rightleftarrows C_6H_{12}$$

lies on the right side of the equation at the lower temperature, but at the higher temperature the energy-consuming process, dehydrogenation, takes precedence. Without the catalyst both reactions proceed unmeasurably slowly; they are accelerated in identical fashion by its presence.

3. REPLACEMENT OF THE OXYGEN IN CARBONYL COMPOUNDS BY HYDROGEN

(*Reduction by Clemmensen's Method*)

Ketones and aldehydes can be deoxidised, usually very readily, with amalgamated zinc and hydrochloric acid; from the groups $>C{=}O$ and $-CH{=}O$ there are formed $>CH_2$ and $-CH_3$.

Preparation of the Zinc Amalgam.—Granulated zinc in *thin* pieces or, still better, zinc foil 0·15 to 0·25 mm. thick and cut into small strips is left for one hour with frequent shaking in contact with an equal weight of 5 per cent aqueous mercuric chloride solution. The solution is then poured off and the metal washed once with fresh water.

(*a*) **Ethylbenzene from Acetophenone.**[1]—Acetophenone (6 g.) and hydrochloric acid (30 c.c., from 1 part of concentrated acid and 2 parts of water) are heated with 15 g. of amalgamated zinc to vigorous boiling on wire gauze in a flask attached to a reflux condenser by means of a ground joint. Boiling is continued for five hours and at the end of each hour 5 c.c. of concentrated hydrochloric acid are added. Then the hydrocarbon produced is distilled in a few minutes with steam, separated from water by means of a small dropping funnel, dried with calcium chloride, and distilled. Boiling point 135°–136°. Yield 3–4 g. The yield can be increased if the acetophenone is added slowly drop by drop.

(*b*) **Dibenzyl from Benzil.**[2]—Benzil (7 g.) and hydrochloric acid (1:1, 100 c.c.) are boiled for five hours under reflux with 30 g. of

[1] E. Clemmensen, *Ber.*, 1913, **46**, 1838.
[2] *Ber.*, 1914, **47**, 683.

amalgamated zinc. Concentrated hydrochloric acid (20 c.c. in all) is poured in from time to time as described under (*a*). When the reaction is over the liquid is decanted from the zinc and cooled. The reaction product solidifies and is collected by filtration and washed several times with water. It is then distilled in a small sausage flask. Boiling point 280°. Melting point 50°–52°. The hydrocarbon can be recrystallised from a little alcohol. Yield 5 g. (almost theoretical).

Benzoin can be reduced to dibenzyl in the same way.

CHAPTER XII

NATURAL PRODUCTS

1. FURFURAL[1]

BRAN (300 g.) is stirred in a three-litre flask with a mixture of 150 c.c. of concentrated sulphuric acid and 800 c.c. of water. About 900 c.c. of the liquid are distilled and the distillate, after neutralisation with sodium carbonate, is saturated with common salt (250 g.). Of this solution 300 c.c. are again distilled and the distillate is extracted with ether after saturation with salt. The extract is dried and the ether evaporated. Then the furfural is distilled. Boiling point 162°. Yield 5-7 g.

The *pentoses*, when boiled with mineral acids, lose three molecules of water and are converted into *furfural*.

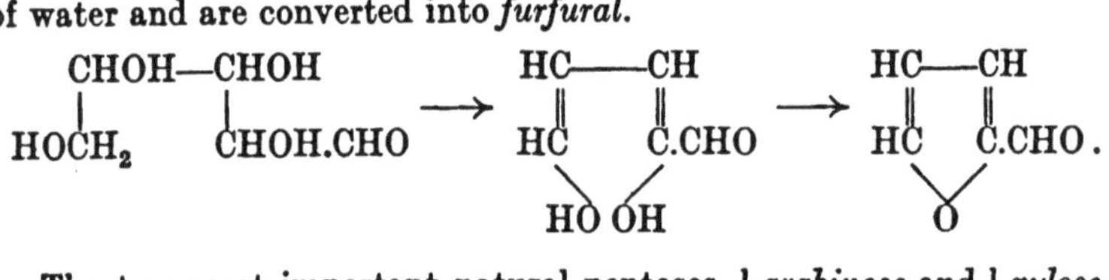

The two most important natural pentoses, l-*arabinose* and l-*xylose*, occur in nature as polymeric anhydrides, the so-called pentosans, viz. *araban*, the chief constituent of many vegetable gums (cherry gum, gum arabic, bran gum), and *xylan*, in wood. From these pentapolyoses there are produced by hydrolysis first the simple pentoses which are then converted by sufficiently strong acids into furfural. This aldehyde is thus also produced as a by-product in the *saccharification* of wood (cellulose) by dilute acids. Furfural, being a " tertiary " aldehyde, is very similar to benzaldehyde, and like the latter undergoes the acyloin reaction (furoin) and takes part in the Perkin synthesis. It also resembles benzaldehyde in its reaction with ammonia (p. 206).

Experiments.—Allow furfural to stand for a short time with five parts of aqueous ammonia; after three hours, separation of the

[1] Stenhouse, *Annalen*, 1841, **35**, 302; Fownes, *Annalen*, 1845, **54**, 52.

product is complete. Recrystallise from alcohol. Melting point 117°. The substance has a structure analogous to that of hydrobenzamide.

Furfural even in dilute aqueous solution gives, almost at once, a precipitate of the phenylhydrazone with phenylhydrazine acetate. Collect the precipitate at the pump and dry. Purify by dissolution in a little ether and careful addition of petrol ether until crystallisation begins. Melting point 97°–98°. Method for quantitative determination of furfural.

Furfural gives two characteristic colour reactions which serve for qualitative detection. With phloroglucinol and hydrochloric acid (1 part concentrated acid, 1 part water) a cherry-red colour is produced on boiling; with a solution of aniline acetate a red colour is produced even in the cold.

Carry out these two tests.

The reaction with aniline salts was explained simultaneously by Zincke and by Dieckmann in 1905. The furane ring undergoes "aminolytic" cleavage and then the anil is formed from the aldehyde.

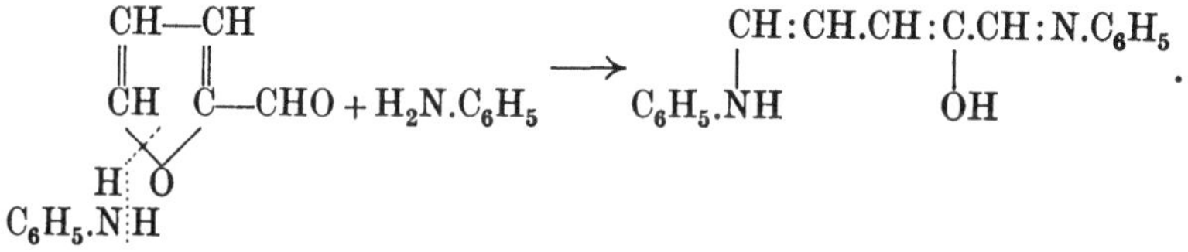

The dyes are salts of the dianil of α-hydroxyglutaconic dialdehyde having the above formula; instead of $C_6H_5.NH.CH{=}CH$ it is equally good to write $C_6H_5.N{=}CH{-}CH_2{-}$; the relationships to glutaconic acid $HOOC.CH_2.CH{:}CH.COOH$ and to its aldehyde then become clearer.

The coloured salts lose one molecule of aniline when heated and are converted into quaternary *β-hydroxypyridinium salts*:

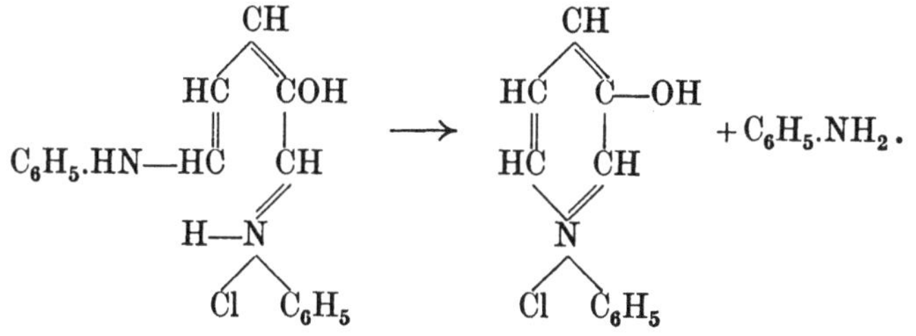

Experiment.[1]—Dissolve 2 c.c. of aniline in 1 c.c. of concentrated hydrochloric acid and add alcohol till the volume is 10 c.c. Then add

[1] Stenhouse, *Annalen*, 1870, 156, 199.

a solution of 1 c.c. of furfural in 8 c.c. of alcohol and warm for a short time. On cooling, the violet dye separates in fine needles. Collect the crystals at the pump and wash with a little alcohol and ether.

2. *d*-GLUCOSE FROM CANE SUGAR[1]

A mixture of 750 c.c. of methylated spirit with 30 c.c. of fuming hydrochloric acid (d. 1·19) is heated to 45°–50° and at this temperature 250 g. of pure, finely powdered cane sugar (castor sugar) are added in portions with continuous shaking. The sugar must dissolve completely. On cooling, the *d*-glucose produced is precipitated as a viscous resin while the *d*-fructose remains in solution. The resinous material is inoculated with a few decigrammes of anhydrous glucose. Crystallisation is promoted by frequent stirring, but nevertheless requires several days to reach completion. The precipitate is now an almost colourless fine crystalline powder which is collected at the pump and at once redissolved in 20–25 c.c. of hot water. Alcohol (120–150 c.c.) is then added to the hot solution until turbidity appears, and the liquid is allowed to cool with stirring and inoculation. The cold mixture is left over night and then filtered at the pump. The precipitate is washed with alcohol and well dried in a vacuum desiccator. Yield 50–60 g. Melting point 146°.

3. HYDROLYSIS OF CANE SUGAR BY SACCHARASE (INVERTASE)

(*a*) **Preparation of the Enzyme Solution.**[2]—Pressed yeast (50 g.) is stirred in a small filter jar by means of a stout glass rod with 5 c.c. of toluene at 30° until the mixture forms a quite thin sludge (about three-quarters of an hour). This sludge, which results from the autolysis of the yeast cells, is diluted with 50 c.c. of water at 30°, and kept at this temperature for one hour. It is then transferred, along with 150 c.c. of water, to a conical flask (capacity 250 c.c.), vigorously shaken with some kieselguhr, and poured rapidly on to a moderate sized Büchner funnel to which slight suction is applied. The residue on the filter is again washed with 50 c.c. of water at 30°. The filtrate is thrown away; it indeed contains a little invertase, but is chiefly a solution of other yeast constituents, which should thus be removed. The residue on the funnel is well beaten up with 50 c.c. of water, a few drops of toluene are added, and the mixture is left for fifteen hours at about 30° to allow the enzyme to be set free.

[1] Soxhlet, *J. pr. Chem.*, 1880, 21, 245.

[2] Willstätter, Schneider, and Bamann, *Z. physiol. Chem.*, 1925, 147, 264.

In order now to remove proteins from the thin sludge it is vigorously stirred with 0·05 N-acetic acid in amount just sufficient to change the colour of methyl red ($p_H = 4$) (test with a sample), and is then filtered as above after shaking with a little kieselguhr if necessary. The filtrate is made neutral to litmus with dilute ammonia and in this condition, protected by a little toluene, can be kept unchanged for several days.

(*b*) **The Inversion.**—Cane sugar (40 g.) is dissolved in 200 c.c. of water in a 250 c.c. measuring flask, 25 c.c. of 10 per cent sodium dihydrogen phosphate solution are added and the mixture is warmed on a large water bath (or in a thermostat) at 30°. Now 10 c.c. of the enzyme solution prepared as described under (*a*) are added and the time at which the latter has run out of the pipette is noted. The flask is filled at once to the mark with water at 30° and shaken up. The first 25 c.c. sample is removed for polarisation and the time is noted in the same way as was done just before. Each sample is run into 5 c.c. of 2 N-sodium carbonate solution in order to stop the action of the enzyme and at the same time to accelerate the "mutarotation" (p. 385). After shaking it with a little animal charcoal the sample is poured through a dry filter and the clear solution is polarised in a 2-dm. tube. Each time the average of three readings is taken.

The main solution in which the reaction goes on is kept at 30°. From it samples are taken for polarisation every twenty minutes during the first hour after the beginning of the experiment, and every thirty minutes during the second hour. During this reaction period that stage of the inversion which is shown by zero rotation is usually passed. This indicates that about 75 per cent of the cane sugar taken has been hydrolysed.

When the rotations are plotted as ordinates against the times as abscissae, the observed points can be connected by a logarithmic curve, which is flatter in the later stages of the reaction and indicates the order as unimolecular. The abscissa of the point in this curve corresponding to a rotation of 0° gives the "zero-rotation time", which is to some extent a measure of the activity of the enzyme solution employed.

The course of the curve already indicates that the logarithmic law is not strictly obeyed. By exterpolation we can obtain the point of intersection of the curve with the Y-axis, and so the rotation at zero time; if one-third of this initial dextro-rotation is plotted (as final laevo-rotation) below the rotation of 0° on the Y-axis, the time when this ultimate rotation is approximately reached can be ascertained graphically. We next test whether the half-value time is constant, *i.e.* whether the moments at which one-half, one-quarter, one-eighth of the total change in rotation have still to occur and similar fractions of the sugar remain unhydrolysed (as indicated by the curve) are really separated by equal intervals of time. Since the diminution in the rotation is directly proportional to the amount of sugar inverted, this diminution is at once a measure of the reaction velocity.

The observed changes of rotation and the corresponding times should be substituted in the following equation for unimolecular reactions, and the constant K should be calculated.

$$K = \frac{\log(a_1 - a) - \log(a_2 - a)}{0{\cdot}4343(t_2 - t_1)}.$$

a_1 and a_2 are the rotations read off from the curve and corresponding to times t_1 and t_2; $t_2 - t_1$ is in each case the interval during which the inversion proceeded (in minutes), a is the (negative) final value of the rotation obtained in the manner indicated above; since it is subtracted it becomes positive in the equation.

4. β-PENTA-ACETYLGLUCOSE AND α-ACETOBROMOGLUCOSE [1]

Powdered anhydrous *d*-glucose (25 g.) is mixed in a mortar with anhydrous sodium acetate (12 g.) and the mixture is then heated in a half-litre flask on the water bath (with frequent shaking) with pure acetic anhydride (125 g.), so that after about half an hour a clear solution forms. After two hours more the solution is poured in a thin stream, with stirring, into a litre of ice-water. The crystalline mass which separates is powdered as carefully as may be, and after some hours, when most of the excess of acetic anhydride is decomposed, is collected at the pump and then kept for a few hours longer under water. Finally it is again collected at the pump, pressed down well, and recrystallised from about 120 c.c. of alcohol. The penta-acetylglucose so obtained is sufficiently pure for further use. Yield 35–40 g.

Acetobromoglucose.—Glacial acetic acid is saturated at 0° with hydrogen bromide [2] and 50 g. of the solution are poured on to 25 g.

[1] Emil Fischer, *Ber.*, 1916, 49, 584; K. Freudenberg, *Ber.*, 1927, 60, 241.

[2] **Preparation of Hydrogen Bromide.**—Red phosphorus (6 g.) is mixed to a sludge with 15 c.c. of water in a distilling flask (capacity 250 c.c.) having the side tube attached high up on the neck. The flask is connected to a U-tube containing loose glass beads covered with moist red phosphorus and a second U-tube containing calcium chloride, or preferably phosphorus pentoxide, is attached to the first. Bromine (20 c.c. = 60 g.) is then *slowly* added to the phosphorus drop by drop from a funnel,

of the peracetylated glucose, which is kept cool in ice. Dissolution is brought about by vigorous shaking, and the solution is left at room temperature for two hours. Then 850 c.c. of ice-water are added with stirring, the liquid is decanted from the precipitate which separates, and the latter, after thorough grinding in a basin with ice-water, is collected at the pump and washed well. The crude product is dissolved in 250 c.c. of cold methyl alcohol, the solution is kept cool in ice, and an equal volume of water is slowly added. In this way the substance is precipitated in the pure crystalline state. It is collected at the pump, washed with water, and dried, first on porous plate and then very thoroughly in a vacuum desiccator over solid potassium hydroxide and concentrated sulphuric acid. Yield 15–20 g.

5. LACTOSE AND CASEIN FROM MILK

Whole milk (2 litres) is diluted with an equal volume of water at 30°–40° and commercial rennet (0·1 g.), dissolved in a few cubic centimetres of water, is added. The mixture is then left at the same temperature until separation of the casein is complete (about two hours). The whey is filtered through a filter cloth, and after the liquid has run off the residue is pressed down well. The casein, which contains a great deal of fat, is ground in a mortar with a little 1 per cent sodium hydroxide solution; 1 to 1·5 litres of sodium hydroxide solution of the same concentration are then poured on to the resulting paste, and the mixture is gently warmed in a porcelain basin until all but the fat dissolves.

When the mixture stands over night in a filter jar the milk fat collects at the surface and can be separated cleanly at the pump. The filtrate is combined with the rest of the casein solution and the substance is again precipitated by the addition of 10–20 c.c. of glacial acetic acid. Finally the precipitate is collected by filtration through linen, washed well with water, and dried in a desiccator. Yield 50–60 g. (For the hydrolysis it is not necessary to dry the casein.)

Lactose.—The whey is concentrated to a small volume on a

---

the tube of which reaches nearly to the surface of the phosphorus. The reaction is very vigorous. The gas which is produced must be quite colourless when it has passed through the U-tube containing phosphorus. The most suitable receiver is a filter flask containing 40 c.c. of liquid glacial acetic acid, protected from atmospheric moisture by means of a calcium chloride tube. The receiver is kept cool in ice-water after part of the hydrogen bromide has been absorbed.

ring burner which causes almost complete precipitation of the albumin. The mixture is filtered through linen, and the filtrate is again concentrated until lactose separates. When the mixture is cold the crude crystalline sugar is collected by filtration through linen on a Büchner funnel and dried. A second portion of lactose is obtained by further concentration of the mother liquor—this time on the water bath. Yield of crude material 70–75 g.

The crude product is dissolved in the minimum amount of hot water (30–40 c.c.) and sufficient alcohol to produce a turbidity (about 100 c.c.) is added to the hot solution.

In the course of a few hours copious crystallisation occurs. Precipitation of the crystals can be initiated and accelerated by scratching with a glass rod. The mixture is left over night before filtration, then the preparation is collected at the pump and washed with alcohol. Yield 60–65 g.

Acid Hydrolysis of Casein.[1]—The casein is boiled for sixteen hours with three times its weight of 25 per cent sulphuric acid in a flask attached by means of a ground joint to a reflux condenser. The sulphuric acid is then removed from the dark-coloured liquid by means of hot saturated barium hydroxide solution and the slight excess of barium ions is precipitated with carbon dioxide. The precipitated material is boiled with 500 c.c. of water. A sample of the barium sulphate precipitate again boiled with water should not react with Millon's reagent (see below). The filtrates are combined and concentrated until crystals begin to separate. After the crystals have been removed from the cold solution by filtration the mother liquor is again concentrated until crystallisation sets in once more, and this process is repeated two or three times until the filtrate gives but a feeble reaction for tyrosine.

The various crops of crystals are combined and repeatedly recrystallised from hot water with addition of animal charcoal. A small amount of l-*tyrosine* is obtained in this way. Melting point of the pure compound 314°–318°.

In addition some *leucine* and *glutamic* acid are obtained; their purification is described in the handbook cited.

Millon's reagent: Dissolve one part of mercury in two parts of nitric acid (d. 1·42) first in the cold and then by warming. Dilute the solution with two volumes of water.

[1] Abderhalden, *Handbuch der biolog. Arbeitsmethoden*, I, 7, p. 19.

6. *d*-GALACTOSE FROM LACTOSE.[1] MUCIC ACID. PYRROLE

Lactose (100 g.) is boiled under reflux for two hours with 250 c.c. of water containing 3 c.c. of concentrated sulphuric acid, and finally for a few minutes with animal charcoal. The sulphuric acid is precipitated from the unfiltered liquid by addition to the well-shaken sugar solution of a hot saturated aqueous solution of the calculated amount of barium hydroxide (about 15 g. of $Ba(OH)_2, 8\,H_2O$). The liquid may not become alkaline. When the solution is free from sulphuric acid (and barium) the precipitated material is removed by filtration at the pump, and after adding 3 c.c. of glacial acetic acid the filtrate is concentrated in a vacuum to 60 c.c. on a bath at 40°–50°. Glacial acetic acid (100 c.c.) is mixed with the syrup thus obtained, while it is still warm, until a clear solution is obtained. When this solution is cold and is scratched with a glass rod or seeded with a few crystals of galactose, the sugar crystallises. Crystals are allowed to separate for one day and are then collected at the pump, washed, first with a little cold glacial acetic acid, then with a little cold methyl alcohol, and finally with ether. Yield 20–25 g. Melting point 165°.

Test the purity of the galactose prepared by determining the specific rotation with a polarimeter. An aqueous solution containing one gramme of substance in 10 c.c. should have a rotation of $+8{\cdot}15°$ in a decimetre tube. Hence $[\alpha]_D^{20°} = +81{\cdot}5°$.

Since galactose exhibits *mutarotation* the attainment of equilibrium is accelerated by addition of a few drops of ammonia solution.

Mucic Acid.[2]—Galactose (25 g.) is mixed with nitric acid (300 c.c.; d. 1·15) and the mixture is concentrated to about 50 c.c. on the water bath with stirring.

The pasty mass obtained on cooling is stirred with 50 c.c. of water, left for some hours, and filtered at the pump. The material on the funnel is washed with a little water. Yield 15–16 g.

Pyrrole from Ammonium Mucate.[3]—Mucic acid (15 g.) is dissolved in a basin in 15 c.c. of 20 per cent ammonia solution and the solution is evaporated to dryness. The ammonium mucate is stirred

[1] M. Heidelberger, *Advanced Laboratory Manual of Organic Chemistry*, New York, 1923, p. 76.

[2] Tollens and Kent, *Annalen*, 1885, 227, 222.

[3] Schwanert, *Annalen*, 1860, 116, 271; see also Khotinsky, *Ber.*, 1909, 42, 2506.

with 20 c.c. of glycerol in a distilling flask and the mixture heated. At 170° the reaction begins and the bulk of the pyrrole distils between 180° and 210°. (If the temperature is raised to 300° a little more pyrrole is obtained.) The substance is dissolved in a little ether, dried, and fractionated. Boiling point of pure pyrrole 131°. Yield 2–3 g.

A splinter of pine wood which has been dipped in concentrated hydrochloric acid is coloured red by the vapour from a boiling mixture of pyrrole and water (*πυρρός*).

7. OCTA-ACETYLCELLOBIOSE[1]

Pure cotton wool (10 g.) is added in portions during a period of half an hour to a well-cooled mixture of acetic anhydride (40 c.c.) and concentrated sulphuric acid (5 c.c.). The temperature of the liquid during this time must not exceed 10°. When the cotton has been thoroughly soaked the flask is removed from the freezing mixture, stoppered, and left, first for some hours in water and then at room temperature for eight days. From time to time the soaked cotton is pressed with a glass rod until a jelly-like and finally a semi-liquid material is obtained. Then the contents of the flask, which are usually interspersed with crystals, are poured with stirring into a litre of ice-water. The precipitate is collected at the pump in a Büchner funnel provided with linen filter cloth, washed carefully with water, dried, and recrystallised from about one litre of ethyl alcohol. Leaflets having melting point 222°. Yield 5–6 g.

Some Remarks on Carbohydrates (with Reference to Sections 2–7)

Two disaccharides have been hydrolysed and in each case the more readily crystallisable sugar has been isolated—d-*glucose* from cane sugar, d-*galactose* from lactose. The hydrolysis of bioses, by cleavage at the oxygen bridge, is biologically accelerated by the catalytic action of specific enzymes (invertase, lactase). The more universal hydrolysis by acids is a unimolecular reaction and its velocity is proportional to the hydrogen ion concentration.

The highly polymerised compound cellulose is broken down into its octa-acetyl disaccharide *cellobiose* by "acetolysis" (Franchimont, Skraup, Ost). The bioses belong to the important group of *glucosides*.

[1] Ost, *Annalen*, 1913, **398**, 332; Freudenberg, *Ber.*, 1921, **54**, 767; Karrer, *Helv. Chim. Acta*, 1921, **4**, 174.

The isomerism of α- and β-glucose is to be attributed to the spatially different arrangement of the H and OH groups attached to the asymmetric carbon atom 1. This atom is asymmetric in the cyclic " lactol " formula (Tollens). The mutarotation of the sugars, *i.e.* the gradual change to the final stationary value of the optical rotation, is to be explained by an equilibrium occurring in solution between the various isomeric forms, which equilibrium is attained more or less rapidly but not instantaneously. Of late the aldohexoses have come to be considered as derived from a ring closely related to that of pyrone and having an oxygen bridge between C_1 and C_5 (Haworth). For this reason they are appropriately called *pyranoses.* The labile γ- or *h*-sugars are supposed to be structural isomers with the bridge between C_1 and C_4 as in the old formula of Tollens ("furanoses"). These are believed to be the forms which are produced during cellular metabolism (fermentation, oxidative degradation of sugars). A corresponding view is held concerning the structure of fructose (cf. W. N. Haworth, *Helv. Chim. Acta,* 1928, **11**, 534).

α-*d*-glucose

β-*d*-glucose

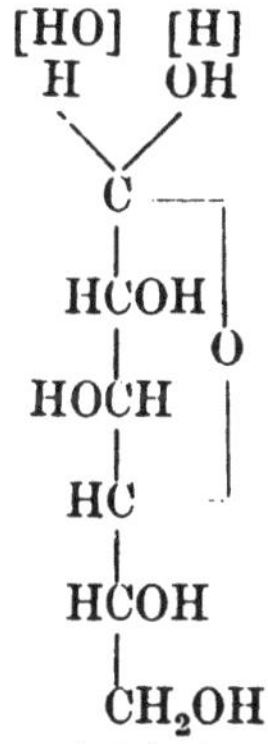

α- and β-*h*-glucose

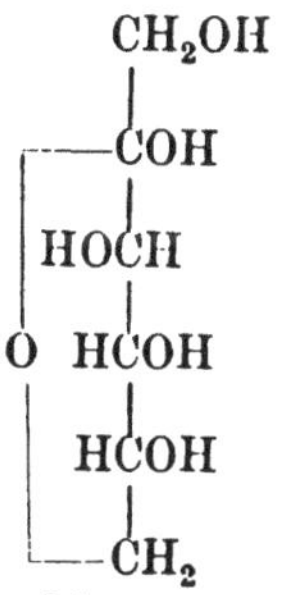

d-fructose

A hexagonal formula is especially instructive :

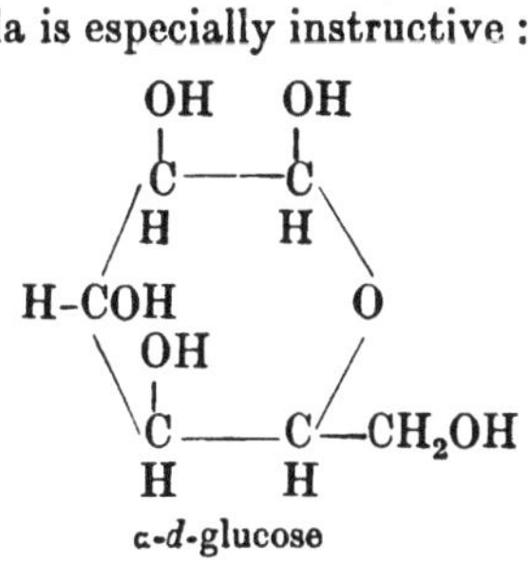

α-*d*-glucose

Corresponding to the two structurally isomeric forms of glucose and other monoses there are also two types of glucosides, each of which exists in two enantiomorphs. All the four ethyl glucosides are known.

2 C

The α-glucosides are especially easily hydrolysed by *yeast* (invertase, maltase), the β-glucosides by *emulsin*, the enzyme of bitter almonds. Accordingly, the biologically important bioses such as sucrose and maltose belong to the α-series, whilst lactose belongs to the β-series. In the same way the biose of amygdalin has been recognised as a β-glucosido-β-glucose, *i.e.* gentiobiose (R. Kuhn, Haworth, Hudson). Helferich has recently succeeded in synthesising this sugar.

The *investigation of the constitution* of the bioses has made considerable progress during the last decade, chiefly through the work of British chemists (Irvine, Haworth, and others). The principle of the method is as follows: the free OH-groups of the biose are fixed by methylation, the octa-methylated molecule is then broken at the glucoside bridge and the structures of the methylated monoses are determined. The formulae which are at present attributed to the most important bioses are given below.

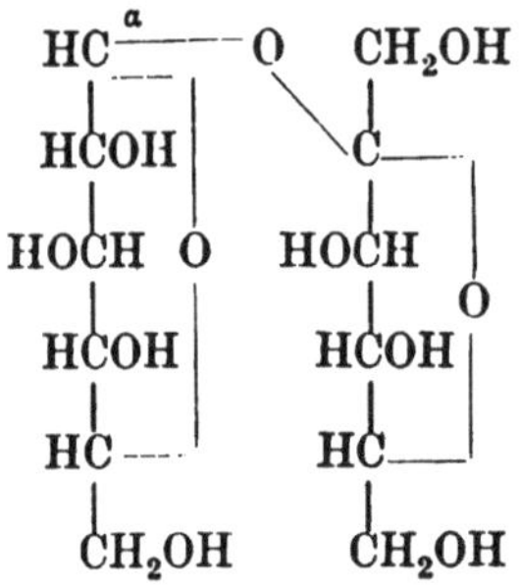

Sucrose

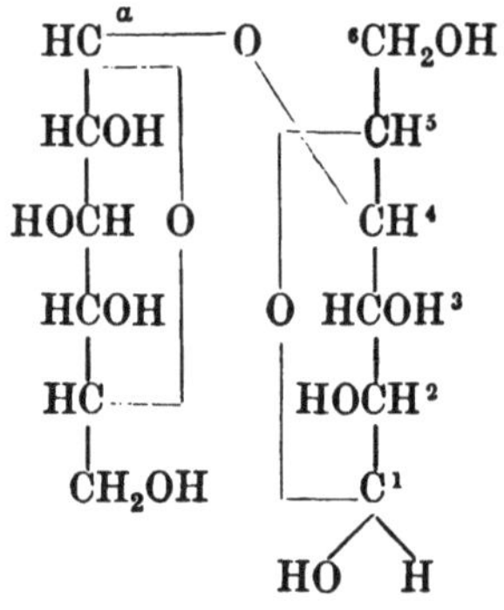

Maltose

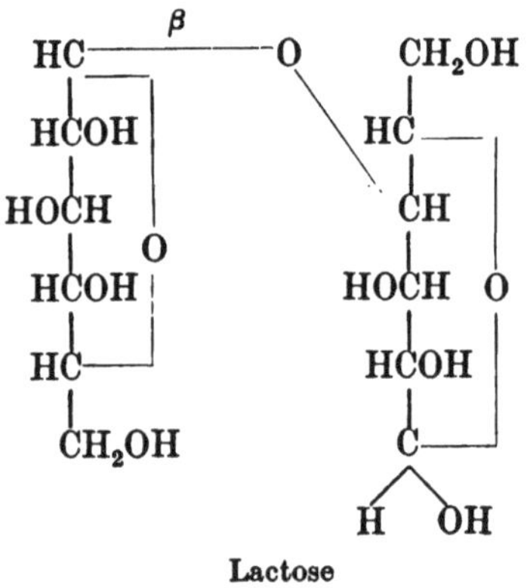

Lactose

According to Haworth *cellobiose* differs from maltose only in having the β-glucosidal configuration. Lactose and cellobiose are built up in identical fashion and differ only in their monose components. *Gentiobiose* is a β-biose; the common oxygen bridge, however, is attached at C_6. The formulae show at the same time the configurations of the constituent monoses.

The fundamental principles governing the spatial isomerism of the simple sugars and the theoretical and experimental work of Emil Fischer cannot be dealt with here. Those who are not already familiar with the subject should at once fill this gap in their knowledge; they are emphatically advised to use space models.

Those bioses in which the aldehydic (cyclo-acetal) carbonyl group is free have a reducing action (on Fehling's solution).

Experiment.—Boil samples of both sucrose and lactose with Fehling's solution.

See p. 288 for information about the production of the osazones of the simple sugars. The osazones of *d*-glucose, *d*-fructose, and *d*-mannose are identical.

The number of *synthetic bioses* is increasing; among natural bioses *gentiobiose* (see above) has been obtained synthetically from *d*-glucose by the action of emulsin (Bourquelot). Attempts at synthesis with the bromoacetyl-sugars do not succeed because the desired OH-group of the second molecule is not attacked. The bromoacetyl-sugars are formed, quite generally, by the action of acetyl bromide on the sugars.

The synthesis of lactose has been described very recently by A. Pictet. Helferich has also been successful in synthetic work of this kind.

Of other recent experiments in the chemistry of the sugars the preparation of the anhydro-sugar *laevoglucosan* $C_6H_{10}O_5$ by the rapid distillation of starch or cellulose in a vacuum may be mentioned (Pictet). This interesting compound possibly has the constitution I.

I 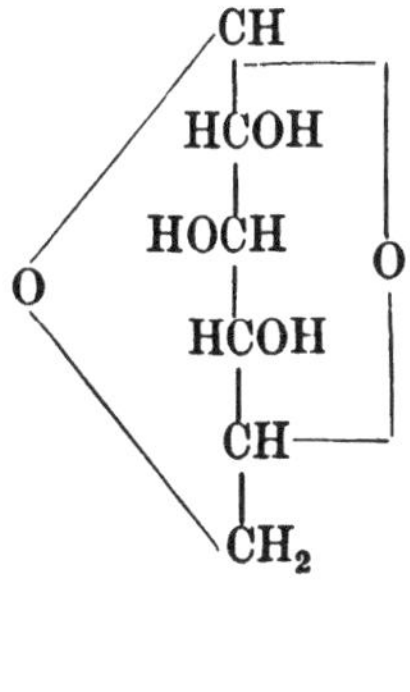

II 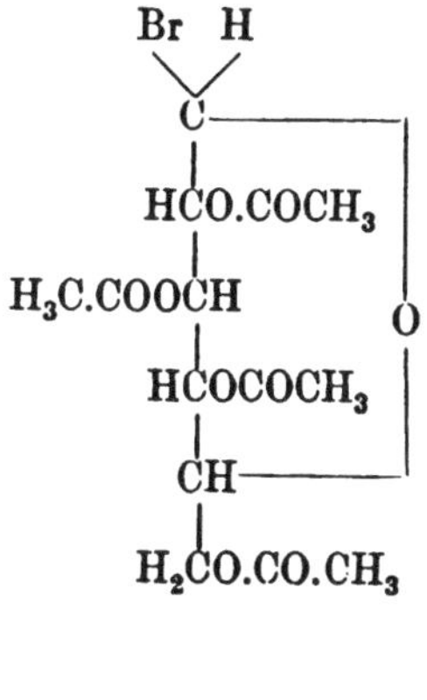

III 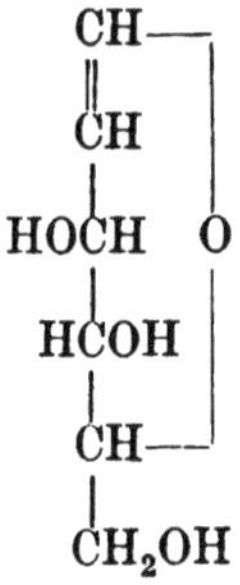 IV 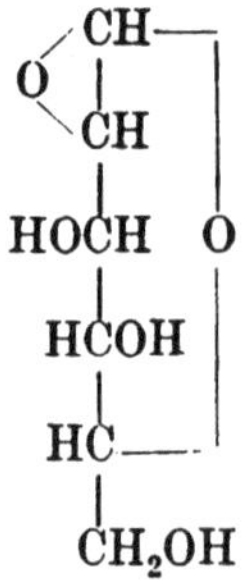

The adjoining formula II belongs to *α-acetobromoglucose* (tetra-acetyl-α-bromo-*d*-glucose), which is also formed from starch (Bergmann) and cellulose (Karrer, Hess) by the action of acetyl bromide. On boiling with 50 per cent acetic acid and zinc dust, the bromine at the first carbon atom is replaced by hydrogen, and at the same time acetic acid is eliminated with the formation of a double bond between carbon atoms 1 and 2. From the resulting compound, the substance III, *glucal* $C_6H_{10}O_4$ (E. Fischer, Bergmann), is formed by hydrolytic elimination of the other three acetyl groups. Prileschaiev's reaction with benzoyl peroxide (p. 99) produces from III the anhydro-sugar IV, of which the 1:2-bridge is very easily broken by water to yield the glycol group. The main product is then *mannose* (Bergmann).

What are the products of oxidation and reduction of the monosaccharides? Mucic acid, of which the preparation from *d*-galactose has been described, is *optically inactive* and *incapable of resolution*, just like *dulcitol*, the corresponding hexahydric alcohol. As the formula shows, the four asymmetric carbon atoms of galactose form two pairs (2, 5 and 3, 4) having the same substituents but opposite arrangement in space, and when the carbon atoms 1 and 6 become alike, inactive forms are produced by intramolecular compensation, as in mesotartaric acid.

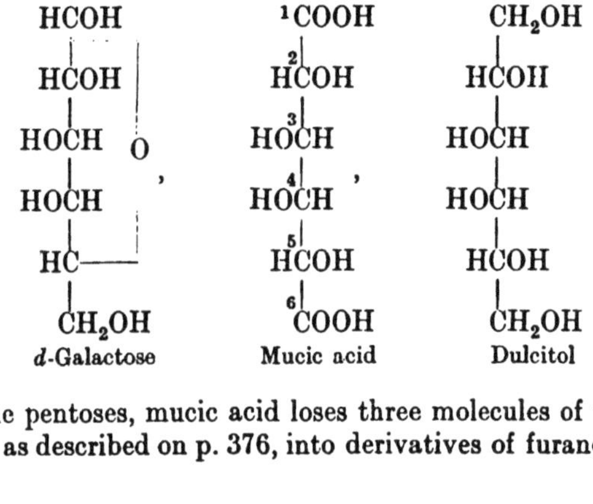

d-Galactose, Mucic acid, Dulcitol

Like the pentoses, mucic acid loses three molecules of water and is converted, as described on p. 376, into derivatives of furane; with concentrated hydrochloric acid the product is furane-α : α′-dicarboxylic acid. Dry distillation yields furane-α-carboxylic or pyromucic acid.

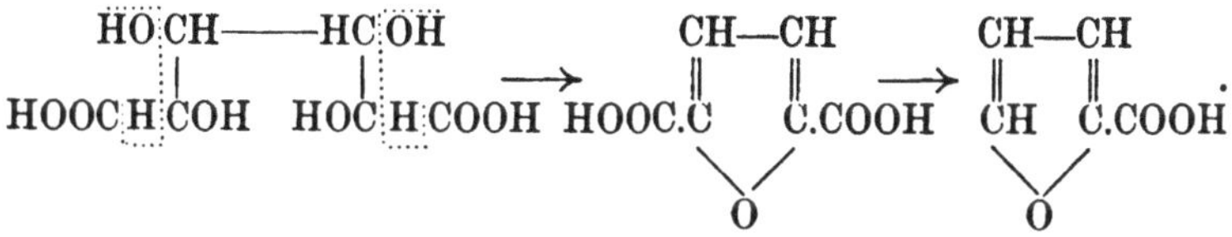

Pyrogenic decomposition of mucic acid in the manner indicated above but in presence of NH_3 yields *pyrrole*, the important unit from which chlorophyll and blood pigments are built up (see p. 395). The oxygen bridge in the furane ring is replaced by NH—a general reaction of hetero-rings containing oxygen.

Monographs: W. N. Haworth, *The Constitution of Sugars*, London, 1929; H. Pringsheim, *The Chemistry of the Monosaccharides and of the Polysaccharides*, New York, 1932, *Zuckerchemie*, Leipzig, 1925, *Die Polysaccharide*, Berlin, 1931; P. Karrer, *Einführung in die Chemie der polymeren Kohlenhydrate*, 1925. For cellulose especially, Kurt Hess, *Die Chemie der Zellulose und ihrer Begleiter*, Leipzig, 1928.

8. SACCHARIFICATION OF STARCH AND ALCOHOLIC FERMENTATION

Saccharification.—Potato starch [1] (100 g.) is made into a thin sludge by stirring with water and then poured slowly into an enamelled jar (capacity 2·5 l.) containing 1500 c.c. of boiling water, which is kept vigorously stirred throughout with a wooden spoon. To the thick homogeneous glassy material so obtained (after cooling to 40°) there is added one-third of a filtered extract, freshly prepared by digesting 15 g. of coarsely ground kiln malt with 100 c.c. of water at 35°-40° for one hour.

The saccharification, which should be carried out at about 40°, is accelerated by stirring and is complete in about an hour. Its course is followed by removing 5 c.c. of the solution a quarter of an hour after the malt extract has been added and determining in this sample, by the method of Willstätter and Schudel,[2] the amount of maltose formed (see below). After thirty minutes more the determination is repeated.

The Sugar Determination.—Dilute the 5 c.c. sample in a measuring flask to 25 c.c. and pour 10 c.c. of this diluted solution into 25 c.c.

[1] If it is desired to start from potatoes (500 g.), they must be heated in a digester under 2–3 atmospheres pressure, since simple boiling till soft does not produce a breakdown sufficient for complete saccharification.

[2] *Ber.*, 1918, 51, 780.

of 0·1 N-iodine solution. Then add 40 c.c. of 0·1 N-sodium hydroxide solution (free from alcohol) and leave for twenty minutes. Make faintly acid with dilute sulphuric acid and titrate back with 0·1 N-sodium thiosulphate solution. One equivalent of iodine corresponds to 0·5 mole of reducing biose, or 1 c.c. of 0·1 N-iodine solution to 17·1 mg. of maltose. What is the course of the reaction ?

When the consumption of iodine is the same in two successive tests and a sample of the mixture is no longer coloured by iodine the saccharification is complete. Usually 75–80 per cent of the starch taken is converted into sugar. The rest of the starch is only broken down into dextrins which are, however, also saccharified in the course of the subsequent fermentation. The volume of the mash is measured in a cylinder, and from the result of the final maltose titration the sugar content is calculated. For the CO_2-determination 10 c.c. are retained (cf. p. 391).

Fermentation.—The solution is next transferred for fermentation to a round-bottomed flask (capacity 3 l.). To set this going 10 g. of baker's yeast made into a dough with water are added after 3 g. of dihydrogen ammonium phosphate have been dissolved in the mash. The flask is now provided with a small two-bulb trap containing a little water and well shaken from time to time; after fermentation has set in it is allowed to proceed to completion for two to three days in a warm room. As the process nears its end the water in the seal hardly moves, or not at all. The alcohol is now distilled through an efficient column (Raschig rings), downward condenser and adapter into a filter flask. Barely half of the liquid is distilled. Then the distillate is transferred to a smaller flask and distillation is twice repeated as before until the final volume collected is about 200 c.c. From the specific gravity determined with a hydrometer at 15° the alcohol content is found.

At 15° the specific gravity of 10 per cent alcohol (by volume) is 0·9857 and that of 30 per cent alcohol 0·9646. Between these two contents the specific gravity decreases nearly linearly by 0·0010 for each 1 per cent by volume, so that from the specific gravity found the concentration of the alcohol obtained can be calculated without the use of a table.

In the aqueous-alcoholic distillate having a volume of about 200 c.c. about 70 c.c. or 56 g. of alcohol will be found. The yield of alcohol approaches the theoretical value and should amount to about 20 per cent more than would be calculated from the maltose determination just carried out, since the latter does not, of course, take into account the additional sugar produced during the fermentation by " diastatic after effect ".

Work out the balance for the whole process, taking into consideration the number of litres of CO_2 formed.

In order to obtain the alcohol in the pure state the final distillate is dropped from a tap funnel into a distilling flask containing 700 g. of quicklime. The flask is connected to a downward condenser and is heated in an oil bath so that the alcohol distils.

Determination of CO_2.—Dilute 10 c.c. of the mash to 25 c.c. in a measuring flask and transfer 10 c.c. of the diluted liquid by means of a pipette to a small distilling flask which is directly connected to a nitrometer. To the solution add about 0·2 g. of yeast made into a cream with a little water and then at once displace the air by passing in carbon-dioxide from above through a tube which does not dip into the liquid. Close the delivery tube by means of stop-cock or spring clip, fill the nitrometer with water saturated with carbon dioxide, and allow fermentation to proceed until the volume of gas evolved ceases to increase. Finally, reduce the volume found to 0° and 760° and compare the result with the yield of alcohol and with the result of the maltose determination, making allowance for the additional saccharification.

The chemical course of alcoholic fermentation, which has already been the subject of investigation for more than a century, has been explained chiefly by the fine work of C. Neuberg. E. Buchner proved that zymase, the enzyme complex of yeast, can be separated from the living cells.

Shortly summarised, the process may be regarded as resulting from a series of successive hydrogenation and dehydrogenation reactions similar to that of Cannizzaro (p. 211), and taking place as follows: *d*-glucose in the form of a phosphoric ester first decomposes into two molecules of *methylglyoxal*.

$$C_6H_{12}O_6 \longrightarrow 2CH_3.CO.CHO + 2H_2O.$$

This aldehyde undergoes "dismutation" to *pyruvic acid* and hydroxyacetone (half and half). The hydroxyacetone undergoes rearrangement to glycerol.

By the action of the enzyme carboxylase, *acetaldehyde* and CO_2 are produced from the pyruvic acid. As soon as the acetaldehyde is formed it reacts, as a result of the influence of the enzyme, with methylglyoxal, which is again dehydrogenated to the acid, while the aldehyde is converted into **ethyl alcohol.**

$$CH_3.CO.CH(OH)_2 + CH_3.\underset{H}{C}{=}O \longrightarrow CH_3.CO.COOH + CH_3.CH_2OH.$$

During the biological oxidation of *d*-glucose in muscle the methylglyoxal—which is the primary product here also—undergoes rearrangement to lactic acid as a result of a wandering of hydrogen within the molecule, a Cannizzaro reaction which corresponds to the benzilic acid rearrangement.

$$H_3C.CO.C\begin{matrix}\diagup OH \\ \diagdown OH\end{matrix}_{H} \longrightarrow H_3C.CHOH.COOH.$$

A small part of the lactic acid is now directly oxidised, but about 5 out of every 6 molecules are used, in a manner at present unknown, for a resynthesis of *d*-glucose (O. Meyerhof).

More detailed information about enzymes will be found in Oppenheimer and Kuhn, *Lehrbuch der Enzyme*, Leipzig, 1927; J. B. S. Haldane, *Enzymes*, London, 1930.

9. *d*-ARGININE HYDROCHLORIDE FROM GELATIN[1]

***d*-Arginine Flavianate.**—Gelatin (100 g.) is boiled vigorously under reflux for eight to ten hours in a half-litre flask with 100 c.c. of 36 per cent hydrochloric acid (d. 1·19). After the liquid has cooled it is diluted to about 500 c.c., made just alkaline with 33 per cent sodium hydroxide solution, and then acidified with 12 c.c. of glacial acetic acid. The liquid is filtered if necessary, and a hot solution of 20 g. of flavianic acid (naphthol yellow *S*, see p. 186) in 100 c.c. of water is added. After a quarter to half an hour the flavianate begins to separate. The mixture is left for one to two days and then filtered at the pump. In order to remove any flavianic acid which is precipitated at the same time, the solid is ground twice with 500 c.c. portions of cold water and filtered at the pump each time. Yield 18–22 g. *d*-Arginine flavianate blackens above 200° and decomposes at 275°.

Monohydrochloride of *d*-Arginine.—The flavianate is suspended in 100 c.c. of hot water in a large mortar and *well* ground with a solution of 40 g. of barium hydroxide dissolved in the necessary amount of hot water. The barium flavianate which is precipitated is collected at the pump while the mixture is still hot, ground again with a hot solution of 8 g. of barium hydroxide in 200 c.c. of water, and once more separated at the pump. A rapid current of carbon dioxide is at once passed into the combined filtrates until the reaction is slightly acid; the barium carbonate which is precipitated is removed by filtration at the pump and washed well with water.

[1] Kossel and Gross, *Z. physiol. Chem.*, 1924, **135**, 167; Felix and Dirr, *Z. physiol. Chem.*, 1928, **176**, 38.

After the filtrate has been concentrated to 100–150 c.c. on the water bath and filtered from more precipitated barium carbonate, it is made acid to Congo red by addition of concentrated hydrochloric acid (3–4 c.c.) and left for a short time. Some more flavianic acid is precipitated from the cold solution on scratching, and is removed by filtration. The filtrate is decolorised by boiling for a short time with animal charcoal and filtered once more. The almost colourless filtrate is made faintly alkaline with ammonia and evaporated to dryness. In this way a mixture of arginine hydrochloride and ammonium chloride is obtained. The mixture is dissolved in as little hot water as possible, and the hot solution is diluted with hot 96 per cent alcohol until a distinct turbidity appears. As the liquid cools the arginine hydrochloride is almost quantitatively precipitated in sheaves of prisms which are recrystallised in the same way. Yield 7–8 g.

d-Arginine monohydrochloride sinters at 218° and decomposes with copious frothing at 235°.

The flavianic acid can be recovered from its sparingly soluble barium salt as follows: Decompose the salt with a slight excess of hot 20 per cent sulphuric acid, filter at the pump while hot, add some concentrated hydrochloric acid to the clear filtrate, and allow the free sulphonic acid to crystallise.

The method given by Bergmann[1] for the preparation of arginine may also be recommended here.

10. CAFFEINE FROM TEA

In the simplified extraction apparatus (Fig. 28 on p. 34) 100 g. of finely powdered tea or tea dust are extracted with 400 c.c. of alcohol. The alcoholic extract is then added to a suspension of 50 g. of magnesium oxide in 300 c.c. of water, and the whole is evaporated to dryness on the steam bath in a porcelain basin with frequent stirring. The pulverulent residue is boiled once with 500 c.c. and then three times with 250 c.c. of water and filtered while hot at the pump.

After 50 c.c. of dilute sulphuric acid have been added to the combined filtrates, they are concentrated to about one-third of their volume, filtered if necessary while hot from the flocculent precipitate which occasionally forms, and then extracted five times with 30 c.c. portions of chloroform.

[1] *Z. physiol. Chem.*, **1926**, **152**, 293.

The pale yellow chloroform solution is decolorised by shaking, first with a few cubic centimetres of sodium hydroxide solution, then with the same volume of water, and is evaporated to dryness. The residue of crude caffeine is recrystallised from a little hot water. Yield 2·0–2·5 g. Soft, flexible, silky needles containing one molecule of water of crystallisation.

In the same way *theobromine* can be isolated from cocoa powder which has previously been freed from fat with ether or petrol ether in an extractor. Prepare *caffeine* from the theobromine by methylation according to the method of H. Biltz.[1]

11. NICOTINE FROM TOBACCO EXTRACT[2]

Commercial tobacco extract (300 c.c., d. 1·8), which can also be prepared by concentration of the faintly acidified dilute extract to be had in any cigar factory, is made strongly alkaline with concentrated sodium hydroxide solution. Steam is passed through the hot solution and the free nicotine bases pass over. About 1·5 l. of distillate are collected, made faintly acid to Congo red with solid oxalic acid (a weighed quantity), and concentrated to a syrup. When the syrup cools *nicotine oxalate* contaminated with some ammonium oxalate separates. The crystalline sludge is transferred to a separating funnel into which is poured rather more potassium hydroxide solution (1:1) than corresponds to the oxalic acid used. Heat is developed, and after some time the crude nicotine rises to the surface as a brown oil, which is separated from the cooled mixture by repeated extraction with ether. The concentrated ethereal solution is dried with a few pieces of solid potassium hydroxide and then the ether is evaporated. The residue is fractionally distilled in a vacuum from a small Claisen flask. Since rubber stoppers are attacked by nicotine air-tight corks are used instead.

By repeated distillation of the higher boiling fraction the pure base is obtained as a colourless liquid of boiling point 114°/10 mm., 120°/14 mm. At atmospheric pressure nicotine boils without decomposition at 240°. The yield varies from 4 to 6 g.

The specimen turns brown very soon if exposed to the air, and must be kept in a sealed glass tube.

From a sample prepare the dimethiodide thus: Dissolve the base in

[1] *Annalen*, 1917, 413, 190. The reaction was first carried out by A. Strecker (*Annalen*, 1861, 118, 170).

[2] Laiblin, *Annalen*, 1879, 196, 130.

a little alcohol and warm the solution with about three parts of methyl iodide. Recrystallise from a little methyl alcohol.

Oxidise some nicotine to nicotinic acid with permanganate.

Study the syntheses of nicotine (Pictet, Karrer). What is tropine, and what alkaloids are derived from it ?

12. HAEMIN FROM OX BLOOD[1]

Glacial acetic acid (3 l.), to which 5 c.c. of saturated brine have been added, is heated to 100° in a four-litre flask on a sand bath or conical (Babo) air bath. During the course of twenty to thirty minutes a litre of defibrinated blood filtered through a linen cloth is dropped from a funnel into the hot solvent without interrupting the heating. The end of the delivery tube of the funnel reaches below the neck of the flask and the blood is not allowed to come into contact with the walls as it flows in ; the temperature should not fall below 90°. After the blood has been added, the liquid is kept gently boiling for a quarter of an hour longer ; the bulk of the haemin will separate in the form of shining crystals. The mixture is cooled to 40°–50°[2] and filtered at this temperature. The haemin is washed successively with 50 per cent acetic acid, water, alcohol, and ether. Dark crystals having a high surface lustre and great purity. Yield 3·5–4·0 g.

In the blood pigment, *haemoglobin*, the coloured component which is separated as haemin in the manner described above is combined with a complicated protein component known as globin. According to Hans Fischer, haemin has the composition $C_{34}H_{32}O_4N_4FeCl$. This compound in the form of the so-called Teichmann crystals serves for the microscopic detection of blood.

It is also the starting point for the fundamental degradation experiments which have provided considerable insight into the constitution of this important substance, although they have not yet completely elucidated it (Nencki, Küster, Piloty, Willstätter, H. Fischer).

When haemin is *reduced* with hydriodic acid it yields a mixture of *substituted pyrroles* and *pyrrole carboxylic acids*, namely :

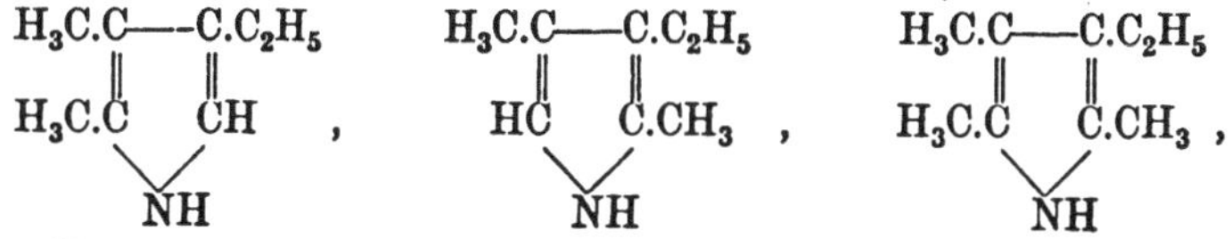

Haemopyrrole Cryptopyrrole Phyllopyrrole

[1] Process of Schalfeiev. Details given by Nencki and Zaleski, *Z. physiol. Chem.*, 1900, 30, 390; Willstätter and Stoll, *Untersuchungen über Chlorophyll*, Berlin, 1913, p. 399.

[2] According to H. Fischer in C. Oppenheimer's *Handbuch der Biochemie*, vol. i. p. 357, 1923.

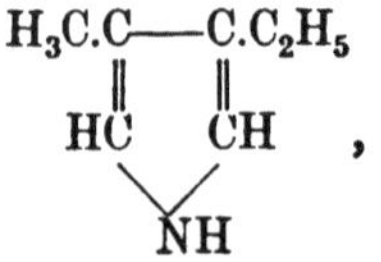

3:4-Methylethylpyrrole

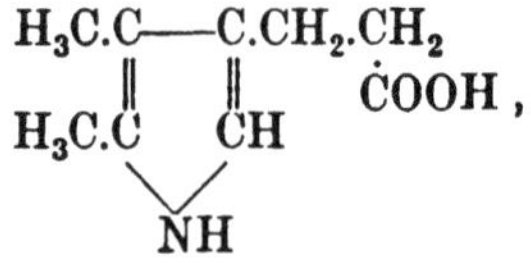

Haemopyrrole carboxylic acid

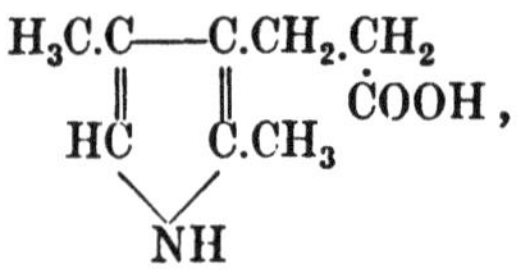

Cryptopyrrole carboxylic acid

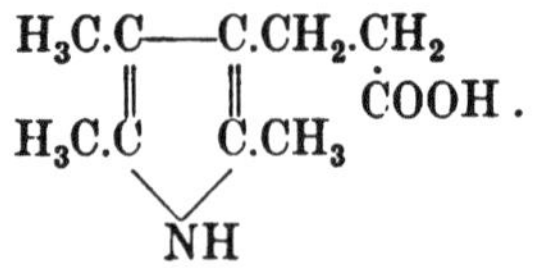

Phyllopyrrole carboxylic acid

All these degradation products have also been obtained synthetically, chiefly by H. Fischer. On *oxidation*, haemin yields derivatives of *maleic imide* (W. Küster) corresponding to the fact that pyrrole itself can be oxidised to this substance.

Thus the imide of methylethylmaleic acid (I) corresponds to the pyrroles, and the so-called *haematinic acid* (II) to the carboxylic acids formulated above:

I 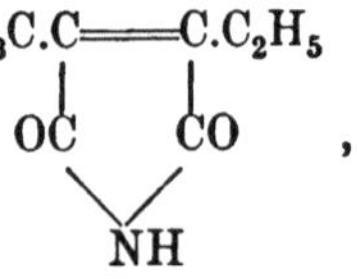

II 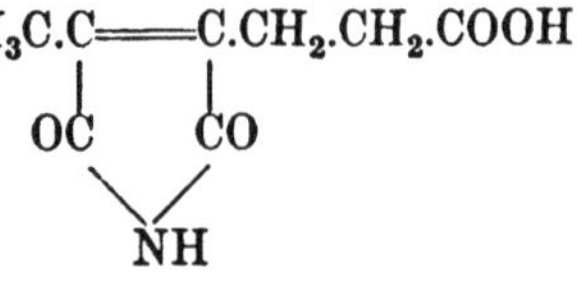

As will be seen later, C—C bonds of substituents in the α-position are broken on hydrogenation and so the various pyrrole units are set free.

The *porphyrins* are pigments of this group, which have been freed from iron; they are closely related to haemin, and iron in complex union can be reintroduced into them. Similar porphyrins are also derived from chlorophyll.

Two natural porphyrins, *uroporphyrin* $C_{40}H_{38}O_{16}N_4$ and *coproporphyrin* $C_{36}H_{38}O_8N_4$, have acquired great importance in connection with the question of constitution. These substances were first isolated by Hans Fischer from the urine and faeces of a patient suffering from deranged porphyrin metabolism; they are also formed in traces during normal metabolism. Uroporphyrin is an octa-, coproporphyrin a tetracarboxylic acid. Both are converted by thermal decarboxylation into the oxygen-free *aetioporphyrin* $C_{32}H_{38}N_4$, a substance which Willstätter had obtained from haemin.

The view formerly held that an analogous degradation product of the porphyrins of chlorophyll might be identical with this aetioporphyrin has not been confirmed (H. Fischer). There is, however, a close relationship between the two substances.

As a result of the brilliant syntheses of aetioporphyrin, coproporphyrin, meso- and deuteroporphyrin (H. Fischer, 1927–28), the structure of aetioporphyrin has been established. The isomers which may arise by varying the arrangement of the CH_3- and C_2H_5-groups in the following formula (suggested by W. Küster) have likewise been synthesised recently by H. Fischer. The readily intelligible synthesis of an aetioporphyrin from *ββ'-methylethylpyrrole* and *formic acid* is here quoted in illustration ; it proceeds according to the equation :

$$4\,C_7H_{11}N + 4\,H_2CO_2 \xrightarrow{+2\,H} C_{32}H_{38}N_4 + 8\,H_2O\,.$$

The two additional H-atoms are doubtless provided by the (excess of) formic acid.

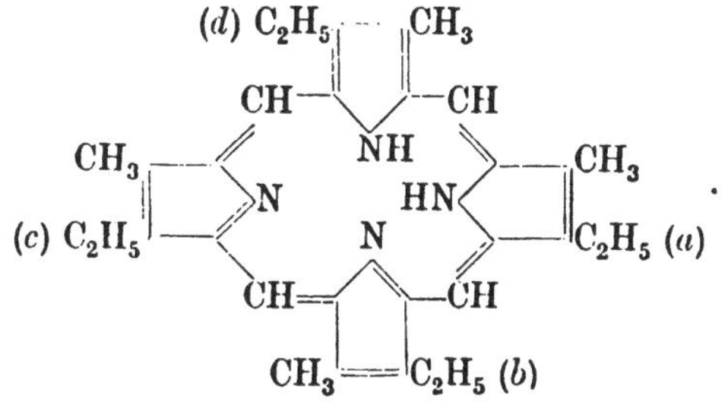

The formula shows very clearly how the four rings are condensed by intercalation of methine groups ; as is very often the case, the pyrrole here reacts in the tautomeric form 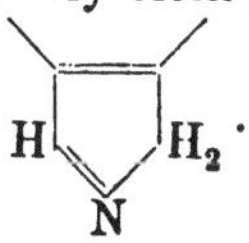 The higher porphyrins are carboxylic acids of the parent substance and derivatives of propionic acid. Hans Fischer has also built up coproporphyrin in an analogous way from β-methylpyrrole-β'-propionic acid. With the *synthesis of haemin* (H. Fischer, 1929) our knowledge of the chemistry of the coloured components of the blood pigment has been made complete. The formula of the iron-free parent substance of haemin, *protoporphyrin*, is obtained by replacing the ethyl groups at *a* and *b* by the propionic acid radicle $CH_2.CH_2.CO_2H$ and the ethyl groups at *c* and *d* by the vinyl group $CH{=}CH_2$.

The starting material for the synthesis was *deuteroporphyrin*. By means of the Friedel-Crafts reaction acetyl groups were introduced at *c* and *d*, the diketone formed was reduced to the glycol (*haemato-*

porphyrin, *c* and $d = CHOH.CH_3$), and the latter converted into protoporphyrin with elimination of water.

In haemin the iron is combined thus: The two hydrogen atoms of the NH-groups of its porphyrin are replaced by two principal valencies of the metal which is also united in complex combination to the two other N-atoms (as in the complex nickel compound of dimethylglyoxime according to Tschugaev). Then the resulting complex cation combines with Cl′ or another univalent anion to form a salt.

For the preparation of *chlorophyll* as a laboratory exercise the processes given on pp. 133 and 138 of the book by Willstätter and Stoll should be considered; see p. 395; footnote 1.

13. THE CHIEF CONSTITUENTS OF OX BILE

Glycocholic Acid.—Fresh ox bile (2·0–2·5 l.) is obtained from a slaughter-house and a small portion is used for isolating the most important of the *conjugated* acids of the bile, namely, *glycocholic acid.* In a 750 c.c. separating funnel 250 c.c. of the bile are vigorously shaken with 150 c.c. of ether; then 25 c.c. of 4 *N*-sulphuric acid are added, and the whole is at once shaken without interruption for three minutes. After a rather long time—usually by the next day—the glycocholic acid has separated in snow-white needles which form a voluminous mass at the liquid interface. The crystals are collected at the pump and washed with a little water and then with ether. For purification they are dissolved in *very little* hot alcohol, and water is added to the solution until turbidity appears.

Occasionally the glycocholic acid does not crystallise from the acidified bile until several days have elapsed. Frequent shaking has an accelerating effect. With summer bile the experiment often fails because the components of such bile crystallise with more difficulty. As far as experience goes, this has never been the case in winter.

Cholic Acid.—Commercial potassium hydroxide (200 g.) is added to fresh ox bile (2 l.) and the mixture is boiled under reflux for eighteen hours. It is best to use a copper or iron flask, but a large round-bottomed flask of Jena glass heated on a sand bath may also be employed. After the alkaline liquid has cooled it is transferred to a filter jar and made just acid to Congo paper by addition of hydrochloric acid (1 part concentrated acid, 2 parts water) which is poured in with constant stirring by means of a stout glass rod. The result of the acidification is that the bile acids and the bile pigment (biliverdin) are precipitated as a green resinous mass. At first this has the consistency of dough, but after being left for twenty-four hours in the salt solution it becomes harder, brittle, and crystalline. The hard material is removed and broken up as far as possible with the fingers under clean water in a basin. Then the crumbly mass is collected at the pump and again washed well with water. The crude acid must now be dried *completely* before further treatment. At least eight days are necessary to accomplish this when the substance is left exposed to the air in a thin layer on filter paper or wood. The object is more rapidly achieved if the adherent water is removed in a vacuum desiccator over concentrated sulphuric acid and solid potassium hydroxide. Ultimately 100–110 g. of pale grey-green substance are obtained, which can be powdered to a fine dust.

In this condition it is transferred to a conical flask into which absolute alcohol (a volume equivalent to two-thirds of the weight of the solid) is poured and the mixture is shaken well so that the whole mass is moistened. Already after one day a sludge of the readily crystallisable alcohol compound of cholic acid has been formed. After standing for two days the thick sludge is collected at the pump on a medium-sized filter plate (5 cm. in diameter), and when the dark green mother liquor has drained almost completely away the remainder of the crystalline material is washed on to the filter plate with a little absolute alcohol. Then the crystals are washed with absolute alcohol until the adherent liquor has been removed and the liquid which passes through is only pale green in colour. In this way (after drying on the water bath) 80–85 g. of cholic acid are obtained, which is already fairly pure. This is equivalent to 70 per cent of the crude precipitate (10 per cent of the weight consists of the alcohol of crystallisation).

For further purification the crystallised acid is boiled for a quarter of an hour under reflux with methylated spirit (one part by volume of spirit for each part by weight of acid), left over night without filtering, then collected again at the pump, washed with alcohol, and finally the almost colourless material is recrystallised from spirit, in which it is dissolved by boiling under reflux. The pure cholic acid separates on cooling in the form of transparent tetrahedral crystals. Melting point 196°. A further quantity of pure substance can be obtained from the mother liquor by concentration. If the acid, partially purified as described, is extracted in a thimble with ethyl acetate, a very fine preparation is obtained. The yield of pure acid amounts to fully 50 g.

The *methyl ester* of the acid can very conveniently be prepared by esterifying in ten parts of methyl alcohol with hydrogen chloride. After boiling for quite a short time, the cold solution, saturated with hydrogen chloride, is poured into a large volume of dilute sodium carbonate solution, and the ester, which is precipitated in solid form, is dried and recrystallised from a little dilute methyl alcohol. Other compounds which may suitably be prepared are *dehydrocholic acid*, *cholatrienic acid*, and *cholanic acid*.

Desoxycholic Acid, Fatty Acids, Cholesterol.—The first alcoholic filtrate obtained during the crystallisation of the cholic acid (see above) is made strongly alkaline to phenolphthalein paper with 2 N-sodium hydroxide solution and concentrated to a syrupy consistency in a porcelain basin on the water bath. The syrup is taken up in 250 c.c. of water, transferred to a separating funnel, cooled, and covered with 250 c.c. of ether. Then 2 N-sulphuric acid is gradually added from a dropping funnel with swirling. From time to time the separating funnel is vigorously shaken in order to dissolve as much as possible of the precipitated acid in the ether. When the aqueous solution has become acid to Congo paper it is run off from undissolved matter and the green ether layer is also poured off. The resinous material is dissolved by shaking with the necessary amount of dilute ammonia, and after 200 c.c. of fresh ether have been added the precipitation is repeated as before. Any resin which now remains is discarded. The combined ether extracts are at once shaken twice with 10 c.c. portions of hydrochloric acid (1 part of concentrated acid, 2 parts of water) so that the pigment is precipitated as a green resin. The clear ethereal solution, which, however, always remains green, is now concentrated to 50 c.c. and left in the cold to allow the acid to crystallise. After twenty-four hours 2·5–3·0 g. of *desoxycholic acid* will have separated in the form of well-crystallised "choleic" acid (see below). The crystals are collected at the pump, washed once with ether, and recrystallised directly from three volumes of glacial acetic acid. From the acetic acid solution, on cooling, the compound of desoxycholic acid with acetic acid ("acetic acid choleic acid") separates in beautiful prisms which melt at 142°. In the simple manner here given only about one-quarter to one-fifth of the total desoxycholic acid is obtained. The other methods of isolation are, however, very troublesome.

In a small round-bottomed flask the ethereal mother liquor is concentrated to a volume of 10 c.c., 50 c.c. of low-boiling petrol ether are added, and the stoppered flask is shaken continuously until a clear solution is obtained. The petrol ether which is now poured out contains fatty acids and cholesterol, whereas the bile acids formerly in solution in the ether have separated as a thick resinous mass. In order to separate the fatty acids and the cholesterol the petrol ether solution is carefully shaken with 50 to 60 c.c. of 2*N*-potassium hydroxide solution. The solution is added in 10 c.c. portions, and each portion is at once removed so as to avoid emulsions.

The extracts containing the soaps are freed from dissolved petrol ether by heating in a beaker on the boiling water bath, and the *fatty acids* are then precipitated with hydrochloric acid; they set to a cake on cooling. The yield is 4-5 g.

When the petrol ether solution from which the acids have been removed is evaporated to dryness 0·2-0·4 g. of cholesterol is left. From a little alcohol it crystallises beautifully in characteristic nacreous scales. Melting point 145°.

The two most important bile acids, *cholic acid* $C_{24}H_{40}O_5$ and *desoxycholic acid* $C_{24}H_{40}O_4$, occur in ox bile in combination, partly with *glycine* and partly with *taurine* as glyco- and taurocholic and glyco- and taurodesoxycholic acids. The linkage between the amino acids and the bile acids is of an amide nature. On hydrolysis the nitrogenous constituents are split off.

Cholanic acid $C_{24}H_{40}O_2$, a monobasic saturated acid containing four hydroaromatic rings, is the parent substance of the two natural acids, which are its trihydroxy- and dihydroxy-derivatives.

Cholesterol $C_{27}H_{45}OH$, an unsaturated secondary alcohol, contains the same ring system as the bile acids and is closely related to them genetically. *Pseudocholestane*, indeed, which is a stereoisomer of the parent hydrocarbon of cholesterol, *cholestane*, can be oxidised to *cholanic acid* by chromic acid with elimination of *acetone* (Windaus).

$$\underset{\text{Pseudocholestane}}{C_{23}H_{39}.CH_2CH\langle{}^{CH_3}_{CH_3}} \longrightarrow \underset{\text{Cholanic acid}}{C_{23}H_{39}.COOH} + OC\langle{}^{CH_3}_{CH_3} .$$

Conversely, pseudocholestane has also been obtained synthetically from cholanic acid.

Ergosterol $C_{27}H_{41}OH$ is a trebly unsaturated alcohol of this group. It is produced in relatively large amounts by fungi, in particular by yeast, and on irradiation it is isomerised to the anti-rachitic vitamin, vitamin-D (Windaus).

Cholic acid, and still more *desoxycholic acid*, excel in their power of combining in stoicheiometrical proportions with organic substances belonging to all kinds of chemical classes. Thus the " choleic " acid of bile, which was formerly regarded as an isomer of desoxycholic acid, is simply an addition compound of eight molecules of desoxycholic acid with one of fatty acid. The high solvent power of solutions of the alkali salts of desoxycholic acid is also connected with this property.

Experiment.—Dissolve 0·4 g. of the acetic acid - desoxycholic acid compound, prepared as above, in 4 c.c. of 2 *N*-sodium hydroxide. Prepare also from about 100 mg. of the fatty acids isolated, by boiling with a few cubic centimetres of dilute sodium hydroxide solution, a soap solution and cool it till it sets to a jelly. Add part of the solution of desoxycholic-acetic acid. The soap dissolves.

In the same way dissolve a few granules of naphthalene in 0·5 c.c. of alcohol, pour the solution into twice its volume of water, and at once (before large crystals form) add a few drops of the desoxycholic-acetic acid solution to the milky suspension of the hydrocarbon. A clear solution is produced.

For further details see Wieland, *Z. physiol. Chem.*, 1916, 97, 1.

Experiment.—Dissolve 0·02 g. of cholic acid in 0·5 c.c. of alcohol and add to the solution 1 c.c. of 0·1 *N*-iodine solution. On cautious dilution with water the blue crystalline iodine compound of cholic acid separates, comparable to "starch iodide ".

Experiment.—Barium Desoxycholate.—Dissolve a small sample of desoxycholic acid in dilute ammonia to a clear solution and add barium chloride solution. The sparingly soluble barium salt is at once precipitated as a tough mass, which crystallises in beautiful rosettes when left over night. Distinction from cholic acid, the barium salt of which is more soluble in water.

Experiment.—Dissolve about 5 mg. of cholesterol in 1 c.c. of acetic anhydride and add a drop of concentrated sulphuric acid. A magnificent play of colours is produced (Liebermann-Burchard cholestol reaction).

HINTS FOR USING THE LITERATURE OF ORGANIC CHEMISTRY

THE multitudinous compounds of carbon, over 200,000 in number, are so well arranged in chemical reference books that it is a simple matter to obtain particulars about any one of them. Students should know their way about the library quite early in their course, at the latest when they begin to prepare substances described in original papers. In order to supply this knowledge some directions are given here.

The foundation-stone of the systematic arrangement of organic compounds is the principle of the " formula index " which was first worked out by M. M. Richter in his *Formel-Lexikon der organischen Verbindungen*, of which the third edition includes the material available up to the year 1910. The compounds are grouped, according to the number of carbon atoms in the molecule, in a numerical series from C_1 upwards. In the separate groups those substances come first which contain only one element in addition to carbon, and they are so arranged that the additional elements follow each other in the order H, O, N, Halogen, S, P, As; then come the compounds with two, three, and more elements besides carbon, arranged as before. For example, if information is wanted about *methyl* p-*chlorobenzoate* a search is made in Richter's *Lexikon* in the volume which contains the compounds of the C_8 series, since the substance has the molecular formula $C_8H_7O_2Cl$. The headings in the upper corners of the pages lead to the group 8 III, *i.e.* to the list of organic compounds which contain three other elements besides eight carbon atoms.

Since the known compounds in the group C_8III are arranged according to the number of atoms of the additional elements (H and O), it is easy to find the compound required.

The *Lexicon* of the carbon compounds corresponds to the great reference work of Beilstein, which will be discussed later. Accordingly, for our compound the *Lexicon* gives, in addition to a brief statement of melting or boiling point, also the place in Beilstein where more detailed information is available. Since Richter's work covers more ground than that of Beilstein, of which the third edition with supplementary volumes only includes the literature up to July 1, 1899, Richter gives references to the original literature for the material which was lacking in Beilstein.

The five-volume work of Stelzner forms the continuation of Richter's formula lexicon. It covers the literature up to the year 1921 and is arranged on the same plan as Richter's, but gives in addition a brief description of the individual substances, their constants and reactions, after the fashion of Beilstein.

The indexing of organic compounds described since 1922 has been taken over by the *Chemisches Zentralblatt*, which follows the system hitherto discussed. The collective indexes of this journal for 1922 to 1924 and for 1925 to 1929 continue the classifying work of M. M. Richter. In these indexes the references are to the *Zentralblatt* itself. Each year of the *Zentralblatt*, from 1925 onwards, contains a formula index covering the annual increment of literature abstracted in this comprehensively planned journal. These formula indexes are amalgamated every few years into a collective index and individually they constitute the reference works for the period which has elapsed since the appearance of the last collective index. Consequently, when it is desired to have complete information from the whole literature about an organic substance, the following works must be consulted at the present day :

1. Richter's *Lexikon*.
2. *Stelzner*, 1911–1921.
3. Collective Indexes of the *Chemisches Zentralblatt*, volumes vi., 1922–1924, and vii., 1925–1929.
4. Formula indexes to the *Zentralblatt* for the years 1930, 1931, etc.

Conscientious study of the literature is indispensable to the organic chemist. After a little practice the trouble of looking up references becomes inconsiderable, and a thorough knowledge of the literature on any particular point is of quite extraordinary value. It may spare the worker the irritation of trying unsuitable preparative methods which he may be tempted to apply because he has neglected to look up a good one hidden in the literature. Likewise, the danger of rediscovering already known compounds can only be obviated by a thorough search through the chain of indexes mentioned.

The references which we obtain in a successful search are of various kinds. The main work of "*Richter*", as already mentioned, first refers us to Beilstein's *Handbuch der organischen Chemie,* which may now be briefly described. The third edition of this work, in four volumes and as many supplementary volumes, gives a brief description of all pure organic compounds prepared up to July 1, 1899, with their physical constants, methods of formation, preparation, and most important reactions. At the same time complete references to the original literature are given. The arrangement is as follows: I. Compounds of the Aliphatic Series. II. Aromatic Compounds. III. Carbonyl Compounds. IV. Heterocyclic Compounds. Each volume contains a subject index. The more detailed classification cannot be given here.

The third edition of "*Beilstein*" is seriously out of date, *i.e.* it contains only a small proportion of the organic compounds known to-day. Unfortunately, the same must be said of the fourth edition (of which so far about two-thirds has been completed), for this only covers the literature to January 1, 1910. Of the twenty volumes which are likely to be necessary, fourteen had appeared in 1931. Of the supplementary volumes covering the literature from the year 1910 onwards, eight volumes of the first series have been published; this series deals with the literature from 1910 to the end of 1919.

The plan of classification and arrangement of the subject matter in the new (fourth) edition is described in its first volume. It does not correspond to any of the formula indexes mentioned. Because of the completeness with which the material from the whole literature is collected, "*Beilstein*" provides an admirably concise account of all the known facts relating to any particular compound. The book constitutes a kind of information bureau for the organic chemist. Almost as important as the descriptive material given are the references to the literature. Whoever turns to "*Beilstein*" for preparative purposes should remember that the ultimate sources of information are the *original publications* to which reference is made. *The student should acquire the habit of going back to these sources in all cases, of always reading the original work* (and of abstracting it). *He should never be satisfied with a mere perusal of books of reference.*

This injunction holds also for those who wish to make the best use of the references to the *Chemisches Zentralblatt.* From the abstracts in the latter the student must likewise turn to the original papers, and look up the journal mentioned at the end of the abstract.

For general information regarding the various classes of substances dealt with in the course of laboratory preparations, the comprehensive textbook of V. Meyer and Jacobson is particularly suitable. Unfortunately, owing to the vast amount of subject matter, some of its parts are unpleasantly out of date.

An excellent reference work for preparations of technical interest is provided by the twelve volumes of F. Ullmann's *Enzyklopädie der technischen Chemie*, arranged alphabetically. Since 1928 the second edition has been rapidly appearing, and seven volumes have now been published.

In the four volumes of *Die Methoden der organischen Chemie* of Houben and Weyl, all the methods used in organic laboratories are systematically collected and described, whilst the comprehensive book by Hans Meyer, *Analyse und Konstitutionsermittlung organischer Verbindungen*, is more concerned with analytical requirements.

PREPARATIONS FROM THE ORIGINAL LITERATURE

As a conclusion to his preparative course the student should undertake *several* preparations, for which he has to look up the appropriate methods in scientific journals, instead of finding them worked out in minute detail in a manual of laboratory instruction. In so doing he will learn, with the aid of the directions given above, how to consult the literature, and how to make use of the chemical library; most of all, however, these more difficult preparations will serve to test his capacity for laboratory work. *When synthesising a compound in several stages it is absolutely essential to test and consolidate each separate stage by means of test-tube experiments* before the whole of the substance is risked. Whoever fails to observe this rule will pay dearly for his neglect by loss of material and time.

The choice of difficult preparations will often be regulated by the requirements and desiderata resulting from the scientific work of the institute; that the interests of the person who carries out the preparation must take precedence in this matter need not be emphasised.

The following list of preparations has proved suitable for the reasons given:

Pinacone, Pinacoline
O-Methylhydroxylamine
Allyl Alcohol
Styrene
Stilbene
Phenacyl Bromide
Biphenyl
Ethylene Oxide
Choline
Dithioglycollic Acid
Oxalyl Chloride
Nitrourea
Nitramide
Ethyl Orthoformate
Fumaric Acid
Glutaric Acid
Adipic Acid
Pimelic Acid
Cyanamide
β-Phenylethyl Alcohol
Pyruvic Acid
Oxalylacetic Acid
Dihydroxytartaric Acid
Dihydroxymaleic Acid
Acetonedicarboxylic Acid
Muconic Acid
Cadaverine
Triphenylamine
Fulminuric Acid
Ethyl Azodicarboxylate
p-Dinitrobenzene
Aldol
Crotonic aldehyde
Phenylacetaldehyde
Benzildioxime
Phenylalanine
Methylcyclohexenone
p-Diketocyclohexane
Diacetyl
Mesitylene
Acetonylacetone
Antipyrine
Carbon Suboxide
Coumarin

Allocinnamic Acid
Xanthone
Acridine
Anthranol
Di-biphenylene-ethylene
β-Nitronaphthalene
Diphenylketene
Camphorquinone
Violuric Acid
Veronal
Phenylnitramine
m-Toluidine
m-Nitrophenol
p-Nitrophenylhydrazine
Fuchsone
Hexahydroxybenzene
Inositol
o-Benzoquinone
Quinol
Hydroxyquinol
Tetraphenylxylylene
Dimethylaminobenzaldehyde
Phenanthraquinone
Biphenylene-ethylene
Diphenylhydrazine
Triphenylhydrazine and Hexaphenyltetrazane
Tolane
Mercury Diphenyl
Dimethylpyrone
1:5-Dibromopentane
Diphenylnitric Oxide
Thioindigotin
Thioindigo Scarlet
Isatin (Sandmeyer's method)
Thymolphthalein
Sodium Triphenylmethyl
Dimethylpyrrole
Mellitic Acid
Methyliminazole from glucose
α- and β-Methylglucosides
dl-Camphor from pinene

Cystine
Hydrolysis of protein (according to E. Fischer and Dakin
Dihydroxyphenylalanine
Glutamic Acid
Glycylglycine
Adrenaline
Mannose
Diacetoneglucose
Dulcitol from galactose
Sylvane from furfural
Guanine
Xanthine
Uric Acid (synthetic)
Camphoric Acid
o-Esdragole from phenylallyl ether
Ionone
Vanillin from iso-eugenol
Tartrazine
Auramine
Tetraphenyl Lead
Boron Triphenyl
Phorone
Ninhydrin
Ethyl Chloro-iminocarbonate
Diphenyldiazomethane
Phenylacetylene
Piperidine
Piperylene
Ethyl Cyclopentanone Carboxylate
Picrolonic Acid
as-Diphenylethylene
Amylene from acetone
Xylose
Glyoxal
Furfuryl Alcohol and Cinnamyl Alcohol (Meerwein's method)
Glutathione

TABLE FOR CALCULATIONS IN THE DETERMINATION OF NITROGEN BY DUMAS' METHOD

Density of Dry Nitrogen

Weight of 1 c.c. of nitrogen in milligrammes under pressures between 700 and 770 mm. at temperatures between 10° and 25°. Calculated on the assumption that, under standard conditions, 1 c.c. of nitrogen weighs 1·2507 mg. (Lord Rayleigh's determination).

p	10°	11°	12°	13°	14°	15°	16°	17°	18°	19°	20°	21°	22°	23°	24°	25°	p
700	1·111	1·107	1·103	1·100	1·096	1·092	1·088	1·084	1·081	1·077	1·073	1·070	1·066	1·062	1·059	1·055	700
702	1·114	1·110	1·107	1·103	1·099	1·095	1·091	1·087	1·084	1·080	1·076	1·073	1·069	1·065	1·062	1·058	702
704	1·118	1·114	1·110	1·106	1·102	1·098	1·094	1·091	1·087	1·083	1·079	1·076	1·072	1·068	1·065	1·061	704
706	1·121	1·117	1·113	1·109	1·105	1·101	1·097	1·094	1·090	1·086	1·082	1·079	1·075	1·071	1·068	1·064	706
708	1·124	1·120	1·116	1·112	1·108	1·104	1·101	1·097	1·093	1·089	1·085	1·082	1·078	1·074	1·071	1·067	708
710	1·127	1·123	1·119	1·115	1·111	1·107	1·104	1·100	1·096	1·092	1·089	1·085	1·081	1·077	1·074	1·070	710
712	1·130	1·126	1·122	1·118	1·114	1·111	1·107	1·103	1·099	1·095	1·092	1·088	1·084	1·081	1·077	1·073	712
714	1·133	1·129	1·125	1·121	1·118	1·114	1·110	1·106	1·102	1·098	1·095	1·091	1·087	1·084	1·080	1·076	714
716	1·137	1·133	1·129	1·125	1·121	1·117	1·113	1·109	1·105	1·101	1·098	1·094	1·090	1·087	1·083	1·079	716
718	1·140	1·136	1·132	1·128	1·124	1·120	1·116	1·112	1·108	1·105	1·101	1·097	1·093	1·090	1·086	1·082	718
720	1·143	1·139	1·135	1·131	1·127	1·123	1·119	1·115	1·111	1·108	1·104	1·100	1·096	1·093	1·089	1·085	720
722	1·146	1·142	1·138	1·134	1·130	1·126	1·122	1·118	1·115	1·111	1·107	1·103	1·099	1·096	1·092	1·088	722
724	1·149	1·145	1·141	1·137	1·133	1·129	1·125	1·121	1·118	1·114	1·110	1·106	1·102	1·099	1·095	1·091	724
726	1·152	1·148	1·144	1·140	1·136	1·132	1·128	1·125	1·121	1·117	1·113	1·109	1·105	1·102	1·098	1·094	726
728	1·156	1·152	1·148	1·143	1·139	1·136	1·132	1·128	1·124	1·120	1·116	1·112	1·109	1·105	1·101	1·097	728
730	1·159	1·155	1·151	1·147	1·143	1·139	1·135	1·131	1·127	1·123	1·119	1·115	1·112	1·108	1·104	1·100	730
732	1·162	1·158	1·154	1·150	1·146	1·142	1·138	1·134	1·130	1·126	1·122	1·118	1·115	1·111	1·107	1·103	732
734	1·165	1·161	1·157	1·153	1·149	1·145	1·141	1·137	1·133	1·129	1·125	1·121	1·118	1·114	1·110	1·106	734
736	1·168	1·164	1·160	1·156	1·152	1·148	1·144	1·140	1·136	1·132	1·128	1·125	1·121	1·117	1·113	1·109	736
738	1·172	1·167	1·163	1·159	1·155	1·151	1·147	1·143	1·139	1·135	1·131	1·128	1·124	1·120	1·116	1·112	738

p	10°	11°	12°	13°	14°	15°	16°	17°	18°	19°	20°	21°	22°	23°	24°	25°	p
740	1·175	1·171	1·166	1·162	1·158	1·154	1·150	1·146	1·142	1·138	1·135	1·131	1·127	1·123	1·119	1·115	740
742	1·178	1·174	1·170	1·165	1·161	1·157	1·153	1·149	1·145	1·141	1·138	1·134	1·130	1·126	1·122	1·118	742
744	1·181	1·177	1·173	1·169	1·165	1·160	1·156	1·152	1·149	1·145	1·141	1·137	1·133	1·129	1·125	1·121	744
746	1·184	1·180	1·176	1·172	1·168	1·164	1·160	1·156	1·152	1·148	1·144	1·140	1·136	1·132	1·128	1·124	746
748	1·187	1·183	1·179	1·175	1·171	1·167	1·163	1·159	1·155	1·151	1·147	1·143	1·139	1·135	1·131	1·128	748
750	1·191	1·186	1·182	1·178	1·174	1·170	1·166	1·162	1·158	1·154	1·150	1·146	1·142	1·138	1·134	1·131	750
752	1·194	1·190	1·186	1·181	1·177	1·173	1·169	1·165	1·161	1·157	1·153	1·149	1·145	1·141	1·137	1·134	752
754	1·197	1·193	1·189	1·185	1·180	1·176	1·172	1·168	1·164	1·160	1·156	1·152	1·148	1·144	1·140	1·137	754
756	1·200	1·196	1·192	1·188	1·183	1·179	1·175	1·171	1·167	1·163	1·159	1·155	1·151	1·147	1·143	1·140	756
758	1·203	1·199	1·195	1·191	1·186	1·182	1·178	1·174	1·170	1·166	1·162	1·158	1·154	1·150	1·146	1·143	758
760	1·206	1·202	1·198	1·194	1·190	1·185	1·181	1·177	1·173	1·169	1·165	1·161	1·157	1·153	1·149	1·146	760
762	1·210	1·205	1·201	1·197	1·193	1·189	1·184	1·180	1·176	1·172	1·168	1·164	1·160	1·156	1·152	1·149	762
764	1·213	1·209	1·204	1·200	1·196	1·192	1·188	1·183	1·179	1·175	1·171	1·167	1·163	1·159	1·156	1·152	764
766	1·216	1·212	1·207	1·203	1·199	1·195	1·191	1·187	1·182	1·178	1·174	1·170	1·166	1·162	1·159	1·155	766
768	1·219	1·215	1·211	1·206	1·202	1·198	1·194	1·190	1·186	1·181	1·177	1·173	1·169	1·165	1·162	1·158	768
770	1·222	1·218	1·214	1·209	1·205	1·201	1·197	1·193	1·189	1·185	1·181	1·176	1·172	1·169	1·165	1·161	770
f (mm.)	5·6	6·0	6·4	6·9	7·3	7·8	8·4	8·9	9·5	10·1	10·8	11·4	12·2	12·9	13·7	14·6	f (mm.)

If the confining liquid is 30 per cent potassium hydroxide solution, the pressure of water vapour over this solution at the temperature concerned (f mm., see above) must be subtracted from the observed barometric pressure after reduction to 0°.

SUBJECT INDEX

(The page numbers which refer to the experiments of the practical course are printed in heavy type.)

THE END

Printed in Great Britain by R. & R. Clark, Limited, *Edinburgh.*

Bei Fragen zur Produktsicherheit wenden Sie sich bitte an:
If you have any questions regarding product safety,
please contact:

Walter de Gruyter GmbH
Genthiner Straße 13
10785 Berlin
productsafety@degruyterbrill.com